OCEAN SCIENCE DATA

OCEAN SCIENCE DATA

COLLECTION, MANAGEMENT, NETWORKING AND SERVICES

Edited by

GIUSEPPE MANZELLA
OceanHis SrL, Torino, Italy

ANTONIO NOVELLINO
ETT SpA - Gruppo SCAI, Genova, Italy

ELSEVIER

Elsevier
Radarweg 29, PO Box 211, 1000 AE Amsterdam, Netherlands
The Boulevard, Langford Lane, Kidlington, Oxford OX5 1GB, United Kingdom
50 Hampshire Street, 5th Floor, Cambridge, MA 02139, United States

Copyright © 2022 Elsevier Inc. All rights reserved.

No part of this publication may be reproduced or transmitted in any form or by any means, electronic or mechanical, including photocopying, recording, or any information storage and retrieval system, without permission in writing from the publisher. Details on how to seek permission, further information about the Publisher's permissions policies and our arrangements with organizations such as the Copyright Clearance Center and the Copyright Licensing Agency, can be found at our website: www.elsevier.com/permissions.

This book and the individual contributions contained in it are protected under copyright by the Publisher (other than as may be noted herein).

Notices
Knowledge and best practice in this field are constantly changing. As new research and experience broaden our understanding, changes in research methods, professional practices, or medical treatment may become necessary.

Practitioners and researchers must always rely on their own experience and knowledge in evaluating and using any information, methods, compounds, or experiments described herein. In using such information or methods they should be mindful of their own safety and the safety of others, including parties for whom they have a professional responsibility.

To the fullest extent of the law, neither the Publisher nor the authors, contributors, or editors, assume any liability for any injury and/or damage to persons or property as a matter of products liability, negligence or otherwise, or from any use or operation of any methods, products, instructions, or ideas contained in the material herein.

Library of Congress Cataloging-in-Publication Data
A catalog record for this book is available from the Library of Congress

British Library Cataloguing-in-Publication Data
A catalogue record for this book is available from the British Library

ISBN: 978-0-12-823427-3

For information on all Elsevier publications visit our website
at https://www.elsevier.com/books-and-journals

Publisher: Candice Janco
Acquisitions Editor: Louisa Munro
Editorial Project Manager: Sara Valentino
Production Project Manager: Debasish Ghosh
Cover Designer: Mark Rogers

Typeset by TNQ Technologies

Working together
to grow libraries in
developing countries

www.elsevier.com • www.bookaid.org

Contents

PART 4   Education

PART 5   Appendix

Contributors

A. Barth
University of Liege, Liege, Belgium

Joana Beja
Flanders Marine Institute (VLIZ), Oostende, Belgium

Abigail Benson
U.S. Geological Survey, Lakewood, CO, United States

T. Boyer
National Centers for Environmental Information, National Oceanic and Atmospheric Administration, Asheville, NC, United States

Jan-Bart Calewaert
Seascape Belgium bvba, Brussels, Belgium; European Marine Observation and Data Network (EMODnet) Secretariat, Ostend, Belgium

C. Coatanoan
Ifremer Centre de Bretagne, Plouzané, Brest, France

Tim Collart
Seascape Belgium bvba, Brussels, Belgium; European Marine Observation and Data Network (EMODnet) Secretariat, Ostend, Belgium

Conor Delaney
Seascape Belgium bvba, Brussels, Belgium; European Marine Observation and Data Network (EMODnet) Secretariat, Ostend, Belgium

Daphnis De Pooter
Commission for the Conservation of Antarctic Marine Living Resources, (CCAMLR), Hobart, TAS, Australia

Federico De Strobel
The Historical Oceanography Society, La Spezia, Italy

S. Diggs
Scripps Institution of Oceanography, University of California San Diego, La Jolla, CA, United States

William Emery
University of Colorado, Boulder, CO, United States

Michele Fichaut
IFREMER/SISMER, Brest, France

Vasilis Gerovasileiou
Hellenic Centre for Marine Research (HCMR), Institute of Marine Biology, Biotechnology and Aquaculture (IMBBC), Heraklion, Greece

Kate E. Larkin
Seascape Belgium bvba, Brussels, Belgium; European Marine Observation and Data
Network (EMODnet) Secretariat, Ostend, Belgium

Dan Lear
Marine Biological Association, Plymouth, United Kingdom

Helen Lillis
Joint Nature Conservation Committee (JNCC), Peterborough, United Kingdom

M. Lipizer
Istituto Nazionale di Oceanografia e di Geofisica Sperimentale — OGS, Trieste, Italy

Eleonora Manca
Joint Nature Conservation Committee (JNCC), Peterborough, United Kingdom

Giuseppe M.R. Manzella
The Historical Oceanography Society, La Spezia, Italy; OceanHis SrL, Torino, Italy

Andrée-Anne Marsan
Seascape Belgium bvba, Brussels, Belgium; European Marine Observation and Data
Network (EMODnet) Secretariat, Ostend, Belgium

Patricia Miloslavich
Scientific Committee on Oceanic Research (SCOR), University of Delaware, College of
Earth, Ocean and Environment, Newark, DE, United States; Departamento de Estudios
Ambientales, Universidad Simón Bolívar, Caracas, Miranda, Venezuela

Gwenaëlle Moncoiffé
British Oceanographic Data Centre, National Oceanography Centre, Liverpool, United
Kingdom

V. Myroshnychenko
Middle East Technical University, Institute of Marine Sciences, Erdemli-Mersin, Turkey

John Nicholls
Norfish Project, Centre for Environmental Humanities, Trinity College Dublin, Dublin,
Ireland

Antonio Novellino
ETT SpA, Genova, Italy

Nadia Pinardi
The Historical Oceanography Society, La Spezia, Italy; Department of Physics and
Astronomy, Università di Bologna, Bologna, Italy

A. Pisano
Consiglio Nazionale delle Ricerche - Istituto di Scienze Marine (CNR-ISMAR), Rome,
Italy

A. Pititto
COGEA, Rome, Italy

Dick M.A. Schaap
Mariene Informatie Service MARIS B.V., Nootdorp, the Netherlands

R. Schlitzer
Alfred Wegener Institute, Bremerhaven, Germany

S. Simoncelli
Istituto Nazionale di Geofisica e Vulcanologia, Sezione di Bologna, Italy

A. Storto
Consiglio Nazionale delle Ricerche - Istituto di Scienze Marine (CNR-ISMAR), Rome, Italy

Nathalie Tonné
Seascape Belgium bvba, Brussels, Belgium; European Marine Observation and Data Network (EMODnet) Secretariat, Ostend, Belgium

C. Troupin
University of Liege, Liege, Belgium

Leen Vandepitte
Flanders Marine Institute (VLIZ), Oostende, Belgium

Anton Van de Putte
Royal Belgian Institute for Natural Sciences, Brussels, Belgium; Université Libre de Bruxelles, Brussels, Belgium

Nathalie Van Isacker
Seascape Belgium bvba, Brussels, Belgium; European Marine Observation and Data Network (EMODnet) Secretariat, Ostend, Belgium

Mickaël Vasquez
Ifremer, Brest, France

Nina Wambiji
Kenya Marine and Fisheries Research Institute, Mombasa, Kenya

Biographies

Giuseppe Manzella received a degree in physics from the Department of Physics, University of Rome "La Sapienza." After some fellowships, and attendance of specialization courses in Europe, he first worked at National Research Council (1982–92) and then was employed as research manager in ENEA (1992–2013). From 1978 he has been active in national, European, and international programs in oceanography. He has worked as expert on marine ecosystem for the Italian Ministry of Research, and the Italian representative to WMO-IOC Joint Committee for Marine Meteorology (JCOMM). He has chaired the Italian Oceanographic Commission from January 2009 to June 2014. He is chairing the Historical Oceanography Society. He is author/co-author of 50 refereed papers published in international journals, co-editor of two books published, and the Topic Editor of the Journal *Earth System Science Data.*

Antonio Novellino received a PhD in Biotechnology and Bioengineering and a MSc in Biomedical Engineering. From 2008 to 2010, he served on the European Commission, JRC – IHCP, as a senior researcher. He is the ETT Research Manager where he coordinates R&D activities (www.ettsolutions.com). He served on the Board of Directors of Consortium Si4Life (www.si4life.com) and on the board of Consortium Tecnomar (SMEs working on maritime and environment technology, www.consorziotecnomar.com). He is serving on the techno-scientific board of the Ligurian Cluster of Marine Technology DLTM (www.dltm.it); the board of Consortium TRAIN (innovation in energy and transport management, www.consorziotrain.org); EMODnet Steering Committee and Technical Working Group; Expert Team on WIS Centres (ET-WISC); and Southern Ocean Observing System Data Management team (SOOS DMSC). He is a member of the EuroGOOS DATAMEQ group for advising on operational oceanography data management procedures. He is the EMODnet physics coordinator (www.emodnet-physics.eu) and CMEMS Dissemination Unit (CMEMS DU) deputy coordinator.

PART ONE

Marine science: history and data archaeology

CHAPTER ONE

A narrative of historical, methodological, and technological observations in marine science

Giuseppe M.R. Manzella[1,4], Federico De Strobel[1], Nadia Pinardi[1,2], William Emery[3]

[1]The Historical Oceanography Society, La Spezia, Italy
[2]Department of Physics and Astronomy, Università di Bologna, Bologna, Italy
[3]University of Colorado, Boulder, CO, United States
[4]OceanHis SrL, Torino, Italy

Introduction

> Our planet is invested with two great oceans; one visible, the other invisible; one underfoot, the other overhead; one entirely envelops it, the other covers about two thirds of its surface.
>
> Matthew Fontaine Maury, *The Physical Geography of the Sea and Its Meteorology*, 1855

The earliest studies of the oceans date back to Aristotle (384 BC—322 BC), but a true methodological approach only began about two millennia after his death. Initially, ocean science derived from the practical arts of navigation and cartography (Henry, 2008). During the 15th century, logbooks and annotated navigation maps began to be collected systematically. Unfortunately, few early travel records have survived due to physical deterioration, loss of logbooks or privacy policies (Peterson et al., 1996). With the Portuguese exploration of new lands and seas, important advances were made in one branch of science in particular: the geography of the sea.

Ocean Science Data
ISBN: 978-0-12-823427-3
https://doi.org/10.1016/B978-0-12-823427-3.00004-9

© 2022 Elsevier Inc.
All rights reserved.

The first methodological and technological approach to observing the sea was established at the meeting of the Royal Society of London on June 14, 1661. The document *Propositions of Some Experiments to Be Made by the Earl of Sandwich in His Present Voyage* (Birch, 1760) defined the guidelines for data collection.

"Diligent observations" were required by Galilei (1564–1642) and were the basis of his experimental method. The concept was underlined, inter alia, in the "Forth Day" chapter, discussing the causes of the tides, in the famous book *Dialogue on the Two Chief World Systems* (Galilei, 1632). The "Propositions" were asking for "diligent observations" and their recommendations were subsequently included in the *Directions for the Observations and Experiments to Be Made by Masters of Ships, Pilots, and Other Fit Persons on Their Sea-Voyages* by Murray and Hooke (1667) on behalf of the Royal Society. Seafarers were asked to

- Observe the declination and variations of the compass or needle from the meridian exactly, in as many places as they can, and in the same place, every several voyage,
- Carry dipping-needles with them,
- Mark carefully the flowings and ebbings of the sea, in as many places as may be,
- Sound the deepest seas without a line, …
- Keep a register of all changes of wind and weather …,
- Observe and record all extraordinary meteors, lightnings, thunders, …
- Carry with them good scales and glass-viols of a pint or so, with very narrow mouths, which are to be filled with sea water in different degrees of latitude, and the weight of the viol full of water taken exactly at every time, and recorded; marking withal the degrees of latitude and longitude of the place, and the day of the month, and the temperature of the weather: and as well of water near the top, as at a greater depth,
- Fetch up water from any depth of the sea.

The "Directions" were accompanied by instructions on the use of methods and instruments. Theoretically, they constituted a systematization of a general request by several European scientists made explicit by Vincenzo Viviani (a pupil of Galileo Galilei), on behalf of the Accademia del Cimento to gather knowledge of the variability of ocean circulation by means of "diligent" observations of the sea (see Pinardi et al., 2018). It is worthwhile to mention that the Accademia del Cimento motto was "by trying and trying again" (e.g., Magalotti, 1667).

A further step toward a more "diligent" observational methodology was made by Ferdinando Marsili (1658—1730) with his famous treatise "Osservazioni intorno al Bosforo Tracio" (Marsili, 1681). For the first time, an appropriate observation strategy was defined in order to understand the effects of density differences on the circulation of water masses (Pinardi et al., 2018; Peterson et al., 1996; Deacon, 1971).

The efforts of the Royal Society to gain greater knowledge of the physical characteristics of the sea met with little success. However, an important contribution came from William Dampier (1651—1715), who was the first person to circumnavigate the globe three times (Dampier, 2012). In Dampier's records, data were not reported as requested in the "Directions," but contained substantial information on winds and currents. The essay "A New Voyage Round the World" (Dampier, 1703) was sent to the Royal Society with the aim to promote "useful knowledge, and of anything that may never so remotely tend to my Countries advantage." During his voyages, Dampier found that currents in the equatorial region were driven by the trade winds (Deacon, 1971).

The westward currents in the north equatorial region were used during the discovery of America and were associated with the Aristotelian conception of sea motion. In actual fact, Aristotle never spoke of a westward sea flow, but this was the scholarly interpretation of a passage in the second book of *Meteorologica* (e.g., Aristotle, 1952):

> The whole Mediterranean flows according to the depth of the sea-bed and the volume of the rivers. For Lake Maeotis (Azov Sea) flows into the Pontus and thus into the Aegean ... In the seas mentioned it (the flow) takes place because of the rivers—for more rivers flow into the Euxine and Lake Maeotis than into other areas many times their size—and because of their shallowness. For the sea seem to get deeper and deeper than Lake Maeotis, the Aegean deeper than the Pontus and the Sicilian Sea deeper than the Aegean, while the Sardinian and Tyrrhenian are the deepest of all. The water outside the pillars of Heracles is shallow because of the mud but calm because the sea lies in a hollow.

The westward flow of the surface currents in the north equatorial region was noted by Pietro Martire d'Angera (1457—1526) in De Orbe Novo, 3rd "Decade," Book 4 of the English translation by MacNutt (Martyr d'Anghiera, 1912):

> It was in the year of salvation 1502 on the sixth day of the ides of May that Columbus sailed from Cadiz with a squadron of four vessels of from fifty to sixty tons burthen, manned by one hundred and seventy men. Five days of favourable

weather brought him to the Canaries; seventeen days' sailing brought him to the island of Domingo, the home of the Caribs, and from thence he reached Hispaniola in five days more, so that the entire crossing from Spain to Hispaniola occupied twenty-six days, thanks to favourable winds and currents, which set from the east towards the west. According to the mariners' report the distance is twelve hundred leagues.

Pietro Martire d'Angera in the same "Decade" (from Latin "decas"—group of 10) analyzed the consequences of this flow in terms of the conservation of water masses and wrote explicitly in Book 6:

The time has come, Most Holy Father, to philosophise a little, leaving cosmography to seek the causes of Nature's secrets. The ocean currents in those regions run towards the west, as torrents rushing down a mountain side. Upon this point the testimony is unanimous. Thus, I find myself uncertain when asked where these waters go which flow in a circular and continuous movement from east to west, never to return to their starting-place; and how it happens that the west is not consequently overwhelmed by these waters, nor the east emptied. If it be true that these waters are drawn towards the centre of the earth, as is the case with all heavy objects, and that this centre, as some people affirm, is at the equinoctial line, what can be the central reservoir capable of holding such a mass of waters? And what will be the circumference filled with water, which will yet be discovered? The explorers of these coasts offer no convincing explanation. There are other authors who think that a large strait exists at the extremity of the gulf formed by this vast continent and which, we have already said, is eight times larger than the ocean. This strait may lie to the west of Cuba, and would conduct these raging waters to the west, from whence they would again return to our east. Some learned men think the gulf formed by this vast continent is an enclosed sea, whose coasts bend in a northerly direction behind Cuba, in such wise that the continent would extend unbrokenly to the northern lands beneath the polar circle bathed by the glacial sea. The waters, driven back by the extent of land, are drawn into a circle, as may be seen in rivers whose opposite banks provoke whirlpools; but this theory does not accord with the facts. The explorers of the northern passages, who always sailed westwards, affirm that the waters are always drawn in that direction, not however with violence, but by a long and uninterrupted movement. Amongst the explorers of the glacial region a certain Sebastiano Cabotto, of Venetian origin, but brought by his parents in his infancy to England, is cited. It commonly happens that Venetians visit every part of the universe, for purposes of commerce. Caboto equipped two vessels in England, at his own cost, and first sailed with three hundred men towards the north, to such a distance that he found numerous masses of floating ice in the middle of the month of July. Daylight lasted nearly twenty-four hours, and as the ice had melted, the land was free. According to his story he was obliged to tack and take the direction of west-by-south. The coast bent to about the degree of the strait of Gibraltar.

Cabotto did not sail westward until he had arrived abreast of Cuba, which lay on his left. In following this coast-line which he called Bacallaos, he says that he recognised the same maritime currents flowing to the west that the Castilians noted when they sailed in southern regions belonging to them. It is not merely probable, therefore, but becomes even necessary to conclude that between these two hitherto unknown continents there extend large openings through which the water flows from east to west. I think these waters flow all round the world in a circle, obediently to the Divine Law, and that they are not spewed forth and afterwards absorbed by some panting Demogorgon. This theory would, up to a certain point, furnish an explanation of the ebb and flow.

Soon after the discovery of the new land mass named "America", one of the most exciting and tragic adventures in the history of seafaring began: the search for the passage from the Atlantic to the Pacific. Martire D'Angera hypothesized that this passage was in Central America, but the idea of a "passage to the East Indies by the North Pole was suggested as early as the year 1527 by Robert Thorn, merchant, of Bristol" (Phipps, 1774, see also McConnell, 1982). The polar passage would have allowed England to shorten the travel time to the Spice Islands, compared to the circumnavigation of South America through the Strait of Magellan or South Africa around the Cape of Good Hope. A chronological history of travel to the Arctic regions and the polar passage between the Atlantic and Pacific Oceans was given by Barrow (1818). Ross (1835) (Fig. 1.1) provided a map showing the possible location of the north-west passage.

This chapter shows how observation technologies and methodologies are important for understanding oceanic phenomena. It provides a general overview of the consequences of the rapid evolution of knowledge and technology's impact on work practices. Our knowledge of ocean science is based on scientific debates that began centuries ago, a cultural aspect that should not be overlooked and should be included in academic courses.

Margaret Deacon (1971) in her *Scientists and the Sea* wrote: Oceanography is a descriptive and environmental science; as such it depends for its existence on the application of knowledge already gained in physical and other sciences. However, observations in the sea are very difficult and expensive, and the data collected cannot be reproduced. Technological and methodological advances were key points of progress in ocean science.

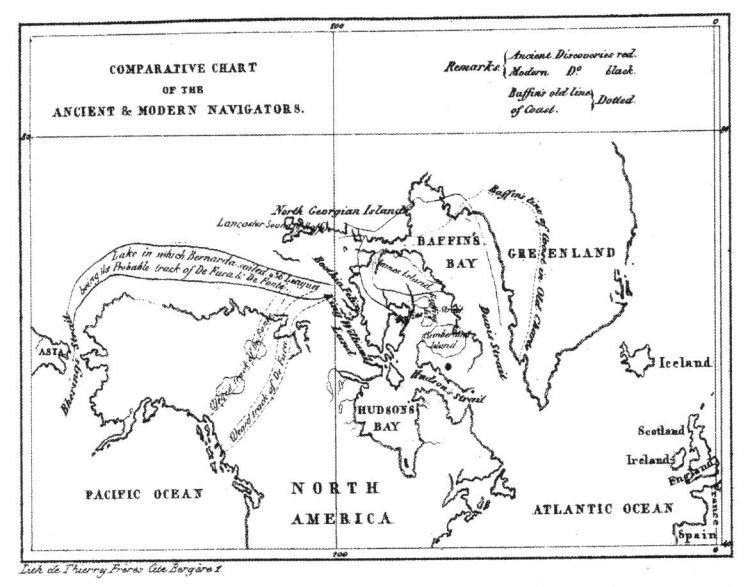

Figure 1.1 The possible location of the north-west passage from Atlantic to Pacific as in the book of John Ross (1835).

The Renaissance brought about an epochal change in human thinking that resulted in the rise of a humanistic culture and major scientific discoveries. The experimental method initiated by Galileo required a procedural systematization which began to take shape in the 17th century. Important cultural and scientific institutions were founded for the advancement of thought and to debate methodologies and technologies. Florence's Accademia del Cimento was founded in 1657 and the Royal Society of London in 1660; both were incubators of ideas on natural sciences.

Methodologies and technologies developed during the 17th to 19th centuries are presented with particular attention to their applications in the northern polar regions. These extreme areas, on account of their oceanographic and meteorological peculiarities, represent interesting case studies for the validity of those methodologies and technologies.

This chapter provides some important historical elements on data collection methods and technologies from the 17th century to the beginning of the 20th century in the science later known as oceanography. In order to evaluate and compare past and present technologies and methods, the data collected in particular areas of the Arctic Sea are presented.

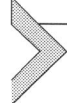

17th century: *Summum frigidum*

> Show us the sensible experience, that the ebb and flow of the sea water is not a swelling, or shrinking of the parts of it element, similar to what we see taking place in the water placed in the heat of the fire, while it for vehement heat becomes rarefied, and rises, and in reducing itself to natural Coldness it reunites, and lowers; but in the Seas there is a true local motion, and so to speak progressive, sometime towards one, sometime towards the other extreme term of the Sinus of the Sea, without any alteration of this element, coming from other accident than from Local Mutation.
>
> Galileo Galilei's speech over the ebb and flow of the sea, 1616; Acts and unpublished memoirs of the Accademia del Cimento, 1780

Speculation on the properties of the oceans during the 17th century was provided by many skilled people ("*virtuosi*"). Galileo's studies (1638) on falling bodies were the basis of many "inquiries" relating to surveys of the sea. Boyle (1627—1691) and Hooke (1635—1703) spent considerable time testing and applying the concept of "gravitation" to ocean studies.

Boyle asked navigators to explore the different aspects of the oceans: with regard to the water are to be considered the sea, its depth, specific gravity, difference of saltness in different places, the plants, insects, and fishes to be found in it, tides, with respect to the adjacent lands, currents, whirlpools, &c (Shaw, 1738). The requirements for these observations were explained in detail in the "Directions for the observations and experiments to be made by masters of ships, pilots, and other fit persons in their sea-voyages" that also contained information on the instruments that should be used routinely for the collection of geographical, atmospheric, oceanographic, and biological data.

The "Directions" were the first step in the creation of a data quality management system:

- essential information describing the sensors and platforms,
- measurement position,
- measurement units,
- processing, date and time information.

During the 17th century, scientists began to define some specific inquiries on natural phenomena (e.g., tides, currents, winds). The diverse interpretations of observations or results of "experiments" made it necessary to adopt precise experimental methodologies. The concept of standards agreed upon by the scientific community and now adopted in everyday practice did not exist then. The "best practices" were defined by one or more highly reputable people (persons of great repute), one of whom was Robert Hooke, who presented the "Method of Making Experiments" to the Royal Society (Derham, 1726). Hooke's experimental method included some specific recommendations:

- After finishing the Experiment, to discourse, argue, defend, and further explain, such Circumstances and Effects in the preceding Experiments, as may seem dubious "or difficult": and to propound what new Difficulties and Queries do occur, that require other Trials and Experiments to be made, in order to their clearing and answering: And farther, to raise such Axioms. and Propositions, as are thereby plainly demonstrated and proved.
- To register the whole Process of the Proposal, Design, Experiment, Success, or Failure: the Objections and Objectors, the Explanation and Explainers, the Proposals and Propounded of new and farther Trials; the Theories and Axioms, and their Authors; and, in a Word, the History of every Thing and Person, that is material and circumstantial in the whole Entertainment of the said Society which shall be prepared and made ready, fairly written in a bound Book, to be read at the Beginning of the Sitting of the said Society.

Sounding: *Nuntius Inanimatus, Esplorator Distantiae*

One of the major problems of the 17th century was the lack of good maps with marine topography for use in the *Art of Navigation, one of the most useful in the World* (Derham, 1726).

The sounding instrument illustrated in the "Directions" was a ball made of waterproofed light wood (e.g., maple), to which an iron or stone weight was tied. When it touched the seabed, the wooden ball came off and rose to the surface (Fig. 1.2). The depth was calculated with tables on the basis of the time taken by the ball to descend and ascend. The "Directions" provided warnings on the weights and dimensions of the different parts of the apparatus.

Hooke gave precise indications on the different components of "instruments for sounding the great depth of the sea", and highlighted two possible technological sources of errors: The first was, that "it was necessary to make

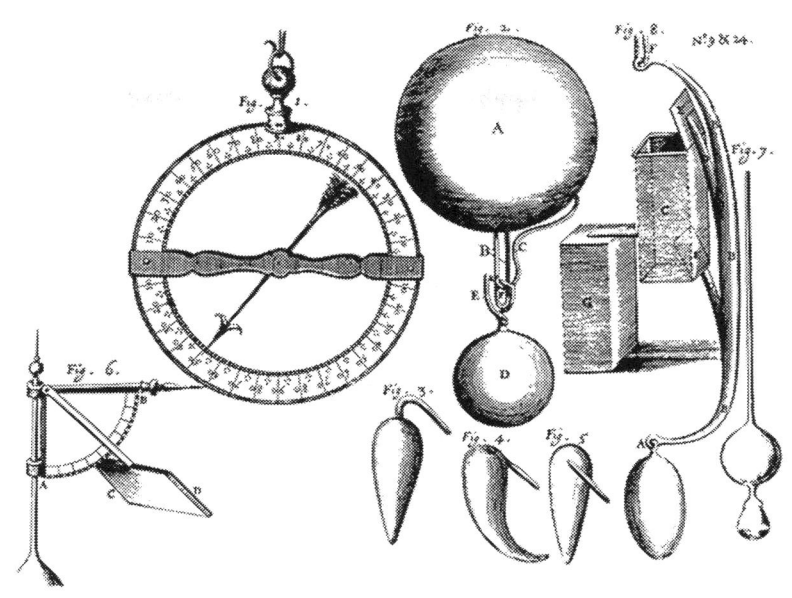

Figure 1.2 Instruments for measurements to be done during voyages, as from "Directions" by Murray and Hooke (1667): Dipping-needle (Fig. 1), Deep sea sounding without a line (Fig. 2) with different forms of weights (Fig. 3, Fig. 4, Fig. 5) substituting the ball D in Fig. 2, Instrument measuring wind strength (Fig. 6), water sampler (Fig. 7). The sounding principle was very simple: a buoyant object attached to a weight that came off in contact with the seabed.

the Weight, that was to sink the Ball, of a certain Size and Figure, so proportioned to the Ball, as that the Velocity of them, downwards, when united, should be equal to the Velocity of the Ball alone, when it ascended in its Return; in Order to which, it required to be prepared with Care, and required also some Charge, it being almost necessary to make it of Lead, of a certain Weight and Figure. The other was, the Difficulty of discovering the Ball at the first Moment of its Return, which was likewise of absolute Necessity; and it was likewise necessary to keep the Time most exactly of its Stay, or Continuance, under the Surface of the Water, by the Vibrations of a Pendulum, held in one's Hand …" (Derham, 1726).

While Hooke acknowledged the error introduced if the ball was not detected immediately upon reaching the surface, he did not realize the difficulty of doing so in anything but a totally calm sea. Many were the complaints as to the difficulties in locating the ball upon its return to the surface.

Hooke was aware of the errors associated with calculations of descent and ascent speeds and of the need to consider the buoyancy of the materials used for the various components of the sounding apparatus. On the contrary, he was confident of the use of the "pendulum clock" described in "Philosophical Experiments" (Derham, 1726). To avoid problems, he proposed a cone-shaped sounding machine (Fig. 1.3) with a small hole to receive water based on external pressure (*Nuntius Inanimatus* or *Explorator Distantiae*). In Hooke's idea, the increasing pressure of sea water at depth would fill the sounding machine in proportion to the actual depth. Therefore, by weighing the content of the water in it after it returned to the surface, it would be possible to have a measurement of the depth of the water.

Whatever the operation of this *Nuntius*, Hooke was sure that the sea temperature would influence the results, as the heat or the cold caused the air contained in the machine to expand or contract. For this reason, he thought of adding a temperature sensitive apparatus. However, there was another important question to answer before evaluating the results of a sounding apparatus, *that is*, "Whether the Gravitation, towards the Center of the Earth, do continue the same, at any Depth; or whether it do increase

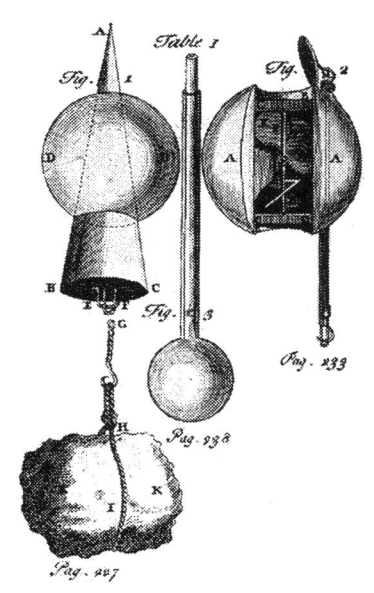

Figure 1.3 The *Nuntius Inanimatus* (on the left) and *Explorator Profunditatis* (on the right) proposed by Hooke (Derham, 1726).

or diminish, according as the Body is posited lower and lower, beneath the Surface of the Sea; for if Gravity do increase, then the Body will move downwards, or sink faster, than at the Top; and if it decreases, it will do the Contrary."

The solution was in the so-called *Explorator Profunditatis*, which consisted of a ball of a selected material with holes allowing the passage of water. The ball had pinions and cogwheels that turned during the descent and during the ascent (Fig. 1.3). The apparatus described by Murray (1912) was composed of two clockwork odometers, one for the descent and another for the ascent. The number of revolutions of the rotors gave values proportional to the depth of the sea.

Esplorator temperature

The history of temperature measurements, from the thermoscope to the thermometer, has been presented in many books (e.g., Knowles Middleton, 2003). Despite still imperfect technology and methodology, temperature measurements revealed some aspects of the marine environment which were analyzed by Boyle, a scientist whose interests ranged from human to natural, chemical and physical sciences. Boyle obtained information on temperature and salt from various sailors and divers and concluded that *sea water is not the summum frigidum*. Therefore, the sea was made up of a surface layer whose temperature was influenced by the atmosphere and a deeper and colder layer (Shaw, 1783). From this information a question arose: why was the deep sea, despite being cold, not frozen? Boyle's conclusion was, "so, I have more than once try'd that salt-water will, without freezing, admit a much greater degree of cold, that is necessary to turn fresh water into ice."

Hooke described a thermometer that was nothing "but a small Bolt-head, filled up with Spirit of Wine, to a convenient Height of the Stem, with a small Embolus and Valve; the Embolus is made so, as to be thrust down the Neck, as the Spirit of Wine shall be contracted by Cold; and the Valve is to let out the Spirit of Wine, when it is again expanded with Heat, in its Ascent". It is important to note that the effect of the pressure on the volume of the Spirit was very well known: "It may, possibly, be thought that the great Pressure, of the incumbent Body of Water, may contribute somewhat to the Contraction, or Shrinking, of the Spirit" (Derham, 1726).

Esplorator Qualitatum

The measurements of sea gravity and saltness were done with a vial of known magnitude having a narrow neck or a graduated glass-tube. The gravity was determined by the weight of the water and the saltness by the weight of substance remaining after evaporation of the water. Water at depth was sampled with a "Square Wooden Bucket" having two valves that remained open during the descent of the sampler and closed in the ascent (Fig. 1.2).

Boyle described various experiments for the calculation of the specific gravity: "We took a vial, with a long and strait neck, and having counter pois'd it, we filled it to a certain height with common conduit-water: we noted the weight of that liquor; which being poured out, the vial was filled to the same height with sea-water, taken up at the surface; and by the difference between the two weights, the sea-water appeared to be about a forty-fifth part heavier than the other."

Having compared the results of different experiments that were giving slightly different results, Boyle deduced that the seawater during the weight operation was "*rarified*" by the effect of the sun. In one experiment Boyle used "distilled rain" as reference, but there were no indications that this water was assumed to be a standard. Boyle gave some values of the gravity of sea water weight using units of measurement from the old English avoirdupois measurement system derived from the Anglo-Norman French "aveir de peis," a derivation of the Latin "habere de pensum."

Specific gravity

The scales used to weigh specific gravity took various forms and Hooke presented some to the Royal Society (Fig. 1.4). The position of the reference weights along the arms (Fig. 1.4a) would provide "the proportionate Weight of those two Bodies" (Derham, 1726). In order to obtain a greater precision, a scale with the beam "in the Form of a Cross, equilibrated upon a sharp Edge in the Center" was proposed, but it is not known if it was actually used (Fig. 1.4b).

Hooke received samples of sea surface water and fresh water, the latter for use as a reference. Unfortunately, there were no indications in the text on the location of the sea sampling location, while the reference fresh water was collected in the Thames River at Greenwich during low tide (which is very likely not completely fresh). The salt content found by Hooke was about 22 parts per 1000, a fairly good value, given the many uncertainties and factors and the use of water from the Thames as reference.

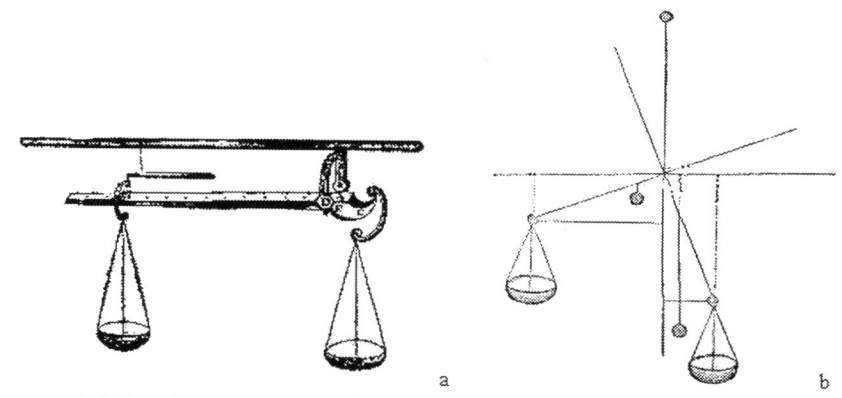

Figure 1.4 Two balances by Hooke. The one on the left is a typical steelyard balance. On the right is a balance proposed by Hooke to improve precision in weight measurements (Derham, 1726).

It can be anticipated that the value of specific gravity measured in the 18th century in Nore, a sandbank in the Thames estuary, ranged from 1000 to 1024.6 and in the North Sea 1000 to 1028.02. These values were provided in the appendix "Account of Doctor Irving's Method of Obtaining Fresh Water from the Sea by Distillation" of "A Voyage towards the North Pole" (Phipps, 1774). One of the many methodologies for the preparation of a reference water is presented in paragraph 18th century: Polar explorations – Distilled water.

18th century: Polar explorations

The usefulness of physical geography is manifest. It teaches us to know the workshop of nature in which we find ourselves, its instruments, its first laboratory, and its attempts.
Immanuel Kant, Physische Geografie (from Augusto Eckerlin edition), 1807

A letter by Stephen Hales (1677—1761) dated June 8, 1751, appeared in the *Philosophical Transaction* (Hales, 1753), which describes a "bucket sea-gage" used by Henry Ellis during his voyage to Hudson's Bay in 1746. This apparatus was used to collect temperature, salinity, and specific gravity information at various depths. The sea-gage "was a common household pail or bucket, with two heads in it; which heads had each a round hole in the middle, near four inches diameter, which were cover'd with valves which open'd upwards; and that they might both open and shut together." The water temperature was measured on board with a mercury thermometer. However, Hales advised users to be very careful since the measurement was altered by contact with air.

Important steps forward in technology and methodology are described in the book *A Voyage towards the North Pole - 1773* (Fig. 1.5), by Phipps (1774). The methodology used during that expedition was based on an intercomparison of measurements made by different people, e.g., longitude was calculated by different people making astronomical observations and timekeepers, and when all the results were reported and compared, corrections made were described in detail.

Temperature

Temperature was measured with Cavendish's overflow thermometers, which were presented to the Royal Society on June 30, 1757 (Fig. 1.6). Cavendish (1704—1783) wrote: "The instrument for finding the greatest heat might be made just like that of Fig. 1. only leaving the top open. It is to be filled with mercury only, as is also the lower part of the ball at top, but not near so high as the end of the capillary tube. The upper part of that ball, being left open, will in a great measure be filled with the seawater, which will be forced into it by the pressure ... The thermometer for finding the greatest cold, if applied to this purpose, must also be left open at top ... the most convenient construction, which occurs to me, is that of Fig. 4" (Cavendish, 1757). The thermometer was filled with mercury (the dark part of the figure) and "spirit of wine" (the gray part).

Soon after the publication of the Cavendish report by the Royal Society, it was noted that "spirits of wine" and other fluids were compressible and that, furthermore, corrections to temperature measurements were necessary. The corrections were presented in an appendix to Phipps' book. The corrected temperature data collected during the Phipps voyage to the North Pole are shown in Table 1.1 (details on Cavendish thermometers and corrections are presented in McConnell, 1982).

The quality of the data can be discussed on the basis of the temperature collected at 780 fathoms (about 1426 m) which, after correction, turned out to be $-3.3°C$, a very low value in light of current knowledge. The corrections to temperatures made by Dr. Irving, a scientific member of Phipps' crew, considered *compression and unequal expansion of spirits.*

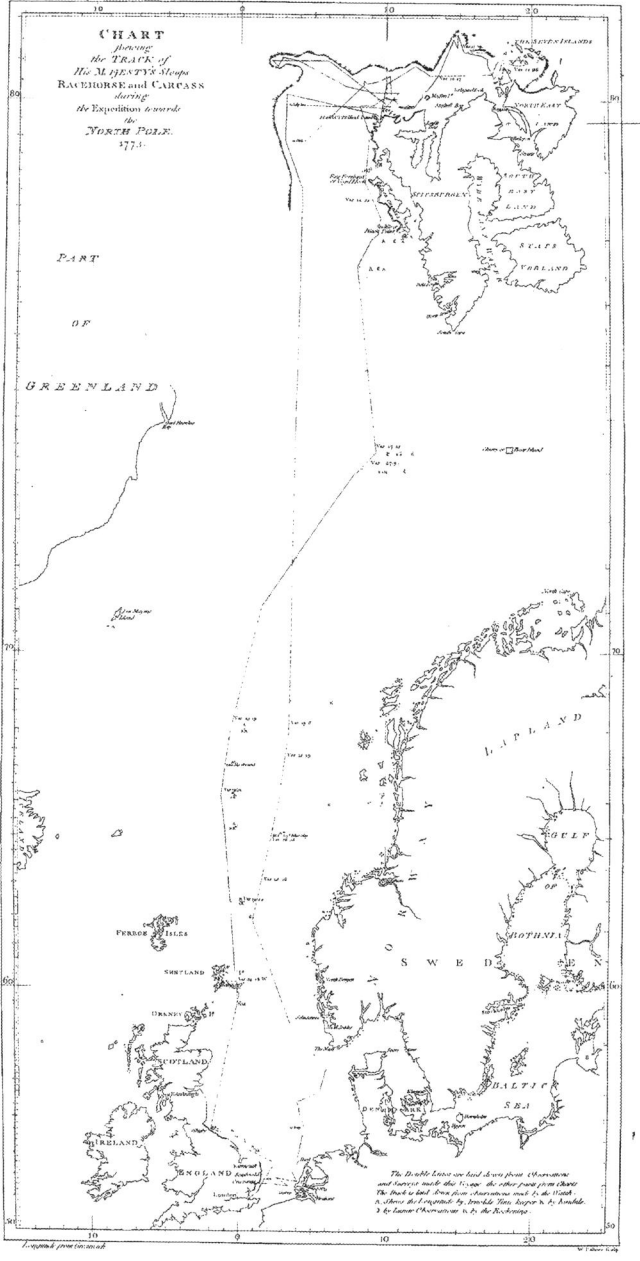

Figure 1.5 Chart of *A Voyage toward the North Pole* by Phipps (1774).

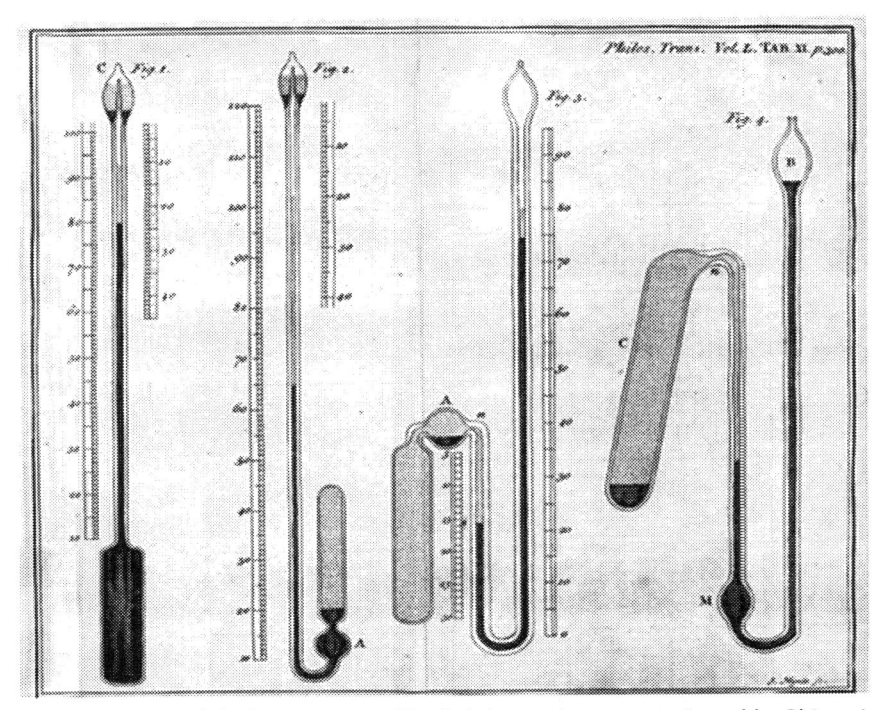

Figure 1.6 Cavendish thermometers. The "minimum thermometer" used by Phipps is presented in "Fig. 4". The gray part was "spirit of wine" and the dark part was mercury. Note the opening on top of "Fig. 4". Cavendish was aware of the effect of pressure on the apparent volume of liquids, causing a shift in reading. Following Cavendish, the pressure exerted on the top causes mercury to pass into the alcohol tank C. Initially C contained "spirit of wine." As the temperature fell, the spirit is contracted and the mercury flows into C where it is trapped. The reading of the mercury in the shorter limb would give a measure of the temperature. More details are in McConnell (1982). (Note the references to figures in the text refer to the numbers next to the thermometers in the figure.). *From Cavendish, C., 1757. A description of some thermometers for particular uses. Phil. Trans. 50, 300–310.*

However, based on some indications provided by Abbe (1888), the temperature should be corrected by more than 0.5°F, probably greater than 2°F, as shown by Fig. 1.7.

To understand the quality of these first observations of the thermal content of the sea in the polar regions, the temperature values collected by Phipps (Table 1.1) are compared with the data collected in recent years (Fig. 1.7). Data above 120 fathoms correspond to the temperature values collected at the beginning of the 21st century, but data for the deepest point are completely out of acceptable ranges. The navigation journal reported

Table 1.1 Temperature data collected during the voyage to North Pole by Phipps in 1774.

Year	Month	Day	Hour	Latitude	Longitude	Depth (fms)	°F corrected	Comments (corrections made by Phipps)
1775	6	20	12	66,9065	−0,9742	780	26	For the position provided on day 19.
	6	30	9	78,1333	9,4742	118	31	
	6	30	14	78,1333	9,4742	115	33	

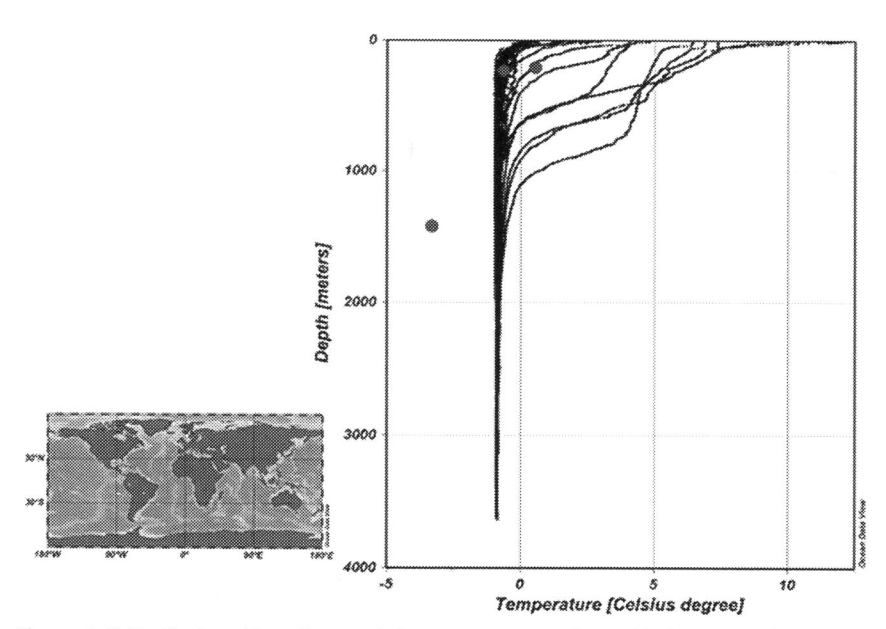

Figure 1.7 Vertical profiles of potential temperatures collected in June in polar regions during the years 2000−05 compared with data collected by Phipps in June 1774 (*blue dots: black dots* in printed version). The vertical profiles were downloaded from SeaDataNet (www.seadatanet.org). The graph has been obtained using Ocean Data View software. *Courtesy Schlitzer, Reiner, 2020. Ocean Data View, https://odv.awi.de.*

that the air temperature was 48.5°F and was calm almost all day; consequently, the low temperature at 780 fathoms was not contaminated by weather events. Any quality problems must be attributed to the measuring device.

The problems in using the thermometers are clearly presented by Camuffo (2002). Members of many societies (e.g., Accademia del Cimento in Florence, The Royal Society of London, and later Societas Meteorologica Palatina in Mannheim) stressed the need to have *a perfectly cylindrical tube … or at least a tube with a constant internal section along its entire length.* Improvements in glassmaking technology enabled scientists and technicians to confirm that the liquid inside the thermometer and the glass both expanded when the heat increased. Fahrenheit used two different liquids, mercury and spirit, to evaluate the law of expansion, obtaining different results. At the end of the 18th century, various volumetric expansions of spirit and mercury were verified, and calibration methods were suggested (Camuffo, 2002).

James Six (1731—1793) invented a maximum and minimum thermometer (Six, 1794) which began to be a commonly used tool during most voyages of exploration. The thermometer was invented in 1782, but the book that described it was published 12 years later, post-mortem. The thermometer contained mercury (the colored or gray part in Fig. 1.8) and spirit

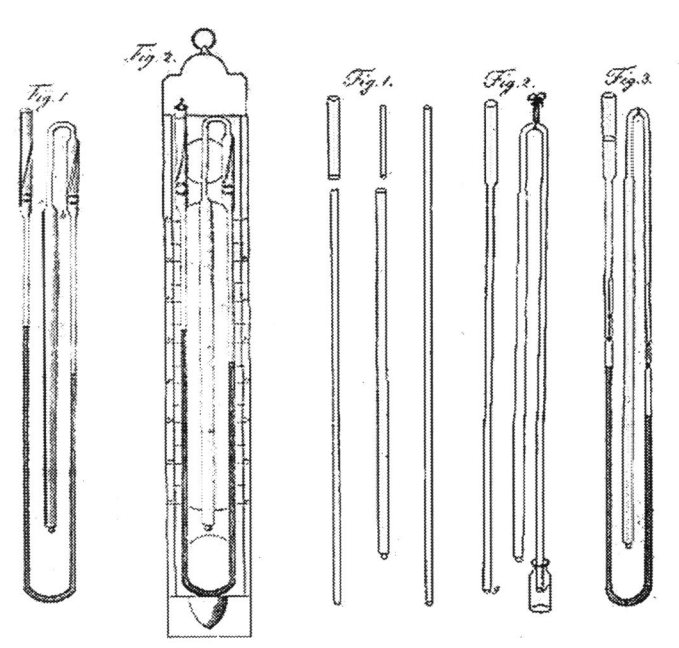

Figure 1.8 The Six's thermometer and the different parts showing the sections of the different parts of it (Six, 1794).

of wine. The expansion of the latter pushed the mercury upwards into the tube on the right. "Within the small tube of the Thermometer, above the surface of the mercury, immersed in the spirit of wine, is placed, on either side, a small index, so fitted as to be moved up and down as occasion may require" (Six, 1794). A magnet was used to restore the position of the metal needle (McConnell, 1982).

Specific gravity and salinity

Specific gravity was measured instead of gravity. A definition of specific gravity was given (among others) by Becket (1775): "that which meant by the term Specific Gravity of bodies, being nothing more than the difference, or comparative weight of those bodies to that of a common water, we might easily find the specific gravity of any fluid, by weighing a quantity of it against an equal quantity of water." In a note, the author provided additional useful information: in hydrostatic calculation, water, as the standard from which all the respective gravities are taken, is reckoned as unity or 1, 10, 100, 1000, &c. as the case requires. The reference liquid selected was distilled water, differently from Marsili, who used rain-water (Marsili, 1681). From a practical point of view, there were many advantages in using distilled water, since it could be obtained by each "weight-keeper," also on board a ship at sea for many months, as was common at that time.

In the 18th century precision balances were introduced in response to scientific as well as commercial needs. They furnished accurate measurements of the specific gravity by defining a standard temperature for the reference water.

Phipps (1744—1792) provided, among other seafarers, information on the salt contained in sea water: "Sea-water contains chiefly a neutral salt, composed of fossil alcali and marine acid (muriatic or hydrochloric acid). It likewise contains a salt which has magnesia for its basis, and the same acid … The mother liquor now remaining, being evaporated, affords a vitriolic magnesia salt, which in England is manufactured in large quantities, under the name of Epsom salt (magnesium sulphate). Besides these salts, which are objects of trade, sea-water contains a selenitic salt (calcium sulphate), a little true Glauber's salt (sodium sulphate), often a little nitre, and always a quantity of gypseous earth suspended (sulphate mineral) by means of fixed air" (Phipps 1774). The measurements of salts in sea water were obtained by dissolving them in alcohol after evaporation of the water.

Distilled water

In *A Voyage toward the North Pole* Phipps (1774) mentioned the participation of experts in various scientific and engineering disciplines, including Dr. Irving who, in an appendix, examined the different methods of obtaining distilled water on board a ship. The distiller was a boiler with openings (the two holes on the back of Fig. 1.9) for cocks. The water was evaporated and forced into tubes that decreased in size and at the end of which distilled water was collected. To clean the tube, steam was forced through for one minute. To ensure maximum purity, the water was distilled until a third of the water originally introduced remained in the boiler.

The text interestingly notes that "The principal intention of this machine, however, is to distil rum and other liquors; for which purpose it has been employed with extraordinary success, in preventing an 'empyreuma' or 'fiery' taste."

Marine zoology

In that historical period, it was normal practice to collect samples of flora and fauna in order to acquire knowledge of the new lands that were discovered. During the voyage, Phipps' crew also recorded biological observations of mammals, fishes, amphibians, insects, etc. Flora and fauna were described and depicted in tables of high artistic value. Examples can be seen in Fig. 1.10.

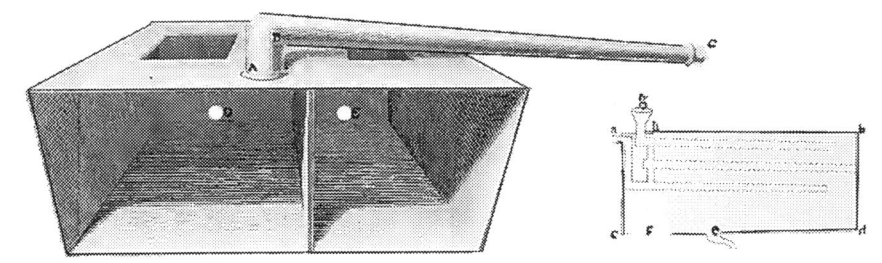

Figure 1.9 The distiller used by Dr. Irving on board the H.M.S. *Racehorse* and *Carcass* during the voyage toward North Pole in 1773.

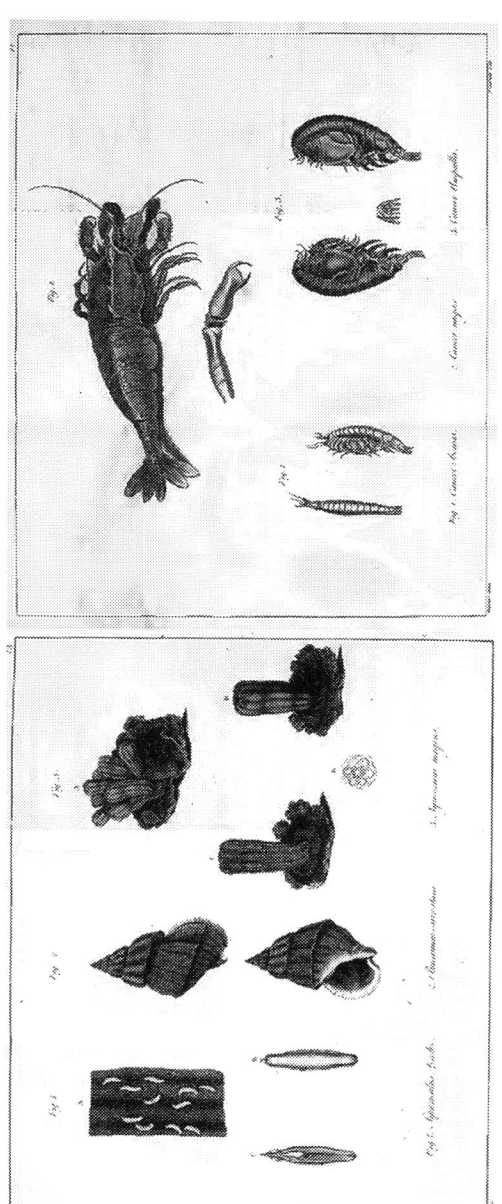

Figure 1.10 Biological observations during the voyage toward the North Pole by Phipps (1774).

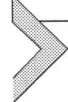

19th century: A century of changes

> The correct analysis of sea-water being a difficult problem, the usual measure of the saltness of the sea, is by its specific gravity; this, though but an approximation to the truth, when the quantity of any particular salt only is considered, gives the saline contents in the gross with tolerable accuracy.
>
> William Scoresby, *An Account of the Arctic regions with a History and Description of the Northern Whale-Fishery*, 1820

The scientific revolution that began in the 16th century saw continuous and increasingly faster advances in mathematics, physics, chemistry, and biology (Preti, 1975). At the same time, there were far-reaching changes in industry, commerce and finance, and, in particular, a surge in the development of commercial relations between Europe and overseas lands that led to the construction of vast and efficient merchant and military fleets (AAVV, 2004). In the mid-19th century, the first submarine telegraph cables were laid. This made bathymetric knowledge increasingly necessary even in the deep sea.

The whaling industry increased significantly during this period. The search for new hunting grounds for the whale fishery led to the exploration of unknown regions, as in the case of William Scoresby Junior, who was cited by Melville (1851) in *Moby Dick or The Whale* ('*No branch of Zoology is so much involved as that which is entitled Cetology,*' *says Captain Scoresby, AD 1820*).

Whaling, an ancient activity that was practiced in the Basque region in the Middle Ages and later moved to the North Atlantic, was carried out in a predatory way. The hunting grounds were depleted considerably in number of animals to the extent that Maury published a map in 1851 showing that the best hunting region was no longer the Atlantic but the Pacific Ocean (https://commons.wikimedia.org/wiki/File:Maurys_whale_chart-1851.jpg; accessed September 2020).

Maury (1806—1873) was an important figure in the history of oceanography. His work *The Physical Geography of the Sea*, dated 1855, marked the boundaries between a geographical description of the seas and oceans and modern oceanography. (Note this book is still in print.)

He promoted the First International Maritime Conference (Houvenaghel, 1990; WMO, 1973), held in Brussels in 1853, "at the invitation of the Government of the United-States of America, for the purpose of concerting a systematical and uniform plan of meteorological observation at sea" (De Groote, 1853). The delegates of Belgium, Denmark, France, Great Britain, the Netherlands, Norway, Portugal, Russia, Sweden, and the United States agreed "on a plan of uniform observation, in which all nations might be engaged in order to establish a concerted action between the meteorologist on land and the navigator at sea."

During the conference, difficulties in concerting comparable and compatible observations were discussed. These difficulties were the variety of scales in use in different countries, the equipment used for observations, and their accuracy. With regard to scales, it was decided that each country could use its own, except for temperature, for which the use of the centigrade scale was agreed, possibly together with the scales of the different countries. The establishment of a *universal system of meteorological observations* was left to future initiatives. With reference to instruments, it was noted that barometers were approximate and gave poor results. It was therefore recommended to accurately determine the errors in them.

It was also noted that the errors for thermometers had been accurately determined. Furthermore, the use of *mercurial thermometers* was recommended. However, the delegates added that the data they produced was of little value, probably referring to their use for navigation. As for wind measurements, the conference decided that the use of anemometers on board ships was a *desideratum*.

The conclusion on instrumentation was important: "In bringing to a conclusion the remarks upon instruments, the Conference considered it desirable, in order the better to establish uniformity, and to secure comparability among the observations, to suggest as a measure conducive thereto, that a set of the standard instruments used by each of the cooperating Governments, together with the instructions which might to given to such Government for their use, should be interchanged."

The conference recommended to carry out the observations reported in Table 1.2. The conference also defined sampling intervals: "at least the position of the vessel and the set of the current, the height of the barometer, the temperature of the air and water should each be determined once a day, the force and direction of the wind three times a day, and the observed variation of the needle occasionally."

Table 1.2 Measurements to be done on board of ships as agreed at the First International Maritime Conference held in Brussels from August 23 to September 8, 1853.

Column 1 Date	Column 2 Hour	Column 3 Latitude Observed	For dead reckoning	Column 4 Longitude Observed	For dead reckoning	Column 5 Currents Directions	Velocity	Column 6 Magnetic variation observed
Column 7 Magnetic variation adopted or used	Column 8 Form and direction of the clouds	Column 9 Part of the sky not obscured	Column 10 Quantity of rain	Column 11 and 12 Winds Direction	 Force	Column 13 Barometer Barometer	 Temperat.	 Reduced to the Temperat. of zero

Column 14 Thermometer for the air	Column 15 Thermometer with the wet bulb	Column 16 Temperature of the water at surface	Column 17 Temperature of the water at certain depth	Column 18 Specific gravity of the water

It was also stated that any additional information reported in logbooks would be of great value.

The Brussels conference was a beginning for international marine meteo-oceanographic cooperation, and was followed by a series of initiatives having the aim to establish a uniform system of meteorological observations (Ballot, 1872). An international coordination and standardization of climatological practices was established during the First International Meteorological Congress held in Vienna in September 1873. The congress was a starting point for the establishment of the International Meteorological Organisation (WMO, 1973), that in 1952 was reestablished as an intergovernmental body: the World Meteorological Organisation (Zillman, 2009).

Deep sea soundings

Difficulties in sounding the deep sea were clearly indicated in this statement by Hjoert (1912): "It has often been said that studying the depths of the sea is like hovering in a balloon high above an unknown land which is hidden by clouds, for it is a peculiarity of oceanic research that direct observations of the abyss are impracticable."

The exploration and study of new lands and oceans sparked an interest in maps describing the trend of the seabed (Fig. 1.11). The methodology for determining the depth of the sea was described by Thomson (1873).

Traditional methodology consisted of a weight attached to a graduated line with strips of variously colored fabric. The distance and the color of the stripes indicated fathoms, tens of fathoms, and hundreds of fathoms, or, for the deep sea, the white stripes were fixed every 50 fathoms, the black every 100 fathoms, and the red every 1000 fathoms. When the weight (*a prismatic leaden block about two feet in length and 80 to 120 lbs in weight*) touched the seabed, an approximate measure of the depth of the sea could be made.

The maximum depth measurement with this system was about 3200 fathoms, beyond which a symbol was used on the bathymetric chart that was $\frac{}{3200}$ meaning *no bottom at 3200 fathoms.*

Deep-sea sounding was done while the ship was moving. When an accurate position was required, as in the case of bathymetric measurements near the coast, the position made reference to some fixed objects on the shore.

The measurements of the depth of the sea with this method were distorted by the currents that inclined the wire, and consequently provided

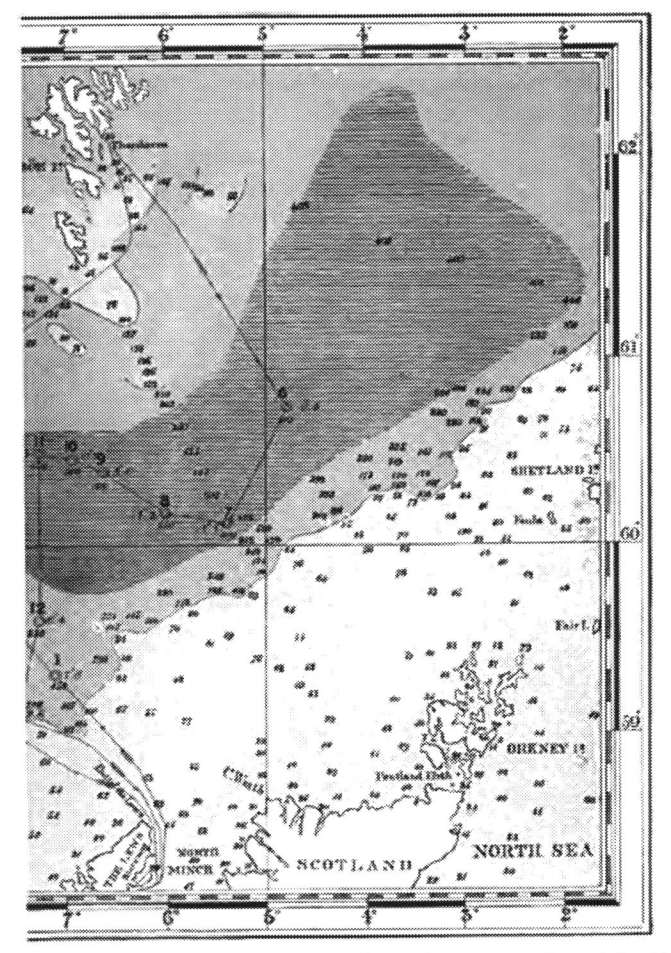

Figure 1.11 Soundings of the North Atlantic from Thomson on board the *Lightning* in 1868 (Thomson, 1873).

measurements higher than the true values. Thomson was aware of this and described another method adopted by the United States Navy. A 32- or 68-pound weight was attached to a fine line and thrown into the water. When the descent speed began to decrease significantly, the wire was cut. The depth of the sea was calculated from the length of the thread left on board the ship. Thomson reported soundings of up to 50,000 fathoms produced by US Navy officers.

During the century, the sounding machine became more sophisticated. Sir William Thomson (Lord Kelvin, 1824—1907) developed a sounding machine using a steel wire instead of a hemp line, thus reducing friction in the water and the weight of the entire apparatus (Fig. 1.12). The Thomson sounding machine took up less space than the previous models, enabled greater speed, and due to less friction, a more perpendicular line and therefore greater precision (McConnell, 1982).

Temperature in the polar regions

In a sea perpetually covered by sea ice, there was initially considerable surprise in finding a sea surface temperature of about $2 \div 3.5°C$ at a latitude of 76—78°N. Measurements made in the polar regions showed that temperature increased with depth. The son of a whaler, William Scoresby Junior (1789—1857), made a significant contribution to the knowledge of the properties of the sea and currents in his book *An account of the Arctic Regions with a History and Description of the Northern Whale-Fishery* (Scoresby, 1820).

The recurrent problem with this type of measurement was to avoid contaminating the temperature at a certain depth with the sea water temperature detected by the thermometer during its rise to the surface. Scoresby devised an apparatus consisting of a fir cask, considered a poor heat

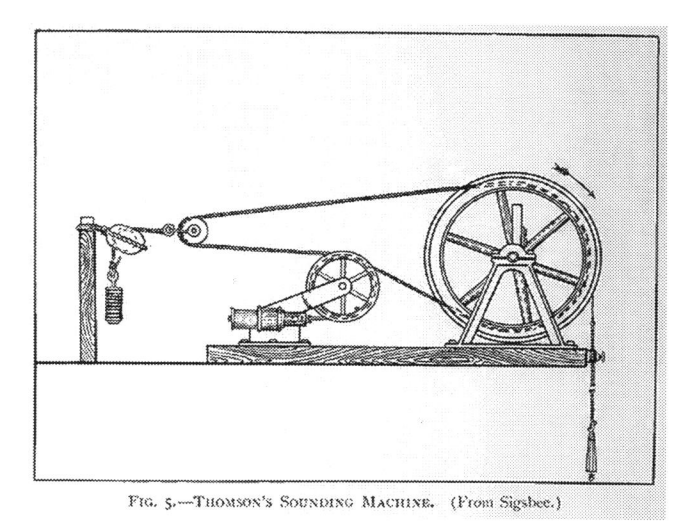

FIG. 5.—THOMSON'S SOUNDING MACHINE. (From Sigsbee.)

Figure 1.12 The Thomson (Lord Kelvin) sounding machine. *From Hjort, J., 1912. The ship and its equipment. In: Murray, J., Hjort, J. (Eds.), Chapter II in the Depths of the Ocean. MacMillan and Co., London, pp. 22—51).*

conductor, containing a thermometer and with two valves at its extremities. The cask was lowered to the desired height and left there for half an hour. During the ascent, the valves closed, maintaining the temperature value. But "… after a few experiments had been made, the wood of the cask became soaked with water; several of the staves rent from end to end; and the apparatus became leaky and useless." This was also a problem with Hooke's wooden balls that became water-logged at depth.

Subsequent experiments were not successful, despite the numerous suggestions of many marine experts including Cavendish; moreover, the use of Six thermometers proved to be problematic as they were extremely fragile.

To avoid any problems, Scoresby built an apparatus called "marine diver" (Fig. 1.13), which had the shape of an octagonal prism with a length

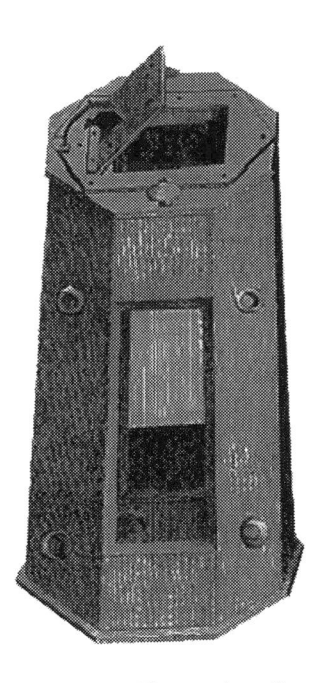

TABLE.

Figure 1.13 The marine diver constructed by Scoresby to measure temperatures at depths below the sea surface (left) and specific gravities and temperatures collected in the Arctic regions. *From Scoresby, W, 1820. An Account of the Arctic Regions with a Description of the Northern Whale-Fishery. Archibald Constable & Co. Edinburg, Hurts, Robinson & co, London.*

of approximately 35.5 cm, a diameter of approximately 13 cm at the top and 15.25 cm at the bottom. On opposite sides there were glass panes approximately 5.5 cm thick. With this apparatus and a reinforced "fir cask" Scoresby collected water samples and measured temperatures at various depths. In a certain sense, these were measurements made with protected thermometers, as the insulating material used for the "marine diver" ensured protection, but it isn't clear that they were protected from pressure.

The temperature values collected in the spring from 1810 to 1816 by Scoresby are shown in Fig. 1.14 and are compared with profiles collected in the same areas in the summer of 2006 and 2007. The differences are very significant and may be due to the annual/seasonal variability and the insulating capacity of the materials of the "marine diver" was most likely far from optimal.

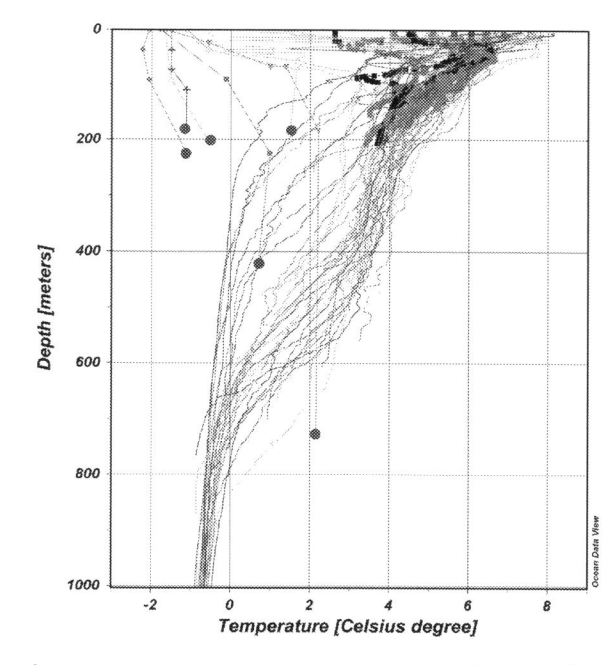
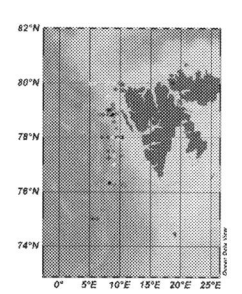

Figure 1.14 Comparison of potential temperature data collected by Scoresby (*lines with blue* [black in printed version] *big dots* at their end) with profiles collected in the same regions during summers 2006–07. These data are free and open in SeaData-Net (www.seadatanet.org). The graph has been obtained with the Ocean Data View software. The potential temperature was computed by using the thermodynamic equation of seawater (IOC, SCOR and IAPSO, 2010). *Courtesy Schlitzer, Reiner, 2020. Ocean Data View, https://odv.awi.de.*

The use of thermometers for the observation of air and water temperatures became part of data collection practice that was widely used, resulting in the recording of measurement errors. However, deep sea temperatures began to be measured accurately when thermometers protected from the effects of pressure came into use. The first to use this type of protected thermometer was probably a certain Captain Pullen in 1857 (Murray, 1912).

Detailed instructions for the measurement of sea temperatures were provided by Abbe (1838–1916). The methodology recommended for the acquisition of sea surface data consisted in a thermometer case to be lowered "to a slight depth below the surface." The case would "allow the water to flow freely through it, but shall then close and bring up from a given depth a sufficient amount of water with the enclosed thermometer, so that no change of temperature can possibly take place before the thermometer is read off" (Abbe, 1888).

Abbe was critical about the usual method "of rising a bucketful from the surface of the water and dipping the thermometer into it for a minute or less." Abbe considered that the errors associated with this latter method were greater than $0.5°F$ (about $0.3°C$).

An interesting presentation of methodologies used to measure the sea surface temperature of the sea from uninsulated buckets has been provided by Folland and Parker (1995). Wooden buckets were recommended as a type to use in instructions derived from the First International Maritime Conference. These instructions were including location of sampling the sea surface temperature, location of the bucket after withdrawal from the sea, time-lapse between withdrawal from the sea and reading the thermometer, stirring and attempts to reduce the influence of the initial temperature of the thermometer and bucket. Correction to sea surface temperature were calculated by Folland and Parker by considering the bucket geometry and physical phenomena such as sensible heat transfer between the bucket or water surface and the ambient air, heat transfer by longwave radiation, influence of solar radiation from monthly 24-hour climatological averages of incident shortwave solar intensity over the sea surface, change in bucket temperature during the time lapse and influence of inserted thermometer on heat transfer rate and water temperature. The sea surface temperature was also affected by the ship speed. The errors associated with sea surface temperature were calculated to be around $0.11–0.30°C$ for data collected at the sea surface in the northern hemisphere in the period 1850–1900. In conclusion, the Folland and

Parker calculations are more optimistic than Abbe's ones, but questions on errors associated with measurements below the sea surface remained unanswered.

Abbe described a protected thermometer for deep sea measurements by means of which "the pressure effect is wholly annulled by adopting a special protection for the bulb. The whole thermometer is placed within a strong bottle or cylinder, which is then partly filled with water or mercury, above which some air remains; the protecting cylinder is hermetically sealed, and when lowered to the ocean depths the external pressure, compressing the cylinder somewhat, causes the water to rise and compress the air slightly. This latter slight increase of pressure is the only one that affects the thermometer bulb."

A functioning maximum and minimum protected thermometer (i.e., providing maximum and minimum temperature measurements in deep water) based on Six thermometers previously adopted by most investigators was described by Negretti and Zambra (Fig. 1.15) in a report to the Royal Society. An article on the instrument was published in the *Chemical News Magazine* (Negretti and Zambra, 1874).

A kind of in situ calibration was done by many scientists who compared the temperature near the seabed with that of the mud in the same place. For example, Ross (1777—1856) "employed, a register thermometer, the indications of which were occasionally compared with the temperature of the mud and earthy fragments of various kinds which he raised from the bottom of the sea, by an appropriate instrument of his own contrivance; as this mud, both from the quantity raised, and from the manner in which it was confined, retained its temperature for a sufficient length of time not to be materially altered on reaching the surface" (Marcet, 1819).

Specific gravity and salinity

In an appendix to the book *A Voyage of Discovery*, Ross (1819a) presented an instrument, called hydraphorus (Fig. 1.16), for the collection of water samples at various depths, "during the voyage, with a view of ascertaining its specific gravity ... This Instrument consists of a copper vessel, the body of which is cylindrical. The upper part, where the machinery is fixed, is square, having on one side a small aperture to admit water. This is covered by a circular plate in which another aperture is made to coincide with the former, when placed opposite the fleur-delis; a cover is fitted to protect this plate, the edge of which being divided into 800 equal parts, the aperture on the

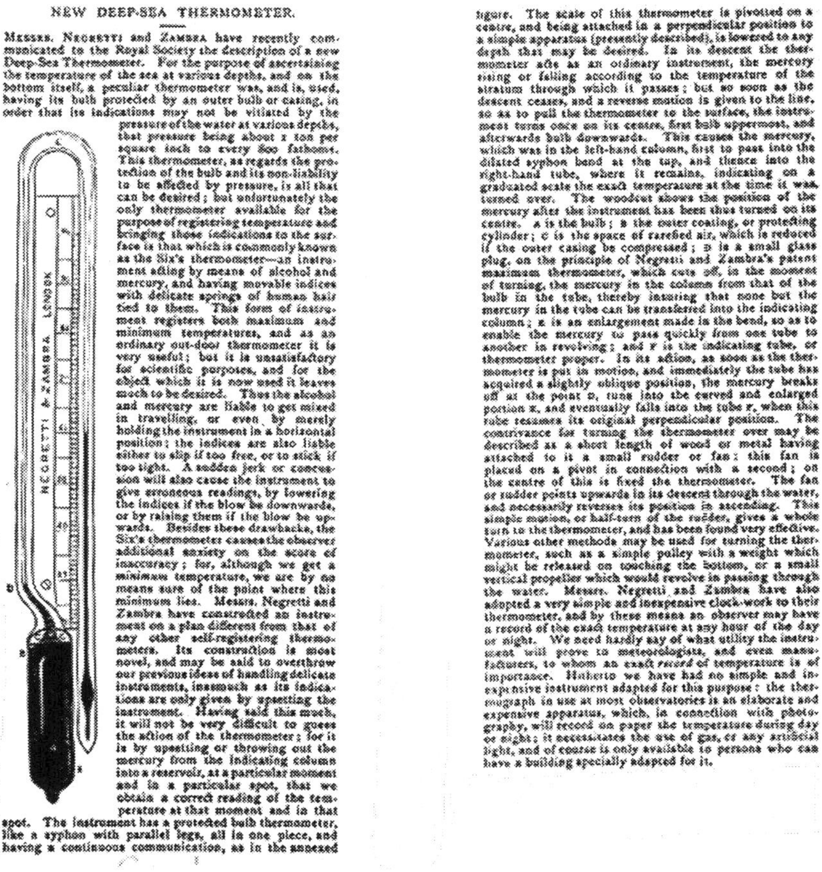

NEW DEEP-SEA THERMOMETER.

Messrs. Negretti and Zambra have recently communicated to the Royal Society the description of a new Deep-Sea Thermometer. For the purpose of ascertaining the temperature of the sea at various depths, and on the bottom itself, a peculiar thermometer was, and is, used, having its bulb protected by an outer bulb or casing, in order that its indications may not be vitiated by the pressure of the water at various depths, that pressure being about 1 ton per square inch to every 500 fathoms. This thermometer, as regards the protection of the bulb against pressure, is all that can be desired; but unfortunately the only thermometer available for the purpose of registering temperature and bringing those indications to the surface is that which is commonly known as the Six's thermometer—an instrument acting by means of alcohol and mercury, and having movable indices with delicate springs of human hair tied to them. This form of instrument registers both maximum and minimum temperatures, and as an ordinary out-door thermometer it is very useful; but it is unsatisfactory for scientific purposes, and for the object which it is now used it leaves much to be desired. Then the alcohol and mercury are liable to get mixed in travelling, or even by merely holding the instrument in a horizontal position; the indices are also liable either to slip if too free, or to stick if too tight. A sudden jerk or concussion will also cause the instrument to give erroneous readings, by lowering the indices if the blow be downwards, or by raising them if the blow be upwards. Besides these drawbacks, the Six's thermometer causes the observer additional anxiety on the score of inaccuracy; for, although we get a minimum temperature, we are by no means sure of the point where this minimum lies. Messrs. Negretti and Zambra have constructed an instrument on a plan different from that of any other self-registering thermometers. Its construction is most novel, and may be said to overthrow our previous ideas of handling delicate instruments, inasmuch as its indications are only given by upsetting the instrument. Having said this much, it will not be very difficult to guess the action of the thermometer; for it is by upsetting or throwing out the mercury from the indicating column into a reservoir, at a particular moment and in a particular spot, that we obtain a correct reading of the temperature at that moment and in that spot. The instrument has a protected bulb thermometer, like a syphon with parallel legs, all in one piece, and having a continuous communication, as in the annexed figure. The scale of this thermometer is pivoted on a centre, and being attached in a perpendicular position to a simple apparatus (presently described), is lowered to any depth that may be desired. In its descent the thermometer acts as an ordinary instrument, the mercury rising or falling according to the temperature of the stratum through which it passes; but so soon as the descent ceases, and a reverse motion is given to the line, so as to pull the thermometer to the surface, the instrument turns once on its centre, first bulb uppermost, and afterwards bulb downwards. This causes the mercury, which was in the left-hand column, first to pass into the dilated syphon bend at the top, and thence into the right-hand tube, where it remains, indicating on a graduated scale the exact temperature at the time it was turned over. The woodcut shows the position of the mercury after the instrument has been thus turned on its centre. a is the bulb; b the outer coating, or protecting cylinder; c is the space of rarefied air, which is reduced if the outer casing be compressed; D is a small glass plug, on the principle of Negretti and Zambra's patent maximum thermometer, which cuts off, in the moment of turning, the mercury in the column from that of the bulb in the tube, thereby insuring that none but the mercury in the tube can be transferred into the indicating column; E is an enlargement made in the tube so as to enable the mercury to pass quickly from one tube to another in revolving; and F is the indicating tube, or thermometer proper. In its action, as soon as the thermometer is put in motion, and immediately the tube has acquired a slightly oblique position, the mercury breaks off at the point D, runs into the curved and enlarged portion E, and eventually falls into the tube F, when this tube resumes its original perpendicular position. The contrivance for turning the thermometer over may be described as a short length of wood or metal having attached to it a small rudder or fan: this fan is placed on a pivot in connection with a second; on the centre of this is fixed the thermometer. The fan or rudder points upwards in its descent through the water, and necessarily reverses its position in ascending. This simple motion, or half-turn of the rudder, gives a whole turn to the thermometer, and has been found very effective. Various other methods may be used for turning the thermometer, such as a simple pulley with a weight which might be released on touching the bottom, or a small vertical propeller which would revolve in passing through the water. Messrs. Negretti and Zambra have also adopted a very simple and inexpensive clock-work to their thermometer, and by these means an observer may have a record of the exact temperature at any hour of the day or night. We need hardly say of what utility the instrument will prove to meteorologists, and even manufacturers, to whom an exact record of temperature is of importance. Hitherto we have had no simple and inexpensive instrument adapted for this purpose: the thermograph in use at most observatories is an elaborate and expensive apparatus, which, in connection with photography, will record on paper the temperature during day or night; it necessitates the use of gas, or any artificial light, and of course is only available to persons who can have a building specially adapted for it.

Figure 1.15 The Negretti and Zambra thermometer. It was an improvement of the Six thermometer which was *very fragile*.

outside can be set to the required position. On the opposite side of the instrument there is a similar plate or wheel, which moves the former; and both are turned by the rotator as the Instrument descends, by the action of the water, the former in a proportion as 1 is to 100. The vanes of the rotator are made to fix in any position, which by actual experiment may be found to be applicable to a graduated wheel; and it is evident, that by placing them in a more vertical or horizontal position, a greater or lesser depth may be obtained during a revolution of the graduated plate; but when it has been once regulated, to agree in a convenient proportion, to these divisions, it will not be necessary to alter the vanes, as the aperture may be easily set to the exact

DESCRIPTION.   FIG. 2.

F — Section of the machinery.
G — Upper part or rope of the instrument.
E — The instrument complete.
No. 5 — Vanes of the rotator.
    6 — Rotator with spiral wheel.
    7 — First large wheel turned by the rotator.
    8 — Small wheel on the same axis, a. No. 7.
    9 — Second large wheel turned by No. 8.
    10 — Swivel to which the rope is attached.
    11 — Spring air valve.
    12 — Aperture in the wheel coinciding with
         one in the cylinder to admit water.
    13 — The ears for attaching additional
         weights.
    14 — Stop cock.
    15 — Rope.

Figure 1.16 The hydraphorus invented by John Ross to sample waters at different depths. *From Ross, J, 1819b. A description of the deep sea clams, hydraphorus and marine artificial horizon. Strahan and Spottiswoode, London.*

depth from which the water is required. At the top of the instrument there is a spring valve, for the double purpose of allowing the air to escape when the water enters, and to let the air enter when the water is drawn off by the stop cock at the bottom, and in the latter case the valve must be moved up by hand."

Many sea water samples were not analyzed by scientists participating in the expeditions. In an article of the "Philosophical Transactions," Marcet (1819) noted this fact by expressing concerns about the loss of data and information: "In the course of few years I became possessed, through the kindness of several friends, of a great variety of specimens of sea water; and I was preparing to examine them, when a most deplorable accident deprived science of the sagacious philosopher from those friendship and enlightened assistance I had anticipated so much advantage. Procrastination and delay were the natural consequence of this misfortune; and I should probably have entirely lost sight of the subject, had not my intention been again directed to it by the late expeditions to the Arctic regions, and the great zeal and kindness of some of the officers engaged in them, in procuring for me specimens of sea water, collected in different latitudes, and under peculiar circumstances, so as to add greatly to the value of those which I previously possessed."

Another problem that was given considerable attention in Marcet's analysis concerned the conservation of sea water specimens. The water collected at depth by a "cylindrical vessel having an opening at the top, and a similar one at the bottom, each closed by a flap or valve opening only upwards, and moving freely upon hinges" was poured into a corked bottle. The sample taken at the desired depth could be contaminated by other water during the ascent of the "vessel" (container). Various improvements to the water collection devices were therefore proposed. Furthermore, the conservation of the samples in corked bottles could be imperfect and therefore cause the evaporation (or contamination) of the water they contained. Marcet remarked that in the records of the Arctic explorers there was a wealth of information about the weather, the state of the sea, and details of navigation, i.e., what we now call "metadata" (Fig. 1.17).

Alexander Marcet (1770—1822) and later John Murray (1841—1914) came to the conclusion that salinity could not be determined by the direct evaporation of sea water. The best solution was to determine salinity by precipitating the components. In 1865 Georg Forchhammer (a Danish chemist) proposed to define a salinity value proportional to hydrochloric acid or chlorine content in sea water, as this was the most abundant and easiest to determine. The reagent for precipitating chlorine was silver nitrate for use in highly acidified sea water (Wallace, 1974).

A little-known Italian chemist, Giulio Usiglio (Sobrero, 2006), carefully analyzed the results obtained by Murray and Marcet, as well as the method proposed by Forchhammer, and suggested a method to find the saline composition using nonconcentrated sea water and then apply the precipitative method (Wallace, 1974).

Friedrich Mohr (1806—1879) was the first to introduce titrimetric analysis. He added a solution of potassium chromate to the sample of sea water, which colored the solution a light yellow. After a process with silver nitrate of known concentration, the solution turned a light pink-hazel color and all the chloride was precipitated (Wallace, 2002).

Marine zoology

Marine zoology observations were made on many expeditions. John Ross (1819a) probed Baffin Bay and Davis Strait for white coral, shells, mud worms, snakes, and shrimps. Ross invented an apparatus called "Deep-Sea Clam" (Fig. 1.18) "to procure substances from the bottom of the sea in

TABLE I. *Specific Gravities of Sea Waters.*

Designation of Seas.	Nos. of Specimens.	Latitude.	Longitude.	Specif. Grav.	OBSERVATIONS.
Arctic Ocean.	1	66,50	68,30W	1025,55	Taken up by Captain Ross, in Sept. 1818, from a depth of 80 fathoms, with Sir Humphry Davy's apparatus. Temperature of the water at 80 fms. 30° ; at 200 fms. 29° ; at 400 fms. 28,5 ; at 670 fms. 25° ; at the surface 33°. Temperature of the air, 36°. Bottle labelled in Capt. Ross's own hand-writing, with all the above particulars.
	2	74,0	——	1025,46*	By Lieut. Parry, from the surface. The ship surrounded by ice in every direction. Temperature of the water 31°, of the air 34° ; 8 July, 1818.
	3	74,50	59,30	1026,19	By Lieut. Parry. Temperature of water 32°, of air 36°.
	4	75,14	4,49E	1027,27	By Lieut. Franklin, from the surface, 10 Sept. 1818.
	5	75,14	4,49	1027,27	By Lieut. Franklin, raised with the cylindrical machine, from a depth of 756 fms. Temperature of the water brought up 36°, of the air 35° ; 10 September, 1818.
	6	75,54	65,32W	1022,7 *	By Captain Ross, from the surface, 4 miles from the land ; 12 August, 1818.
	7	75,54	65,32	1025,9	By Capt. Ross ; from a depth of 80 fms. with Sir H. Davy's machine. Soundings 150 fms. 12 August, 1818.
	8	76,32	76,46	1024,05*	By Capt. Ross ; from the surface. Soundings 109 fms. 22 August, 1818.
	9	76,32	76,46	1026,22	By Capt. Ross ; from a depth of 80 fms. Temperature 30,5° ; 22 August, 1818.
	10	76,33	——	1026,64	By Lieut. Parry ; with Sir H. Davy's machine, from a depth of 80 fms. Temperature of water 32° ; 21 Aug. 1818.
	11	79,57	11,15E	1026,7	By Lieut. Franklin, from a depth of 34 fms. Temperature of the sea at the surface 30,3°, at 34 fms. 33,2° ; of the air 35,2°.
	12	80,26	10,30	1022,55*	By Lieut. Franklin, 13 July, from the surface ; ship beset with ice ; 12 leagues from the Coast of Spitzberg. Temperature of the surface, 32,5°, of air 36°.
	13	80,26	10,30	1027,14	By Lieut. Franklin ; from the bottom, depth of 237 fms.
	14	80,26	10,30	1027,15	By Lieut. Franklin : from the bottom, depth of 237 fms. with Dr. Marcet's machine. Temperature of the bottom 35,5° ; 13 July, 1818.
	15	80,28	10,20	1026,8	By Lieut. Franklin ; from bottom, depth of 185 fms. surface being frozen. Temperature of the bottom 36½°, surface 32½° ; 15 July, 1818.
	16	80,29	11,0	1026,84	By Lieut. Franklin ; from a depth of 305 fms. being the bottom. Temperature of the air 36°, of the surface of the sea 32,2° ; 16 July, 1818.

N. B. The specimens marked * in the three first tables, cannot be taken into account in calculating the mean specific gravity of the waters of the ocean, their saline contents being much diminished either by the vicinity of large masses of ice, or of great rivers, which reduce them much below the average standard of density of sea water.

Figure 1.17 A table of water samples analyzed by Marcet and coming from many different voyages in the Arctic regions.

deep water." The instrument was able to take samples of substances at any depth and preserve the mud temperature on the bottom, allowing a comparison with the water temperature near the seabed.

The "marine diver" built by Scoresby for temperature measurements was also used to collect small fishes at different depths. The upper valve was replaced by a wire gauze, so that any animal that entered the apparatus during its descent was brought back to the surface together with the sea

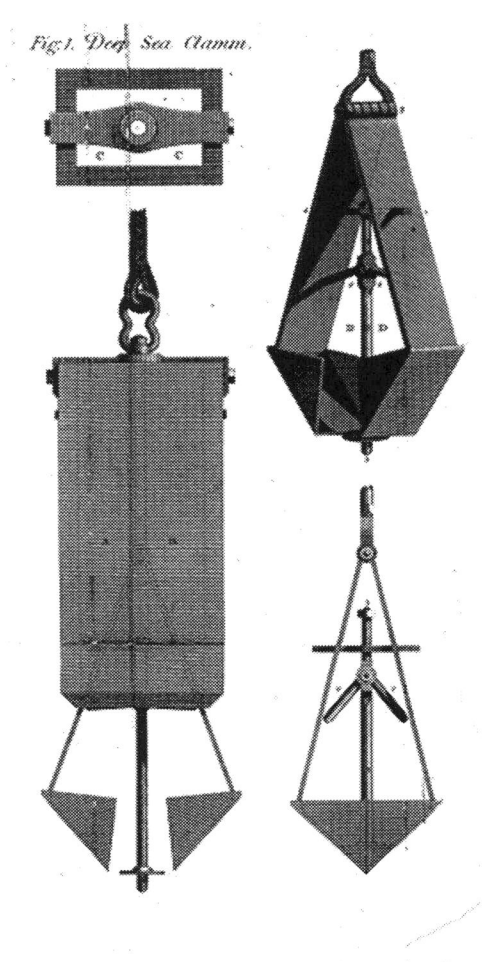

Figure 1.18 Ross' deep sea clam sounder and sampler (Ross, 1819b).

water. Examples of jellyfish and other animals observed by Scoresby are shown in Fig. 1.19. As he was the son of a whaler, he was very attentive to whale food.

"Deep-Sea Dredging Experiments" were conducted from 1868 to 1870 to explore the distribution of living things in the sea. The main questions behind the experiments were:
– the bathymetric limits of life
– laws governing geographic distribution

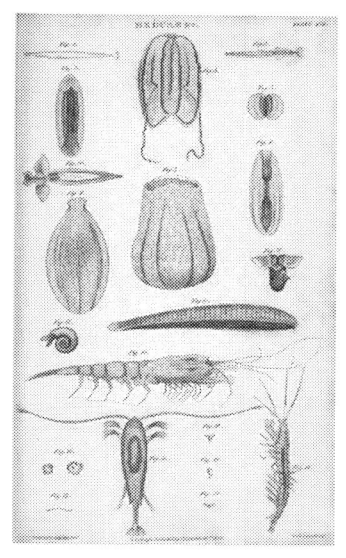

XVI. Figures of Medusæ and other animals, constituting the principal food of the whale.

1, 2. [Natural size.]

3, 4, 5, 6, 7 and 8. Medusæ, described in Vol. I. p. 548–550.

9. An orange-coloured animal, possibly of the same genus. I. p. 550.

10. Clio borealis or C. limacina. I. 544.

11, 12. Clio helicina. I. 543.

13. Cancer boreas. I. 542.

14. Squilla.

15. Beautiful little animal brought up by the marine-diver I. 545.

16, 17, 18, 19 and 20. Minute medusæ and animalcules. I. 545.

Figure 1.19 Figures of Medusae and other animals, constituting the principal food of the whale. On the right the description given by Scoresby (1820).

- representative species and zoological provinces
- influence of pressure, temperature, absence of light on life at great depths

The experiments were conducted in the Arctic regions, across the Atlantic, and in the Mediterranean. The main problems that needed to be solved to carry out the experiments lay in the possibility of dredging the bottom of the sea and of "send down water bottles and registering instruments to settle finally the question of zero animal life" (Thomson, 1873).

The first step in sampling the seabed was obviously to determine the depth of the sea. This was done with the method described in section 'Deep-sea sounding'. Thomson (1873) described many sampling tools, from the simple "Cup Lead" (Fig. 1.20a), for use in shallow water, to Brooke's sound apparatus (Fig. 1.20b), to the Bull-dog probing machine (Fig. 1.20c), a modification of John Ross' Deep-Sea Clam (McConnell, 1982).

Thomson also described other devices adapted to dredge the seabed. Some were simple "oyster dredges," or modifications of these with scrapers on both sides of the apparatus. In this way, regardless of which part touched the bottom, dredging was always possible. One of these apparatuses was

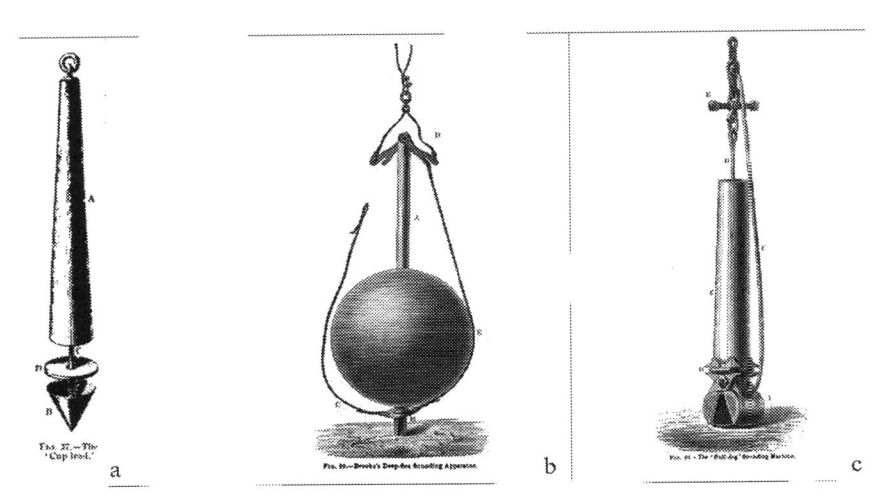

Figure 1.20 Sounding apparatus described by Thomson (1873). The Cup Lead was ending with an iron cup that was penetrating the mud or the sand at the sea bottom. As soon as the line to which the apparatus was attached was pulled up, sand or mud entered the cup and, theoretically, remained protected there by the shape of the instrument. The Brook deep sea sounding was composed by a hollow iron bar inside a shot. Touching the bottom, the tube filled with sand or mud and when it was pulled upwards it released itself from the round-shaped weight. The modified Deep-Sea Clam functioning like marine buckets used today.

introduced in 1838 by Robert Ball, a naturalist whose dredge (Fig. 1.21) was described in a section of the *Zoologist: A Popular Miscellany of Natural History* (Hepburn, 1847).

The other apparatus shown in Fig. 1.21 was designed to dredge the deep sea. The ship's engine was used to descend and retrieve the barge. An "accumulator," consisting of a weight anchored to the ship, supported the sampling apparatus and guaranteed the amortization of the impacts that could occur when the apparatus hit the seabed.

Surface currents

William Scoresby Junior was a true marine science enthusiast, and he was also the author of a description of surface currents derived from his own observations and from the drift of his ship or of shipwrecks. In his book, he also reported that a sinking area of polar waters (the black circle in Fig. 1.22) lay between Norway and Greenland.

Maps of surface currents were produced by many authors on the basis of information on the drift of bottles thrown into the sea by sailors. The bottles

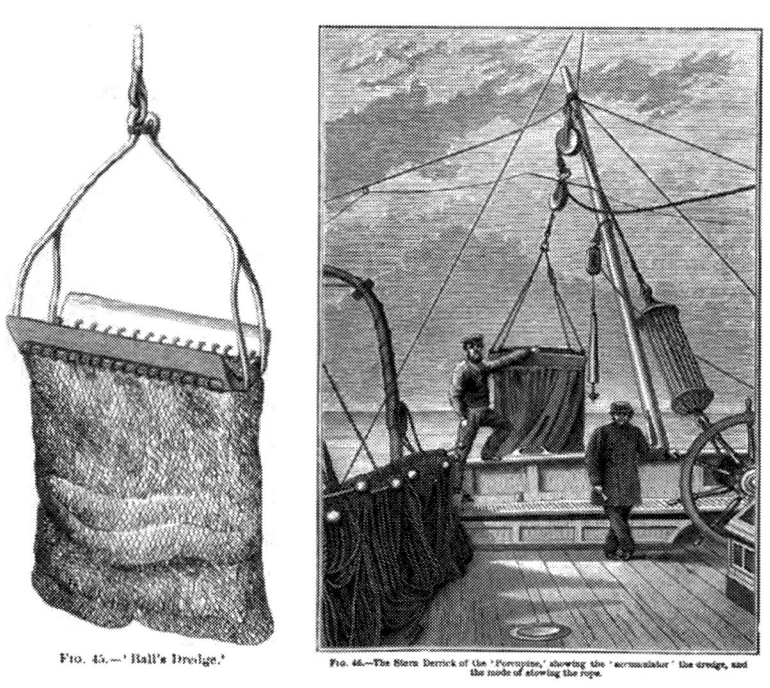

Fig. 45.—'Ball's Dredge.'

Fig. 46.—The Stern Derrick of the 'Porcupine,' showing the 'accumulator' the dredge, and the mode of stowing the rope.

Figure 1.21 Some dredging systems described by Thomson. The Ball's dredge (on the left) was a modification of the usually used oyster dredge. The deep sea dredging apparatus (on the right).

contained a document with the date and location of the release point. When collected by other sailors, the finding date and location were reported in logbooks. One of the maps derived from this measurement methodology was made by Maury (1855) and is shown in Fig. 1.22.

Around 1840, Matthew Fontaine Maury, a United States Navy lieutenant, conceived the idea of collecting information from logbooks to map winds and ocean currents with the aim of reducing ship travel time. He came upon this idea while he was serving as the director of the Depot of Charts and Instruments where he was assigned after a carriage accident that broke his legs and made it impossible for him to go to sea. This agency collected all of the ship logs and Maury realized that there was a lot of information there on winds and currents that he could share with the clipper fleet sailing from the east coast to San Francisco. Initially his results were shunned but when one clipper ship captain used them and cut 30 days off his transit, they became very popular. The popularity of Maury's analysis was such that

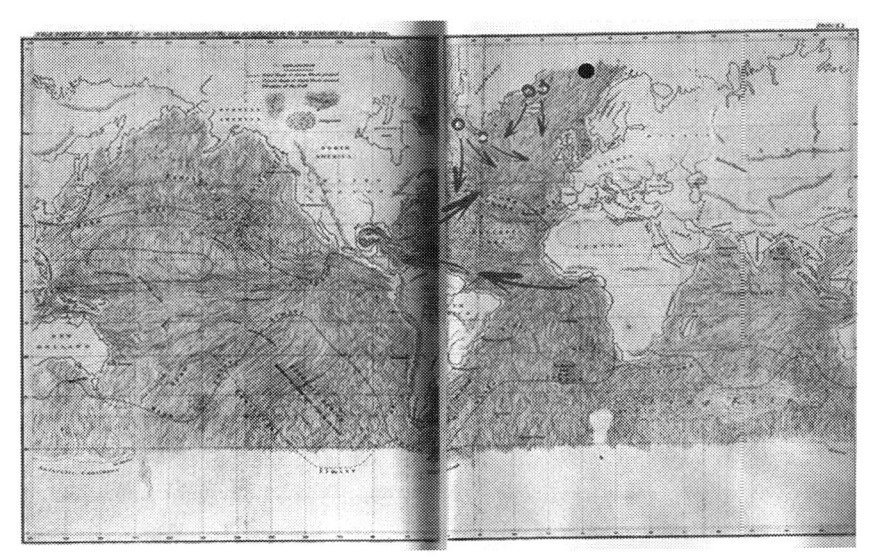

Figure 1.22 Map of surface currents derived from the indications in the book *An Account of the Arctic Regions with a Description of the Northern Whale-Fishery* by Scoresby (1820). The underlying map is by Maury (1855). The arrows indicate the currents as derived from descriptions in Scoresby's book or from indication of movements of wrecks (indicated by ships inside blue [dark gray in printed version] *circles*). The black *circle* indicates the area of waters' sinking. The Maury's map shows surface drifts and areas of whale catches.

he was later invited to visit Europe to share his methods. One of the best-known results is the book published in 1855, *The Physical Geography of the Sea and Its Meteorology*, in which Maury recorded observations during many voyages. The book was a popular success, but it also contained many errors in the interpretation of the observations. For example, Maury linked the Gulf Stream to the trade winds, an idea that was contested by Herschel (1867), who attributed the current to *an impulse that acts horizontally on surface waters* within a closed area. Herschel explained the concept with the analogy of billiard balls that slide on a plane following an impulse induced by the blow of the cue. Maury also wrongly attributed the salinity of the ocean to an undiscovered "salt fountain" somewhere in the ocean not realizing how the salts leached from the land and accumulated in the ocean after they were deposited by the rivers.

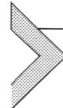 **Farthest north**

> And you are to understand, that although the finding a passage from the Atlantic to the Pacific is the main object of this Expedition, and that the ascertaining the Northern boundary of the American Continent is the next, yet that the different observations you may be enabled to make, with regard to the magnetic influence, as well as such other observations as you may have opportunities of making in Natural History, Geography, &c. in parts of the globe so little known, must prove most valuable and interesting to science.
>
> William Edward Parry, *Journal of a Second Voyage for the Discovery of a North-West Passage from the Atlantic to the Pacific*, 1824.

Farthest North is the book, written by Fridtjof Nansen (1861—1930), that describes the audacious voyage toward the North Pole from 1894 to 1896 (Nansen, 2020). On board of the topsail schooner *Fram* ("Forward" in Norwegian), the 13 crew members collected data on weather, drift currents, bathymetry. This ship was a unique attack on the rush to the pole. It was made out of green wood and had a rounded bottom so that when frozen into the ice the ship would rise up and not be crushed by the ice. The rudder could be raised into the ship and the propeller was shrouded for protection. The expedition was an attempt to reach the geographical North Pole by means of the east–west Transpolar Drift Stream (e.g., Damm et al., 2018). Nansen had surmised from evidence of wood from ship wrecks found on the coast of Greenland that the ice was flowing from east to west with a slight northward component. He planned to sail north–eastward as far as possible and then use a dog sled to reach the North Pole. He took along a number of Siberian huskies for this purpose. When it became evident that *Fram* was as far north as possible, Nansen departed the ship together with Hjalmar Johansen. The ice conditions were terrible with ridges and rafting, and the two explorers never did reach the North Pole and they had to return south to Franz Josef Land where they survived until a British Expedition found them and returned them to Oslo just a few days before the *Fram* and its crew also returned home to Oslo. The expedition was one of the most fascinating human adventures and was regarded as a *pioneer undertaking* (Nansen, 1900). The *Fram* was later used by Amundsen and his party in their successful trip to be the first to the South Pole.

During the expedition, Petterson insulated bottles were used to sample Arctic waters at various depths (https://folk.uib.no/ngfso/The_Norwegian_Sea/TNS-0330.htm; accessed August 2020). The complete set of measurements done during the *Fram* expedition to North Pole is available in the National Oceanic and Atmospheric Administration NOAA—National Snow and Ice Center NSDIC portal (https://nsidc.org/data/g02120; accessed August 2020). A representative temperature profile in the Arctic region has been provided by Nansen in his book *Farthest North* and is shown in Fig. 1.23.

The Nansen temperature data was collected during 2 days with a thermometer attached to a Petterson water bottle. The profile obtained with the data is compared with two temperature profiles collected with XCTD (eXpendable Conductivity—Temperature Depth) probes in September 1997 and available in the US National Oceanographic Data Center—World Data Center (NODC-WDC). Nansen modified the Petterson water sampling bottle to create the reversing Nansen bottle which became a standard in collecting routine hydrographic measurements.

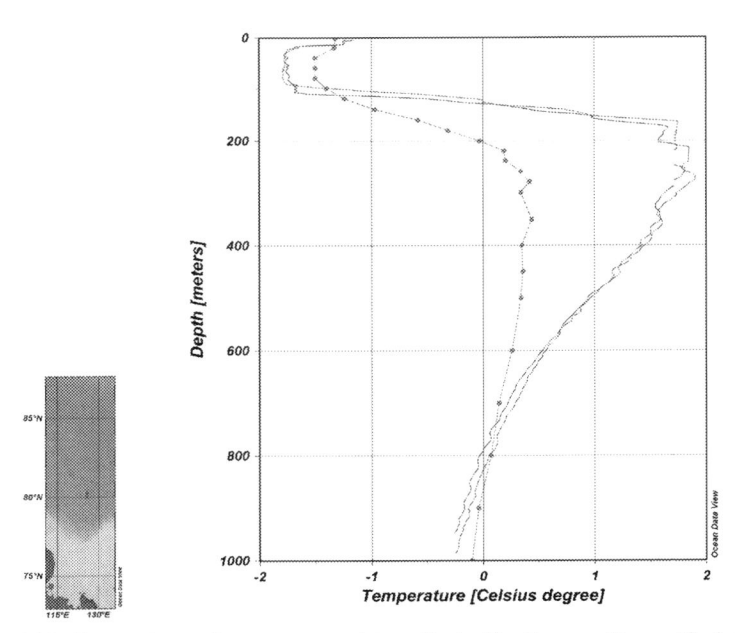

Figure 1.23 Comparison of temperature data collected by Nansen (*lines* with *dots*) with profiles collected in the same regions during September 1997. These data are free and open in NODC-WDC (https://www.nodc.noaa.gov/access/index.html). The graph has been obtained with the Ocean Data View software. *Courtesy Schlitzer, Reiner, 2020. Ocean Data View, https://odv.awi.de.*

Finally, there is a relatively good agreement between data collected at the end of the 19th century and those collected about 100 years later.

From physical geography of the sea to oceanography

> Nature presents itself to meditative contemplation as a unity in diversity.
> Alexander von Humboldt, *COSMOS: A Sketch of the Physical Description of the Universe*, 1844

Two apparently contrasting but complementary aspects coexisted in studies of the seas and oceans. Observations of currents and winds on the sea surface were obviously useful for navigation and, therefore, a geographical description facilitated "the art of navigation." At the same time, ideas and theories were presented on the relationships between winds and currents, currents and temperatures, and the consequences of density stratification.

These two aspects in the study of the sea were noted by Margaret Deacon (1971), who compared the work of Edmund Halley, a scientist, and the actions of William Dampier, a pirate, navigator, and explorer who wrote about the sea from a purely empirical and utilitarian point of view.

Papers on the physical geography of the sea (which included maps of winds, currents, and tides) were published during the 19th century (Murray, 1912; Deacon 1971). The meaning of the term "physical geography," a branch of "descriptive geography," was well explained, among others, by Herschel (1867) who wrote that it is "the business of a perfect 'descriptive geography' to exhibit a true and faithful picture, a sort of daguerreotype, without note or comment. Such comment, or at least one of such comments, it is the object of Physical Geography to supply." He added that any comment should be based on knowledge of the general laws of physics.

In his general considerations about winds and currents, Herschel specified the role of currents in the dispersion of material of terrestrial origin: "From meteorology we learn to refer the great system of aquatic circulation, which transfers the waters of every one region of the ocean, in the course of time, to every other, to the action of our trade-winds, and their compensating currents, the anti-trades; themselves the results of solar action in combination with the earth's rotation on its axis. By the oceanic currents thus arising, the material carried down by rivers, or abraded by the action of the waves (increased in their efficiency by the extent of sloping beach produced by the rise and fall of the tides), is carried off and dispersed abroad, or, it may be, collected by subsidence in deep and comparatively motionless hollows, or in eddy-pools."

Alexander von Humboldt (1769—1859) had a wider concept of physical geography: "Physical geography is not limited to elementary inorganic terrestrial life, but, elevated to a higher point of view, it embraces the sphere of organic life, and the numerous gradations of its typical development. Animal and vegetable life. General diffusion of life in the sea and on the land; microscopic vital forms discovered in the polar ice no less than in the depths of the ocean within the tropics" (Humboldt, 2005).

In the second half of the 19th century, scientists turned their attention to the physical structure of the world and the interrelationships between different natural phenomena. Alexander von Humboldt's ambition was to explain the entire physical world in his work *Cosmos: Essay for a Physical Description of the World* (Humboldt, 2005).

This "new geography," which appeared at the end of the century, placed cartography and discoveries in a more specialized and different context with respect to the new science: oceanography.

Helland-Hansen (1877—1957), in Chapter 5 "Physical Oceanography" of the book *The Depths of the Ocean*, clarified the temporal origin of the term *oceanography*: "In the middle of last century the idea of "physical oceanography"' did not exist, but in the course of a few decades it has become a widespread branch of knowledge, with a copious literature and bulky textbooks" (Helland-Hansen,1912).

The term "oceanography" appears to have been coined by John Murray (Rehbock and Benson, 2002), and in the 19th century *the development of the modern science of oceanography* began (Murray, 1912). The various passages were highlighted in a brief historical excursus by this scientist, who summarized them in three main elements (McConnell, 1982):

- scientific observations that have given a great boost to the study of marine biology;
- surveys in marine laboratories and international initiatives on the biological aspects of marine sciences and fisheries;
- physical exploration of the deep seas and their bathymetries, also in connection with the need to lay submarine telegraph cables.

In the second half of the 19th century, important institutes were established in Europe and in the United States for the study of ocean life (see Chapter 2: Data services in ocean science with a focus on the biology). An International Commission for Scientific Investigations of the North Sea was established at the end of the century with the participation of Great Britain, Germany, Holland, Belgium, Russia, Denmark, Sweden, and Norway. It should be noted that at that time the fields of marine biology, marine geology and marine chemistry were fairly well established leaving the term of oceanography to apply primarily to physical oceanography. This was basically the status in the early 20th century. It is important to realize that the emphasis in oceanography was the collection of new data from dedicated research ships and their analysis. The first group to involve dynamical analyses was the Geophysical Institute at the University of Bergen, Norway, which was founded in 1917.

The birth of modern oceanography

> Proud of our splendid, strong ship, we stand on her deck watching the ice come hurtling against her sides, being crushed and broken there and having to go down below her, while new ice-mass tumble upon her out of the dark, to meet the same fate.
>
> Fridtjof Nansen, *Farthest North*, 1897

Advances in oceanography were achieved with new instruments and increasingly diligent fieldwork, and also due to the development of mechanical tools. Steam winches began to be used from the end of the nineteenth century and, in calm weather, it was possible to operate different apparatuses simultaneously, as in the case of the Norwegian steamship "*Michael Sars*". Hjort (1912) provided an enlightening sketch of the "steamer" that had the ability to carry out physical investigations, as well as fishing and biology experiments: "On the starboard side there are two small winches, the forward one of 3 horsepower and the aft one of 1 horsepower. The forward winch (2), by means of a long axle drives a big reel with 6000 metres of wire, 3.5 mm. in diameter, for the hydrographical instruments and the Lucas sounding machine (6 and 5), and it can also be used to drive the big centrifuge (10) by means of a hemp line. By a similar arrangement the aft winch drives two drums with 2000 metres of wire, 3 and 4 mm. in diameter, for the vertical nets and hydrographical work in moderate depths" (Fig. 1.24).

Pelagic fish were caught with trawl nets; otter trawls were used to catch fish on the seabed, as well as plankton nets and other net systems that eventually made it possible to move from marine zoology to marine biology studies. Deep sea surveys made it possible to produce maps of physical properties and deposits of organic and nonorganic origin on the seabed.

New dredging apparatuses were built to take samples of the seabed containing animals to be classified, measured, counted, weighed, and subjected to chemical analysis. To obtain a good measurement of the depth of any instrument, the hemp lines marked at regular intervals were substituted by meter-wheels, communicating "with a clock-work arrangement with dials and hands, by means of which the length of wire run out can always be read off correct to within a metre" (Helland-Hansen, 1912).

At the beginning of the 20th century, scientists built several new tools. The Petterson-Nansen bottles had a very high insulating capacity and were equipped with a closure system activated with a messenger invented by Meyer, a scientist from Kiel, Germany, and with inversion thermometers (Fig. 1.25). Nansen observed that when "a water-sample is drawn up in an insulating water-bottle from a depth of 1000 metres, the temperature of the

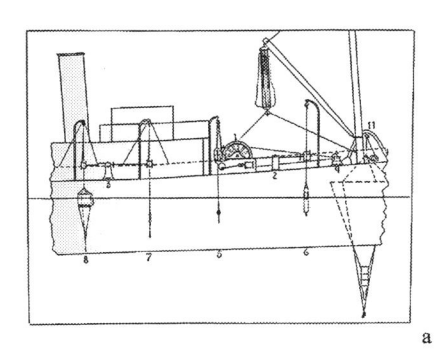 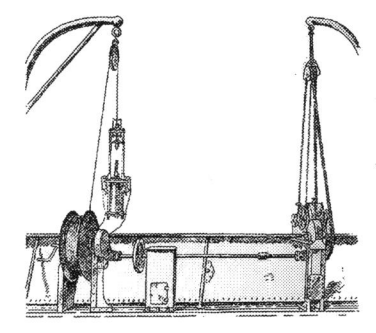

a   b

Figure 1.24 A sketch of the different apparatus on board the Norwegian steamer *Michael Sars*. Explanation in the text. In 1.24b on the left a Petterson-Nansen water bottle is represented, while on the right there is a sounding machine. A sketch of the different apparatus on board the Norwegian steamer 'Michael Sars'. Explanation in the text. In 1.24b on the left a Petterson — Nansen water bottle is represented, while on the right there was a sounding machine.

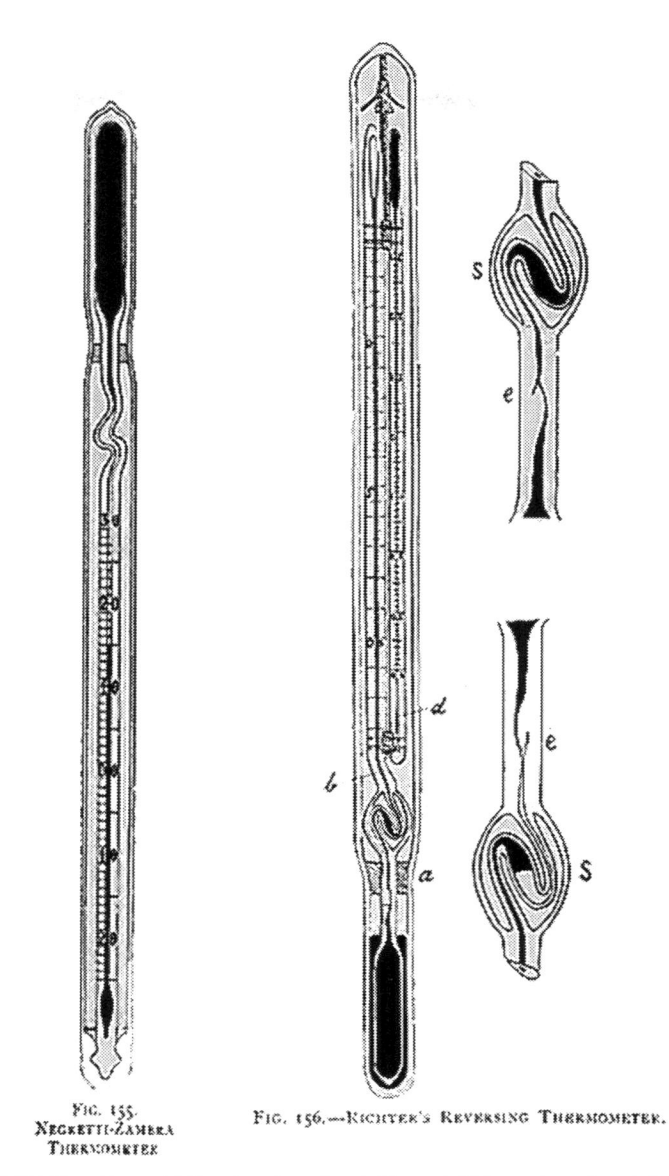

FIG. 155.
NEGRETTI-ZAMBRA
THERMOMETER

FIG. 156.—RICHTER'S REVERSING THERMOMETER.

Figure 1.25 The Negretti-Zambra (on the left) and Richter (on the right) reversing thermometers. Note both of these thermometers are completely enclosed in glass, making them protected from the ocean's pressure. *From Helland-Hansen, B., 1912. Physical oceanography. In: Murray, J., Hjort J. (Eds), Chapter V in the Depths of the Ocean. MacMillan and Co., London, pp. 210—306.*

water sample sinks a little." He also discovered that this temperature difference varies in different seas, for example, it was 0.06°C in the Norwegian Sea and 0.17°C in the Mediterranean.

The Negretti–Zambra reversing thermometers, widely used by early 20th century oceanographers, occasionally malfunctioned: sometimes the "mercury broke off not exactly at the narrowing" or there was an overflow of mercury "during the process of hauling up." An improvement was due to "C. Richter of Berlin, who altered the breaking-off arrangement so as to render it quite reliable, and formed the tube in such a way that no superfluous mercury could enter it during the ascent" (Helland-Hansen, 1912 — Fig. 1.25).

During the *Michael Sars* expedition in the North Atlantic, Helland-Hansen highlighted the difference between "the temperature (measured in situ), and the temperature that the water would acquire on account of the reduction of pressure if it were raised to the surface. The latter temperature has by the author of the present chapter been called the potential temperature, a term used in meteorology." Lord Kelvin studied the problems related to sea water pressure on volumes and temperatures and provided a formula by which changes in temperature were calculated. In order to determine the precision, both Negretti–Zambra and Richter reversing thermometers were used in about 600 *double determinations*, and differences of about 0.001°C were found, when using the Lord Kelvin corrections.

It is important to note that in the Richter thermometer there are essentially two thermometers with one having the "pig-tail" and one being a straight thermometer. The straight thermometer is known as the auxiliary thermometer and is used to correct the reversing thermometer reading for the warming of the thermometer that takes place at the surface. The procedure for reading reversing thermometers is that two people read each thermometer twice and the readings are averaged together for the final value. This reduces the errors introduced by humans reading thermometers. Realizing that the temperature would be affected by the pressure, unprotected versions of the Richter thermometers were introduced that were open at the top to allow the ocean to exert its pressure upon the thermometer. As a result, by properly processing the protected and unprotected thermometer readings it was possible to compute a very accurate estimate of the depth at which the sample had been collected.

Figure 1.26 Ekman current meter. *Courtesy Enrico Muzi, The Historical Oceanography Society.*

Ekman, a Swedish student of Nansen's at the Bergen Geophysical Institute, built a current meter (Svensson, 2002), which was used for years by physical oceanographic communities (Fig. 1.26). It consisted of a propeller oriented by a vane and a mechanism that recorded the number of revolutions, a compass, and a recorder that provided a statistical indication of the current directions. Inside the current meter, metal balls in a reservoir fell, one at a time, onto sectors of the compass. The number of balls in each sector provided an indication of the direction of the main currents. For the German *Meteor* expedition (1925–27) Ekman built a repeating current meter that could be used when the *Meteor* was anchored in the middle of the Atlantic Ocean. Alfred Merz also built a modified version of Ekman's current meter.

An improved color titration method allowed the calculation of salinity in less than five minutes with an accuracy of one centigram of salt per kilogram of sea water (Fig. 1.27; that is a fairly low accuracy when compared with modern electronic methods). Precise instructions for the determination of salinity were given in the 'Bulletin de l'Institut Océanographique: Instructions pratiques sur la détermination de la salinité de l'eau de mer par la méthode de titrage de Mohr-Knudsen' (Thomsen, 1954). The density of the water was calculated by means of "Knudsen tables" after measuring the temperature and determining the salinity.

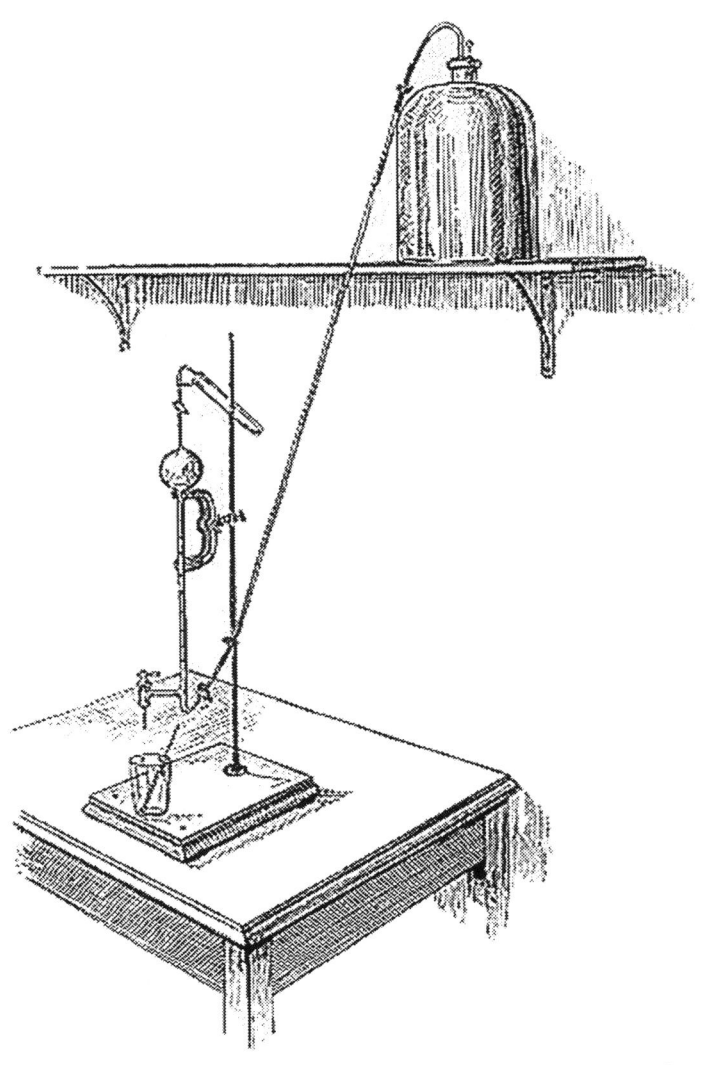

Figure 1.27 The titration apparatus for the determination of salinity. The reservoir above the burette was containing silver nitrate. The water sample was within the sphere of the burette. *From Hales, S., 1753. A letter to the president, from Stephen Hales. Phil. Trans. XVLII. For the years 1751 and 1752.*

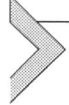

Crossing the north-west passage

From the days when the Phoenician sailors groped along the coasts of the Mediterranean, in the early dawn of civilisation, up to the present time, explorers have ever forged their way across unknown seas and through dark forests—sometimes slowly, and with centuries of intermission, at other times with giant strides, as when the discovery of America and the great voyages round the world dispersed clouds of ignorance and prejudice even in reference to the globe itself.

Roald Amundsen, *The North West Passage*, 1908

The age-old dream was fulfilled by Roald Amundsen (1872—1928), who gave an idealistic aura to the quest for the passage to the north-west: "It is in the service of science that these numerous and incessant assaults have been made upon what is perhaps the most formidable obstacle ever encountered by the inquisitive human spirit, that barrier of millennial, if not primaeval ice which, in a wide and compact wall, enshrouds the mysteries of the North Pole" (Amundsen, 1908).

In the spring of 1903, Amundsen began preparations for the journey through the north-west passage. He chose a 45-meter fishing vessel (the Gjöa), embarking enough food and equipment so that the seven crew members could survive for 5 years. They set sail on June 16, 1903, from Christiania (Oslo) and, after 25 days, the explorers arrived near Cape Farewell, Greenland. During the night of July 25, the Gjöa was anchored in Godhaven, in the western part of Greenland.

The journey continued north and on August 8 the Gjöa entered Melville Bay, where the journey was slowed by icebergs and ice. On August 20, the seven men of the Gjöa reached Lancaster Sound and 2 days later Beechey Island, where the tragic fate of Franklin's lost Polar expedition aboard two ships, H.M.S. *Erebus* and H.M.S. *Terror*, had begun years earlier. This tragedy has inspired writers (e.g., Cussler and Cussler, 2009) and directors (e.g., *The Terror* [TV series]) and solicited many archaeological researches (e.g., Hickey et al., 1993).

On September 14, Amundsen and his men began preparing their first winter quarters in Gjöahavn (Fig. 1.28) in the southern part of King William Island, where they made a fortuitous encounter with the indigenous Eskimo

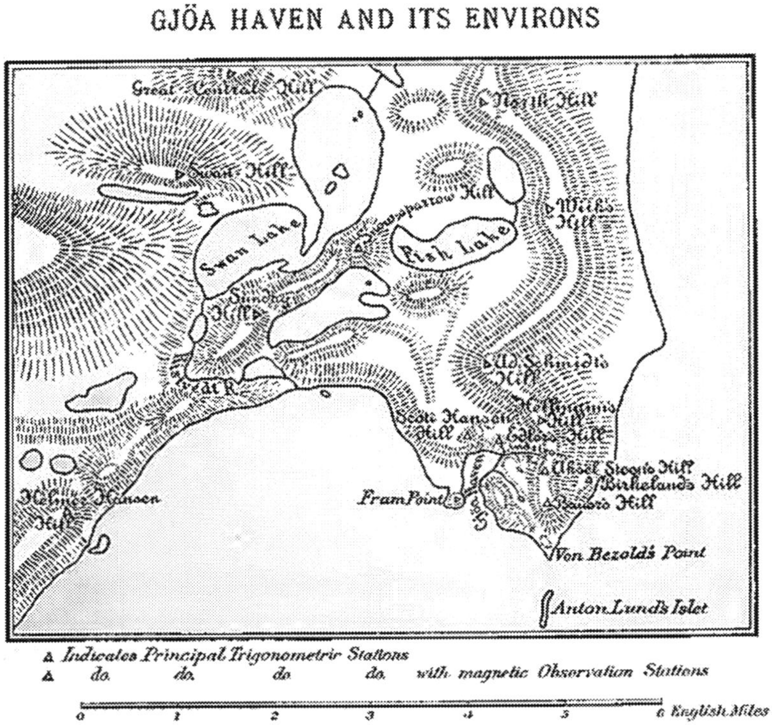

GJÖA HAVEN AND ITS ENVIRONS

Figure 1.28 The position of Gjoahavn winter quarter south of King Williams Island. *From Amundsen, R., 1908. The North West Passage. E.P. Dutton Company, New York.*

people from whom the seven men learned how to survive in an Arctic environment with temperatures as low as $-76°F$ ($-60°C$). The Norwegian team also learned how to use a dog sleigh pulled by sled dogs for transportation and animal skins for protection from the bitterly cold winter weather.

The summer of 1904 was dedicated to exploring and mapping the area and to the measurement of magnetic fields and meteorology. The Norwegians spent their second winter in Gjöahavn together with their Eskimo friends.

In the summer of 1905, the explorers continued their journey on the Gjöa westward through the Canadian Arctic Archipelago, reaching Herschel Island off the coast of Yukon in August (Fig. 1.29). The dream had come true and the north-west Passage had been fully explored!

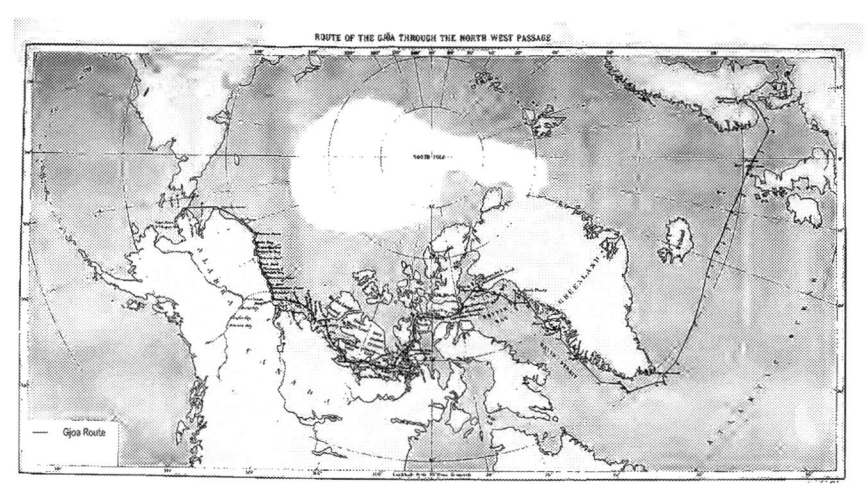

Figure 1.29 The Gjöa route from Christiania (Oslo) to Alaska's Pacific coast.

The deep sea, the global ocean, the physics of the straits, the dense formation of water, the functioning of the marine ecosystem now became new challenges.

Lesson learned

I am confidently of opinion, that the greatest Credit belongs to those learned Men who have forced their Way through all Difficulties, and have preferred the Profit of instructing to the Grace of pleasing.

Pliny, *Natural History*, 77 AC, translation by Philemon Holland, 1601

Oceanography began to be a science when a viable set of observations collected from learned people from different countries became available to scientific communities around the world. These early datasets grew out of practitioners of physical geographers who applied their interests to the ocean. From the outset, it required collaboration among observers, scientists, and laboratories. It soon became clear that a fundamental requirement in collaborative marine studies is the comparability of the data obtained from the participating research groups, regardless of the country of origin.

An official starting point for effective international collaboration was marked by the Congress held in Brussels in 1853, motivated by the works of Maury. Under the aegis of King Leopold, the nations interested in trade (Belgium, France, Great Britain, Norway, Russia, Sweden, the Netherlands) agreed to establish a uniform system of observations. Other countries subsequently offered their collaboration: Austro-Hungarian Empire, Brazil, Chile, Kingdom of Sardinia, Prussia, Spain, the free cities of Hamburg and Bremen, as well as the Papal States (Manzella, 1992).

The first phase in the history of oceanography was an empirical knowledge resulting from surveys of navigators and explorers. Physical geography was followed by the study of flora, fauna, and geological surveys, and finally a marine science was established with the foundation of marine laboratories and stations, university curricula, and national research programs.

We can reasonably believe that the beginning of "best practices" stems from investigations that were established in the 17th century by *persons of great repute*, such as Boyle and Hooke (e.g., Shaw, 1738).

Best practices

Phipps (1774), Scoresby (1820), and collaborators identified one of the actions to be performed before and during data collection, i.e., a knowledge of the measuring apparatus and related work methodology. Dr. Irving, a member of Phipps' scientific team, was familiar with the functioning of Cavendish thermometers, the nonlinearity of the responses, and the compression effects. Scoresby, after experiences with various devices, invented the "marine diver" that, to a certain extent, protected the thermometer from the effects of compression. To collect animals from the bottom of the sea, Ross (1819a,b) invented the Deep-Sea Clam.

Intercomparison of instruments was another element of attention by many scientists. During the voyage toward the North Pole, Irving conducted experiments with the Cavendish and Fahrenheit thermometers at different temperature values (called *heats* by the author) and found significant differences. However, the consistency of the comparison cannot be assessed, as no details have been provided (Phipps, 1774). It should be remembered that these early instrument comparisons were very new because there were no established standards against which any of the observations could be measured.

It is not easy to define what "heat" in the ocean is. Since the 17th century, "heat" was empirically measured with physical parameters such as the height of the mercury column in a glass tube. In principle, the concept of temperature is linked to the thermodynamic properties of bodies (e.g., Rumer and Ryvkin, 1980). The temperature can therefore be considered as an empirical measure of the level of thermal energy. The advection/diffusion of "potential temperature" introduced by Helland–Hansen (1912), has been interpreted as advection/diffusion of heat, as in (for example) MacDonald et al. (1994), where the heat flux was calculated by multiplying the potential temperature flux per unit area. McDougall (2003) demonstrated that the potential enthalpy flux can be called "heat flux" and introduced the "conservative temperature" a variable proportional to the potential enthalpy.

The concepts underlying the methodologies and technologies are therefore necessary to make the right measurements and since the 17th century *Persons of Best Repute* have recommended some specific actions to ensure the correctness of the experiments (observations), identify and deal with errors and omissions, and finally document experimental activities (paragraph 2).

The expertise of scientists and staff with a hands-on knowledge of tools and data has been fundamental to ensure the highest possible quality of the data with the technologies existing at that time.

Importance of standards

Chapter 4 of this book discusses the data quality dimensions. Three main pillars form the basis of collaborative studies, as part of the lesson learned from the history of oceanography:

- Common standards: there is a minimum level of quality control to which all oceanographic data must be subjected. The data must be qualified by further information relating to the measurement methods and subsequent data processing compared against some well-established standards.
- Consistency: the data collected by different groups should be as consistent as possible.
- Reliability: to serve the research community and others, the data must be reliable and this can be best achieved if the data has been analyzed against internationally accepted standards.

These requirements lead to the definition of common units, certified reference materials and analysis techniques, including sampling, pretreatment of samples, transport, and storage. From a historical point of view, the first attempt to define several standard procedures to be adopted internationally was made during the First International Maritime Conference, held in Brussels in 1853.

More effective resolutions were agreed during the International Conference for the Exploration of the Sea, held in Stockholm, 1899 (Trybom, 1900). In particular, it was stressed that "international co-operation is the best way of arriving at (scientific) satisfactory results." It was important to have common "meters" for measurements of the many properties of sea water. The agreements included hydrographic and biological works "in the northern parts of the Atlantic Ocean, the North Sea, The Baltic and adjoining seas." The establishment of a central office responsible for, among other things, *the control of the apparatus to ensure the uniformity of methods*, was recommended.

In Appendix II of the conference recommendations, Nansen stated the need for standards (Fig. 1.30) and, in particular, he recommended that there should be a central laboratory to produce and distribute standard sea water.

On the basis of this recommendation, a standard sea water called "Normal Water" was subsequently produced initially by the Hydrographic Laboratory in Charlottenlund (Denmark) and after by the Danish Biological Station housed in Charlottenlund Slot (e.g., Charlottenlund Palace), from 1935 (Thomson and Emery, 2001; Thomsen, 1954). This "standard" was also called "Copenhagen Water."

APPENDIX II.

In connection with the central bureau there should be a central laboratory, where, amongst other things, the following work might be carried out :—

1. The various methods for determining salinity, temperature, gases, plankton, etc., of the sea should be carefully tested, in order that standard methods may be fixed.

2. The various apparatus and instruments now used for hydrographical and biological research should be examined in order to settle which are the most trustworthy. Experiments may also be made to improve the apparatus and instruments, or to construct new and better ones.

3. Instruments and apparatus used in the investigations should be approved and tested at certain intervals at the central laboratory.

4. The water-samples sent by the workers of the participating states should be analysed and examined at the central laboratory, from which also samples of standard water should be provided. (See A. IV.).

Figure 1.30 The recommendations for standards agreed during the International Conference for the Exploration of the Sea, held in Stockholm, 1899). The appendix was written by Fridtjof Nansen.

Conclusions

Today many marine observation and data collection initiatives are promoting the free sharing of data with the main objectives of supporting research, making data available to the public, promoting innovation, and supporting the blue economy. This is, in part, a consequence of the abundance of data available today with satellites transferring data from a variety of autonomous platforms resulting in a plethora of data that greatly exceeds the number of interested oceanographers today ready to analyze these data.

The use of data requires well-defined quality assessment and quality control procedures, reference methods, and interlaboratory exercises. All these elements are essential components of oceanographic data management.

The reuse of data normally involves the use of data collected for purposes that vary from researcher to researcher. Reuse therefore requires the integration and harmonization of data from different sources (see Chapter 4 of this book). This is particularly important for the production of historical series (Fichtinger et al., 2011).

The need for standards has been addressed as one of the most important actions in collaborative studies (e.g., Tenopir, 2011). Use of internationally agreed upon standards allows the creation of products "by knitting together data from different sources, ensuring continuity and coherence across borders and across different disciplines" (Shepherd, 2018).

The comparison of measurements made by different laboratories and different campaigns in many cases led to the identification of discrepancies that indicate inconsistencies in the preparation of calibration standards. Intercalibrations, certified reference materials, clean room techniques, qualified personnel, etc., were not sufficient to guarantee good data quality. Sampling strategies and methodologies for data collection, sample pretreatment, transport and storage have been identified as an integral part of quality assurance and good measurement practices.

Data quality control aims to provide users with information on the errors they contain (IOC and CEC, 1993; Consortium for Ocean Leadership, 2013). Fichtinger et al. (2011) analyzed problems related to the combination and integration of data from different sources. The authors stressed that data integration also requires their harmonization, and they defined an abstract data management scheme derived from data and product quality specifications. The complete abstract scheme includes data evaluation to ensure relevance, reliability and fitness for purpose, adequacy, comparability, and compatibility (see Chapter 4 for further details).

The Scientific Committee for Ocean Research (SCOR) is today the most relevant international organization for the development of marine observation norms and standards. Within the SCOR, dedicated working groups are developing and making available algorithms for calculating oceanographic parameters. Among the most important results of the SCOR working groups worth mentioning are the thermodynamic equation of state of sea water and the numerous studies on the harmonization of data that underpin the development of best practices.

The United States Integrated Ocean Observation System (IOOS) has instituted an authoritative Quality Assurance/Quality Control of Real-Time Oceanographic Data (QARTOD). The procedures include information on sensors and measurement methodologies to ensure the best use of the data. The QARTOD documents and INSPIRE data specifications (https://inspire.ec.europa.eu/data-specifications/2892) are based on the same concepts and schemes that allow for easy interoperability.

To conclude this chapter, it would be fair to say that today's practices on the use of scientific data of the ocean are the natural evolution of the debates that began centuries ago. The methodological and technological changes have been very rapid over the past century, and there is the risk of not understanding how and why we have reached our current situation. The introduction of electronic methods and satellite data transfer has revolutionized the data collection platforms making many of them autonomous. The necessarily brief presentation of this chapter aims to provide some bibliographic elements that students and scholars can use to deepen their understanding of the themes they prefer.

Acknowledgments

Public and private libraries and collections have provided the materials for this chapter, including the libraries of the Italian National Research Council (CNR), the Italian National Agency for New Technologies, Energy and Sustainable Economic Development (ENEA), and the National Institute of Geophysics and Volcanology (INGV). During their activities, the authors had the opportunity to consult the libraries of many institutes and, in many cases, they were struck by the beauty of the hand-coloured drawings contained in antique biology books. Reproduction work carried out by both the public and private sectors is of inestimable cultural importance for the dissemination of knowledge. Reiner Schlitzer and his team have kindly allowed the use of Ocean Data View in this chapter.

References

AAVV, 2004. La storia. La biblioteca di Repubblica, vol. 10, pp. 115−160. Chapter 3.
Abbe, C., 1888. Treatise on meteorological apparatus and methods. In: Annual Report of the Chief Signal Officer of the Army to the Secretary of War for the Year 1887. Washington Government Printing Office. Appendix 46.

Amundsen, R., 1908. The Northwest Passage. E.P. Dutton Company, New York.

Aristotle, 1952. In: Lee, H.D.P. (Ed.), Meteorologica. With an English Translation. Harvard University Press, Cambridge.

Ballot, B., 1872. Suggestions on a Uniform System of Meteorological Observations. Royal Dutch Meteorological Institute, printing office 'The Industry', Utrecht.

Barrow, J., 1818. A Chronological History of Voyages into the Arctic Regions; Undertaken Chiefly for the Purpose of Discovering a North-East, North-West, or Polar Passage between the Atlantic and Pacific: From Earliest Periods of Scandinavian Navigation, to the Departure of the Recent Expeditions, under the Order of Captains Ross and Buchan. John Murray, Albemarle Street, London.

Becket, J.B., 1775. The Use of the Hydrostatic Balance Made Easy. Bristol: Printed for the Author; - Sold by G. Robinson, London; and R. Cruttwell, in Bath.

Birch, T., 1760. The History of the Royal Society of London for Improving of Natural Knowledge from its First Rise. Supplement to the Philosophical Transactions, vol. 1 (Consulted on February, 2020). https://books.google.it/books?id=stptfhQDyas C&pg=PA29&lpg=PA29&dq=Propositions+of+some+experiments+to+be+made +by+the+Earl+of+Sandwich+in+his+present+voyage.

Camuffo, D., 2002. Calibration and instrumental errors in early measurements of air temperature. In: Camuffo, D., Jones, P. (Eds.), Improved Understanding of Past Climatic Variability from Early Daily European Instrumental Sources. Kluver Academic Press, pp. 297−330.

Cavendish, C., 1757. A description of some thermometers for particular uses. Phil. Trans. 50, 300−310.

Cussler, C., Cussler, D., 2009. Arctic Drift. Penguin Putnam Inc.

Damm, E., Bauch, D., Krumpen, T., Rabe, B., Korhonen, M., Vinogradova, E., Uhlig, C., 2018. The Transpolar Drift Conveys Methane from the Siberian Shelf to the Central Arctic Ocean, vol. 8, p. 4515. https://doi.org/10.1038/s41598-018-22801-z. www. nature.com/scientificreports.

Dampier, W., 1703. A new voyage round the world. Printed for James Knapton, London.

Dampier, W., 2012. Memoirs of a Bucaneer: Dampier's New Voyage Round the World. Dover Publications, Inc., Mineola, NY, 1697.

De Groote, C., 1853. First International Maritime Conference Held at Brussels in 1853 for Devising an Uniform System of Meteorological Observations at Sea. Official Report. Wikisource (Consulted on Feb. 29, 2020). https://en.wikisource.org/wiki/ First_International_Maritime_Conference_Held_for_Devising_an_Uniform_System_of_ Meteorological_Observations_at_Sea.

Derham, F.R.S., 1726. Philosophical Experiments and Observations of the Late Eminent Dr. Robert Hooke S.R.S. And Geom. Prof. Gresh and Other Eminent Virtuoso's in His Time. Royal Society, p. Pp403.

Deacon, M., 1971. Scientists and the Sea, 1600 − 1900. A Study of Marine Science. Academic Press, London.

Folland, C.K., Parker, D.E., 1995. Correction of instrumental biases in historical sea surface temperature data. Quat. J. R. Meteorol. Soc. 121, 319−367.

Fichtinger, A., Rix, J., Schäffler, U., Michi, I., Gone, M., Reitz, T., 2011. Data harmonisation put into practice by the HUMBOLDT project. Int. J. Spat. Data Infrastruct. Res. 6, 234−260.

Galilei, G., 1638. In: Discorsi e dimostrazioni matematiche intorno a due nuove scienze. Ludovico Elzeviro editor, Leida.

Galilei, G., 1632. In: Landini, G.B. (Ed.), Dialogo di Galileo Galilei Dove ne i congressi di quattro giornate si discorre sopra i due Massimi Sistemi del Mondo Tolemaico e Copernicano (Firenze).

Hales, S., 1753. A letter to the president, from Stephen Hales. Phil. Trans. XVLII. For the years 1751 and 1752.

Helland-Hansen, B., 1912. Physical oceanography. In: Murray, J., Hjort, J. (Eds.), Chapter V in the Depths of the Ocean. MacMillan and Co., London, pp. 210–306.

Henry, J., 2008. The Scientific Revolution and the Origins of Modern Science. Palgrave McMillan, New York, p. 165.

Hepburn, A., 1847. In: Newman, V.E. (Ed.), Description Of Ball's Dredge. The Zoologist: A Popular Miscellany of Natural History. John Van Voorst, London, pp. 1847–1849.

Herschel, J.F.W., 1867. Physical Geography of the Globe. Adam and Charles Black, Edinburgh.

Hickey, C.G., Savelle, J.M., Hobson, G.B., 1993. The route of Sir John Franklin's third arctic expedition: an evaluation and test of an alternative hypothesis. Arctic 46 (1), 78–81.

Hjort, J., 1912. The ship and its equipment. In: Murray, J., Hjort, J. (Eds.), Chapter II in the Depths of the Ocean. MacMillan and Co., London, pp. 22–51.

Hjort, J., Bjerkan, P., Bonnevie, K., Brinkmann, A., Broch, H., Chun, C., Dons, C., Hoek, P.P.C., Nordgaard, O., Sars, G.O., Woltereck, R., 1912. In: Murray, J., Hjort, J. (Eds.), General Biology. Chapter X in the Depths of the Ocean. MacMillan and Co., London, pp. 660–786.

Houvenaghel, G., 1990. La Conference Maritime de Bruxelles, en 1853: premiere conference oceanographique international. Tijdschrift voor de geschiedenis der geneeskunde, natuurwetenschappen, wiskunde en techniek 13, 1.

Humboldt, von A., 2005. COSMOS - A Sketch of the Physical Geography of the Universe, vol. 1 eBook #14565. http://www.gutenberg.org/ebooks/14565.

IOC and CEC, 1993. Intergovernmental Oceanographic Commission and Commission of the European Communities. Manual of Quality Control Procedures for Validation of Oceanographic Data. Manual and Guides 26. Prepared by: CEC: DG-XII, MAST and IOC: IODE sc-93Nvs-19. 1993 UNESCO.

IOC, SCOR and IAPSO, 2010. The International Thermodynamic Equation of Seawater – 2010: Calculation and Use of Thermodynamic Properties. Intergovernmental Oceanographic Commission, Manuals and Guides N, vol. 56. UNESCO.

Knowles Middleton, W.E., 2003. A History of Thermometer and its Use in Meteorology. Johns Hopkins University Press, Baltimore.

Magalotti, L., 1667. Saggi di naturali esperienze fatte nell'Accademia del Cimento sotto la protezione del Serenissimo Principe Leopoldo di Toscana e descritte dal segretario di essa accademia. Cocchini Giuseppe all'Insegna della Stella, Firenze.

Manzella, G., 1992. La Dinamica Degli Oceani. Nuova Eri Edizioni Rai, Torino, p. 151.

Marcet, A., 1819. On the specific gravity, and temperature of sea waters, in different parts of the ocean, and in particular seas; with some account of their saline contents. Phil. Trans. Roy. Soc. Lond. 109, 161–208 (1819).

Marsili, L.F., 1681. Osservazioni intorno al Bosforo Tracio overo Canale di Costantinopoli. Rappresentante in lettera alla sacra real maestà di Cristina Regina di Svezia. In: Roma, Per Nicolò Angelo Tinassi, p. 1681.

Martyr D'Anghera, P., 1912. De Orbe Novo, translation. In: MacNutt, F.A., Putnam's Sons, G.P. (Eds.), New York and London. Volume One of Two, p. 448.

Maury, M.F., 1855. The Physical Geography of the Sea. Harper and Brothers Publishers, New York, p. pp283.

McConnell, A., 1982. No Sea Too Deep. Adam Hilger Ltd, Bristol.

MacDonald, A.M., Candela, J., Bryden, H.L., 1994. An estimate of the net heat transport through the Strait of Gibraltar. In: La Violette, P.E. (Ed.), Seasonal and Interannual Variability of the Western Mediterranean Sea, Coastal and Estuarine Studies. American Geophysical Union, pp. 13–32.

McDougall, T.J., 2003. Potential enthalpy: a conservative oceanic variable for evaluating heat content and heat fluxes. J. Phys. Oceanogr. 33 (5), 945—963. https://doi.org/10.1175/1520-0485(2003)033<0945:PEACOV>2.0.CO;2.

Melville, H., 1851. Moby Dick, Lit2Go ed. Retrieved March 18, 2020, from: https://etc.usf.edu/lit2go/42/moby-dick/.

Murray, J., 1912. A Brief historical review of oceanographical investigations. In: Murray, J., Hjort, J. (Eds.), Chapter I in the Depths of the Ocean. MacMillan and Co., London, pp. 1—21.

Murray, R., Hooke, R., 1667. Directions for the observations and experiments to be made by masters of ships, pilots, and other fit persons in their sea-voyages. Phil. Trans. 1666—1667, 2. https://royalsocietypublishing.org/doi/10.1098/rstl.1666.0009.

Nansen, F., 1900. Preface. In: Nansen, F. (Ed.), The Norwegian North Polar Expedition 1896 — 1986. Scientific Results, vol. 1. printed by Brogger A.W., Christiania.

Nansen, F., 2020. Farthest North. Gibson Square Books, London.

Negretti, H., Zambra, J., 1874. New deep-sea thermometer. Chem. News, Crookes, W. (Ed.), vol. XXIX, pp. 201—202.

Peterson, R.G., Stramma, L., Kortum, G., 1996. Early concepts and charts of ocean circulation. Prog. Oceanogr. 37, 1—115.

Phipps, C.J., 1774. A voyage towards North Pole. In: J. Nourse. London.

Preti, G., 1975. Storia del pensiero scientifico. Oscar Studio Mondadori.

Pinardi, N., Özsoy, E., Latif, M.A., Moroni, F., Grandi, A., Manzella, G., De Strobel, F., Lyubartsev, V., 2018. Measuring the sea: Marsili's Oceanographic Cruise (1679—80) and the Roots of Oceanography. J. Phys. Oceanogr. 48, 845—860. https://doi.org/10.1175/JPO-D-17-0168.1.

Rehbock, P.F., Benson, K.R., 2002. Introduction. In: Bengtsson, K.R., Rehbock, P.F. (Eds.), Oceanographic History — the Pacific and beyond. University of Washington Press, pp. ix—xii.

Ross, J., 1819a. A Voyage of Discovery. Longman, Hurst, Rees, Orme and Brown Editors.

Ross, J., 1819b. A Description of the Deep Sea Clams, Hydraphorus and Marine Artificial Horizon. Strahan and Spottiswoode, London.

Ross, J., 1835. Narrative of a second voyage in search of a North-West Passage, and of a residence in the Arctic regions during the years 1829, 1830, 1831, 1832, 1833. Webster 854.

Rumer, Y.B., Ryvkin, M.S., 1980. Thermodynamics, Statistical Physics and Kinematics. MIR Publishers, Moscow.

Scoresby, W., 1820. An Account of the Arctic Regions with a Description of the Northern Whale-Fishery. Archibald Constable & Co. Edinburg, Hurts, Robinson & co, London.

Shaw, P., 1738. The Philosophical Works of the Honourable Robert Boyle Esq. In: Innis, W., Manby, R., Longman, T. (Eds.).

Six, J., 1794. The Construction and Use of a Thermometer. In: Blake, J., Wilkie, G., Wilkie, T. (Eds.), p. pp24. London.

Shepherd, I., 2018. European efforts to make marine data more accessible. Ethics Sci. Environ. Polit. 18, 75—81. https://doi.org/10.3354/esep00181.

Sobrero, A., 2006. Memorie scelte di Ascanio Sobrero pubblicate dall'Associazione chimica industriale di Torino con discorso storico-critico ed annotazioni di I. Guareschi. E-book Progetto Manuzio, Liberliber.

Svensson, A., 2002. Swedish oceanographic instruments up to 1950. In: Bengtsson, K.R., Rehbock, P.F. (Eds.), Oceanographic History — the Pacific and Beyond. Proceedings of the Fifth International Congress on the History of Oceanography, Scripps Institution of Oceanography, La Jolla, California, June 1993. University of Washington Press, pp. 358—361.

Tenopir, C., Allard, S., Douglass, K., Aydinoglu, A.U., Wu, L., Read, E., Manoff, M., Frame, M., 2011. Data sharing by scientists: practices and perceptions. PloS One 6 (6), e21101. https://doi.org/10.1371/journal.pone.0021101.

Thomsen, H., 1954. Instructions pratiques sur la détermination de la salinité de l'eau de la mer par la méthode de titrage de Mohr-Knudsen. Bulletin de l'Institute Océanographique N. 1047.

Thomson, W.C., 1873. The Depth of the Sea. MacMillan & C., London.

Thomson, R.E., Emery, W., 2001. In: Emery, W., Thomson, R.E. (Eds.), Data Analysis Methods in Physical Oceanography. Elsevier Science.

Trybom, F., 1900. International conference for the exploration of the sea, Stockholm, 1899. J. Mar. Biol. Assoc. U K 6 (1), 101–114.

Wallace, W.J., 1974. The Development of the Chlorinity/salinity Concept in Oceanography. Elsevier Oceanographic Series, vol. 7. Elsevier Scientific Publishing Company, Amsterdam, p. 227.

Wallace, W.J., 2002. The history of the chemical determination of salinity. In: Bengtsson, K.R., Rehbock, P.F. (Eds.), Oceanographic History — the Pacific and beyond. University of Washington Press, pp. 362–368. Proceedings of the fifth International Congress on the History of Oceanography, Scripps Institution of Oceanography, La Jolla, California, June 1993.

WMO — World Meteorological Organisation, 1973. One Hundred Years of International Co-operation in Meteorology (1873 — 1973). A Historical Review. WMO — No. 345. https://library.wmo.int/index.php?lvl=notice_display&id=7069#.Xzjy7TXOPIU. (Accessed August 2020).

Zillman, J.W., 2009. A history of climate activities. World Meteorol. Organ. Bull. 58 (3), 141–150.

PART TWO

Data services in ocean science

CHAPTER TWO

Data services in ocean science with a focus on the biology*

Joana Beja[1], Leen Vandepitte[1], Abigail Benson[12],
Anton Van de Putte[2,3], Dan Lear[4], Daphnis De Pooter[5],
Gwenaëlle Moncoiffé[6], John Nicholls[7], Nina Wambiji[8],
Patricia Miloslavich[9,10], Vasilis Gerovasileiou[11]

[1]Flanders Marine Institute (VLIZ), Oostende, Belgium
[2]Royal Belgian Institute for Natural Sciences, Brussels, Belgium
[3]Université Libre de Bruxelles, Brussels, Belgium
[4]Marine Biological Association, Plymouth, United Kingdom
[5]Commission for the Conservation of Antarctic Marine Living Resources, (CCAMLR), Hobart, TAS, Australia
[6]British Oceanographic Data Centre, National Oceanography Centre, Liverpool, United Kingdom
[7]Norfish Project, Centre for Environmental Humanities, Trinity College Dublin, Dublin, Ireland
[8]Kenya Marine and Fisheries Research Institute, Mombasa, Kenya
[9]Scientific Committee on Oceanic Research (SCOR), University of Delaware, College of Earth, Ocean and Environment, Newark, DE, United States
[10]Departamento de Estudios Ambientales, Universidad Simón Bolívar, Caracas, Miranda, Venezuela
[11]Hellenic Centre for Marine Research (HCMR), Institute of Marine Biology, Biotechnology and Aquaculture (IMBBC), Heraklion, Greece
[12]U.S. Geological Survey, Lakewood, CO, United States

Introduction

> Past, present, and future coming together …
> If Aristotle were still alive today, he would have made an excellent ambassador for modern marine biodiversity research.
> http://www.lifewatch.be/en/2016-news-aristotle (LifeWatch Belgium, n.d).

The human desire to learn about, catalog, and understand the diversity of life goes back to ancient times. Biodiversity knowledge arose from the

---

* All authors have equally contributed to the conceptualization and writing of the text. The first two authors (Joana Beja and Leen Vandepitte) were responsible for the general coordination, writing, final reviewing and editing. Anton Van de Putte has additionally created the graphics included in this text.

Ocean Science Data
ISBN: 978-0-12-823427-3
https://doi.org/10.1016/B978-0-12-823427-3.00006-2

© 2022 Elsevier Inc.
All rights reserved.

67

necessity to identify and recognize edible plants, to know where to find them, understand that they appear in seasonal cycles, that some animal species seasonally migrate, how and where large animals are best found, so that times of food scarcity could be overcome. Knowledge on the healing effects of plants, insights in migration patterns, and the understanding and use of mariculture were and are still common practices in indigenous communities. This knowledge is traditionally handed over from generation to generation, not in writing form as is done by modern science, but by songs, drawings, tales, dances, etc (e.g., Nicholas, 2018; Inglis, 1993; Grenier, 1998). The recognition of Indigenous Knowledge (IK) as fundamental for sustainable development has been enshrined in international initiatives, as is the case of principle 22 of the United Nations (1992), Agenda's 21 Chapter 26 (1992), or article 8j of the Convention on Biological Diversity (CBD, https://www.cbd.int/).

Although this biodiversity knowledge has always been part of human life, it was Aristotle (384–322 BCE) who defined the scientific method and pioneered the study of living organisms. Aristotle was the first to systematically observe and describe biological diversity and developed the first classification for animals. He brought existing names together in a zoological treatise and created new names where none existed. Aristotle spent a large part of his life studying marine species, with over 40% of the animals in his zoological works being found in the marine environment. Remarkably, about 30% of the fish and 70% of the marine species' names cannot be found in earlier documents. Aristotle is, therefore, often referred to as "the father of marine biodiversity." His works were the first written documents on zoology and marine biodiversity, and they were heavily relied on by early taxonomists until the 16th–18th centuries, most notably Linnaeus and Cuvier (Voultsiadou et al., 2017).

With the formalization of the binomial nomenclature and the international acceptance of this standard, Linnaeus became the "father of modern taxonomy" (Calisher, 2007). In the establishment of his system, Linnaeus was greatly inspired by the works of Aristotle, which is reflected in the variety of Greek names he used (Voultsiadou et al., 2017).

Meanwhile, the cataloging and documenting of marine biodiversity was being done with varying degrees of precision and detail, linked to when and why they were collected. In fishing journeys, for example, notes were taken on the fauna and flora observed and fishers kept a record of their catches to be able to visit the more favorable fishing grounds in future trips.

Charles Darwin's work represented a huge step forward in the documentation and understanding of (marine) biodiversity, starting with his zoology notes taken during the HMS Beagle voyage (1831—36), Darwin (1838—43), his work on barnacles (Van Wyhe, 2007), and the intellectual link between biodiversity and ecosystem processes, culminating in his findings on species divergence and theory on natural selection (Van Wyhe, 2007 and references therein; Hector, 2009).

In parallel with Darwin's work, the first marine station was founded in 1843 by Pierre-Joseph van Beneden, in Oostende, Belgium. At this station, there was a continuous and dedicated research on marine biodiversity, which was addressed from various perspectives and by a wide range of marine scientists (Breyne et al., 2010). Historical marine research stations were also founded in other countries, including among others: in 1872, the Station Biologique de Roscoff, founded by Henri de Lacaze-Duthiers in Roscoff, France (Anderson, 2016); in 1875, the Stazione Zoologica Anton Dohrn, founded by Anton Dohrn in Naples, Italy (http://www.szn.it/index.php/en/who-we-are/our-history); in 1882, the Station Marine d'-Endoume, founded by Antoine-Fortuné Marion in Marseille, France (Marion, 1897). By the end of the 19th century, marine stations were well established within Europe (Dolan, 2007), but their rise was not constrained to this continent and their proliferation could be seen in other countries. In Australia, the Marine Biological Station at Camp Cove was built in 1881, under the lead of Russian scientist Nikolai Nikolaevich de Miklouho-Maclay (https://www.harbourtrust.gov.au/en/see-and-do/visit/marine-biological-station/#mod-12426 and https://www.harbourtrust.gov.au/en/see-and-do/visit/marine-biological-station/). The University of Tokyo established the first Japanese station, Misaki Marine Biological Station, in 1886, with part of its design based on the Italian Anton Dohrn's station (Inaba, 2015 and reference therein). The United States of America followed shortly after, with the Marine Biological Laboratory at Woods Hole (Massachusetts) being established in 1888 (https://www.mbl.edu). The National Institute of Oceanography and Fisheries (NIOF) in Egypt started in phases from 1918 during the reign of King Fuad, where the Royal Hydrobiological Institute in Alexandria was established. In 1952, it changed its name to the Institute of Hydrobiology and Fisheries and is currently located at Keyed Bey on the Eastern Harbor of Alexandria, Egypt. In 1928, the Egyptian Government commissioned the establishment of a Marine Biological Station at Hurghada on the Red Sea under the leadership of the British scientist Cyril Crossland (1930—38). The Station was associated with the King Fuad I University, Faculty of Sciences (currently Cairo University). It was renamed Farouk I Institute of Oceanography in 1947, and in 1954 finally changed to the Institute of Oceanography of the Red Sea - NIOF (https://niof-eg.com/). Tunisia also established a marine station, the Oceanographic Station of Salammbô in Carthage, in 1924, under the management of Mr. H. Heldt, with the support by the Oceanographic Museum of Monaco funded by the Prince Albert of Monaco (Bristol, 1926). In 1956 when Tunisia gained independence, this station was upgraded to a complete museum, currently named the National Institute of Marine Sciences and Technology. Further south, in Mozambique, the Inhaca Marine Biology Research Station was established in the 1930s.

Following the proliferation of marine stations, several national and international initiatives began. In 1902, the International Council for the Exploration of the Sea (ICES) was established, following the international conferences held in 1899 (in Stockholm, Sweden) and 1901 (in Kristiania, Norway) (Went, 1972, see also Chapter 1). It was the first example of an international scientific collaboration aimed at improving our understanding of the marine ecosystem and its response to human exploitation and, in particular, pressure on fish stocks (Engesæter, 2002). In 1931, the marine community witnessed the start of the most geographically extensive marine monitoring program and biological time series in the world led by Sir Alister Hardy, who initiated what was to become the Continuous Plankton Recorder (CPR) Survey in the North Sea. Since then, the survey has conducted more than seven million nautical miles of towing across the world's oceans, regional seas, and in freshwater lakes while maintaining a consistent methodology. In 1932, under the guidance of the International Geophysical Year (IGY), the second edition of the Polar Year was organized. This initiative collected a significant amount of data, the need for a World Data Center (WDC) was then recognized and created under what would later become the World Meteorological Organization (WMO) (https://en.wikipedia.org/wiki/International_Polar_Year).

Due to the World War II, much of the data collected and scientific analyses from the IGY initiative were lost. This defining moment led the IGY organizing committee to require that "all observational data shall be available to scientists and scientific institutions in all countries" (Odishaw, 1959). This gave rise, in 1957, to the actual establishment of the WDC system. Three WDCs were hosted in the (a) United States, (b) Soviet Union, and (c) one subdivided among countries in Western Europe, Australia, and Japan (Odishaw, 1959). These WDCs would, decades later, in 2008, be superseded by the International Council for Scientific World Data System (ICSU-WDS) while retaining the original goals to preserve quality assured scientific data and information, facilitate open access, and promote the adoption of standards. 10 years after this, in 2018, the ICSU and ISSC (International Social Science Council) merged into ISC, the International

Science Council (https://council.science/about-us/a-brief-history/; https://en.wikipedia.org/wiki/International_Geophysical_Year).

In 1960, during the Intergovernmental Conference on Oceanographic Research, held in Copenhagen, a recommendation to make every effort to publish oceanographic data and distribute them to all interested institutions was ratified by the member states (Wolff, 2010). Following that, and in response to one of the greatest early international interdisciplinary oceanographic research efforts, the Intergovernmental Oceanographic Commission (IOC) was established with the goal to coordinate the global ocean research program "the International Indian Ocean Expedition" (1959−65), where 45 research vessels, under 14 different flags explored through pioneering voyages of discovery the ecological mysteries of the Indian Ocean and defined a new age in ocean research. One year later, in 1961, the International Oceanographic Data and Information Exchange program (IODE) was born, with the purpose of facilitating the exchange of oceanographic data and information among participating member states, and of meeting users' needs for data and information products (https://www.iode.org/; Hofmann and Gross, 2010). All these initiatives covered several Earth sciences disciplines, including oceanography, also known as oceanology, which is the study of the physical, chemical, and biological aspects of the ocean.

Contrasting with early scientific history when progress was slow, the achievements accomplished in the past few decades have resulted in an exponential increase in data and information sharing. Up until the late 1980s to the early 2000s, scientists relied on printed resources to process their data and publish the outcomes of their studies. These resources were only available either through their institutions or university libraries. Communication with peers was slow and conducted through the tedious "snail mail." This process took until the 1990s to drastically change, with the rise of email and the internet, and it took until the late 1990s and early 2000s for digital submissions to journals to become the norm.

The build-up to the third millennium (Anno Domini) brought a digital revolution, with new insights and many opportunities and possibilities to push marine biological science forward. A major global endeavor to increase the knowledge on marine biodiversity was the Census of Marine Life (CoML) project. CoML was a 10-year-long international effort, initiated in 2000, to assess the diversity, distribution, and abundance of marine life globally. The main goal was to answer the most timeless questions "what lives in the sea?", "what lived in the sea?" and "what will live in the sea?". To achieve this, CoML implemented a two-way approach: (1) explore and collect new data through a variety of field programs in the decade-long effort and (2) synthesize and model existing data collected over centuries with data obtained during the decade to identify the baseline for measuring future changes in the biology of Earth's oceans (Alexander et al., 2011). The Census stimulated the development of the marine biodiversity discipline by tackling these issues globally and engaging circa 2700 scientists from about 80 countries around the globe.

In parallel with the start of CoML, Carlo Heip initiated in Europe a major long-term Network of Excellence (NoE), aimed at building long-lasting alliances between European marine researchers, their institutes, and universities and developing central data systems to service the marine scientific community (Heip et al., 2009). Carlo Heip's plans and vision were brought together in the Marine Biodiversity and Ecosystem Functioning EU Network of Excellence (MarBEF), running from 2004 to 2009 and closely linked with CoML's activities. Throughout the duration of CoML and MarBEF, several marine biology systems came into existence and still stand today offering scientists and other users online access to a wide variety of resources and services (see also Section Use cases and stories).

Over the past 2 decades, many initiatives began to promote making data more easily available online and making sure that different systems can connect with each other. The European Strategy Forum on Research Infrastructures (ESFRI) is a strategic instrument created to develop European scientific integration and to strengthen their international outreach. Several of them (partially) relate to marine biological sciences, e.g., LifeWatch (https://www.lifewatch.eu), European Life-Science Infrastructure (ELIXIR; https://elixir-europe.org/), European Marine Biological Research Center (EMBRC; https://www.embrc.eu/), and the European part of the global Long-Term Ecosystem Research (eLTER; https://www.lter-europe.net/).

We have witnessed the rise and the sometimes confusing proliferation of national and international networks, but they all progressively converge toward adoption of shared key concepts, e.g., several phases of the Data Life Cycle, data management best practices (e.g., use of Data Management Plans [DMPs]), community standards, and guiding principles like FAIR (Findable, Accessible, Interoperable, and Reusable, Wilkinson et al., 2016). The adoption of these practices is facilitated by cross-domain community efforts like the Research Data Alliance (RDA; https://rd-alliance.org/; Treloar, 2014) and the Earth Science Information Partners (ESIP, https://www.esipfed.org/).

Data Services can improve data accessibility but equally important in this process are the sharing and easy access to guidelines and best practices (Pearlman et al., 2019) on how to collect but also store, process, and quality control data. The requirement by funding bodies to compile a DMP, and the adoption of common controlled vocabularies to improve the consistency, accuracy, and machine readability of information stored alongside the data values, is key to enabling long-term data reusability and interoperability, as included, for example, in Jones et al. (2020), Steinhart et al. (2012).

While the data community continuously looks to improve access and reusability of historical and recently acquired data, technology evolves and pushes the limits. Engineers and scientists are discovering new ways to increase knowledge about marine biodiversity using new technologies, e.g., imagery, acoustics, animal tracking, or eDNA (environmental DNA— organismal DNA that can be found in the environment, USGS, https:// www.usgs.gov/special-topic/water-science-school/science/environmental-dna-edna?qt-science_center_objects=0#qt-science_center_objects). New measurement techniques bring new challenges for the data management community including issues related with large data volume, multiplicity of file formats and levels of processing, continuity and comparability with existing techniques, and capturing essential (minimum) information about these new techniques to optimize data reuse (De Pooter et al., 2017). All these require that the data community is able to adapt their methodology, infrastructure, and services in order to take best advantage of these developments.

Pressure is mounting for data to be reusable regardless of why it was first collected. The future of scientific knowledge relies on this and is important for making sound policy decisions and developing strategic resource management plans. This type of pressure creates unique challenges that require diversification of tools and services to address the needs and requirements of different user groups and stakeholders. In addition, there is an urgent need for building cohesive, synergetic, and interoperable networks across national, continental, and domain boundaries that are backed-up by trusted governance processes and traceable provenance. Examples can be found in Borgman et al. (2012) and Lear et al. (2020).

The implementation of the Essential Ocean Variables (EOVs) framework has increased the collaboration across marine disciplines, providing communities, managers, and policymakers with data on our changing oceans. However, the implementation of biological EOVs faces challenges in order to further improve their usability and impact (Miloslavich et al., 2018a; Duffy et al., 2019). Widespread use of the EOV framework for biological observations will likely be addressed over the next decade (Miloslavich et al., 2019a), which will be especially relevant given the exponential growth of data and knowledge as we implement more automated technologies (e.g., imaging and genomics). The next decade will also allow for tremendous progress to be achieved through the expansion of what has been accomplished previously, and through dealing with the challenges that remained at the end of the current decade.

With the start of the second United Nations (UN) Decade of Ocean Science for Sustainable Development, beginning in 2021 and initiated 50 years after the International Decade of Ocean Exploration (IDOE, 1971—80), a new era in ocean research will be underway. Whereas the first decade was aimed at generating the science we wanted, to better understand our oceans and seas, this second Decade is focused on generating "the science we need for the ocean we want" (IOC, 2020). The second decade will provide a "once in a lifetime" opportunity to create a new foundation, across the science—policy interface, to strengthen the management of our oceans and coasts for the benefit of humanity. In parallel, at the heart of the 2030 Agenda for Sustainable Development by the UN, the 17 Sustainable Development Goals (SDGs) are defined as a basis to achieve peace and prosperity for people and the planet. Knowledge on marine biodiversity is critical to people's health and that of our planet (https://www.un.org/sustainabledevelopment/oceans/). Achievement of SDG 14 "Life below water—conserve and sustainably use the oceans, seas and marine resources for sustainable development" (https://sdgs.un.org/goals) requires that biodiversity data are open, shared, and interoperable with other biological data and, more importantly, between disciplines.

The development of data management and e-science facilities that support interpretation of observations and improved biodiversity change modeling will be key in the coming decade and will strengthen the links between biodiversity researchers, biological data centers, and all related disciplines through intensive cross-border and cross-discipline interoperability (Fig. 2.1 exemplifies the biological timeline as described in the previous paragraphs).

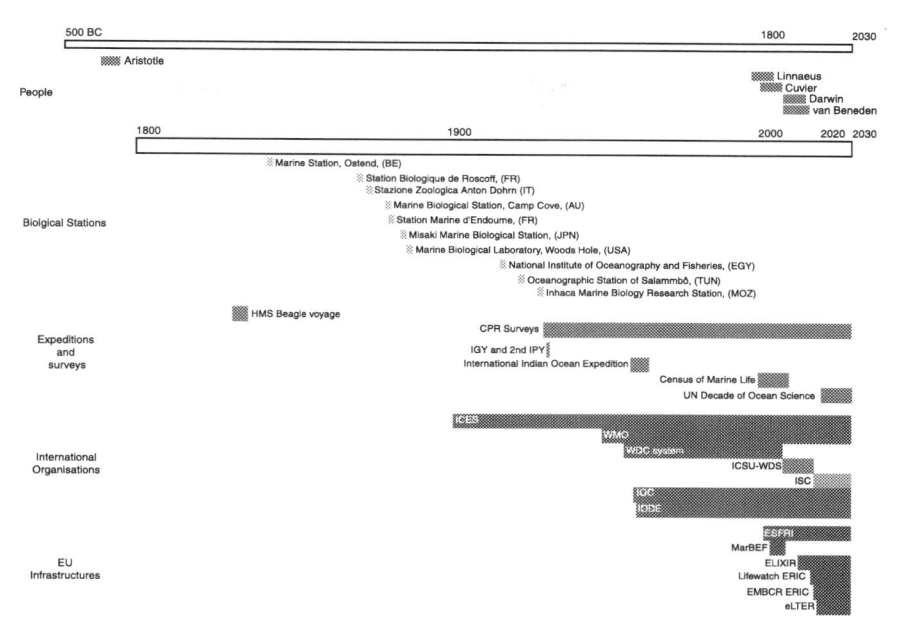

Figure 2.1 Biological data timeline.

Historical data

> When a knowledgeable old person dies, a whole library disappears.
>
> **An old African proverb.**

> Data are just summaries of thousands of stories—tell a few of those stories to help make the data meaningful.
>
> **Chip and Dan Heath.**

Introduction

Why are historical (or "archaeological") data important? There are many reasons, not least of which is that these data provide the historical context for present observations, thereby facilitating the process of establishing reference conditions for monitoring and management. Historical data (defined as data collected before the 1950s) may be employed in reconstructing and modeling past and present conditions, and they are invaluable in determining predictive future trends and shifts in species distribution ranges, regional species extinctions, biological invasions, and human activity impacts on the environment and biodiversity (McClenachan et al., 2012; Faulwetter et al., 2016; Mavraki et al., 2016). Historical data are vital for marine areas that are vulnerable to an ever-increasing number of human activities and pressures like the European Seas or those around South-east Asia, among others.

Historical datasets often contain descriptions of species that are important for taxonomic awareness; the first description of a species has legal priority for the naming of that species. Loss of data can have fatal consequences. It may become impossible to retrieve those data that were collected from a certain area over a certain period and occurrences that are not known within the framework of historical enquiry, that were never recorded (in archived documents) or are simply outside of human awareness (natural events never witnessed by humans) (Faulwetter et al., 2016). These data are therefore not known to be missing, represent unknown "gaps" in our knowledge, or they were lost over time due to rudimentary archiving procedures or the fact that they never made it to an archive. These unknown missing elements constitute gaps in knowledge that create an invisible impact and may influence overall awareness. This may occur with knowledge of actual species occurrences over time being unreliable, or levels of biodiversity being inaccurately reported despite an assumption of awareness and accuracy (Mavraki et al., 2016). Consequently, the loss of data equates to the loss of unique resources and ultimately to the loss of our natural wealth (Faulwetter et al., 2016; Mavraki et al., 2016). The European Commission has championed the principle of "collect once, use many times," referring to all data ever collected (https://emodnet.ec.europa.eu/en/about_emodnet). By reusing historical data in a comparative framework, scientists can gain ever more accurate perspectives to guide current and future research (McClenachan et al., 2012; Faulwetter et al., 2016).

In a typically more nuanced and standardized approach, historical data are characterized by several marked differences from "survey data" and from "scientifically collated information." Perhaps, the primary differences lie in the combined qualitative and quantitative nature of these data. Can historical data ever enjoy the same level of accuracy and reliability as, for example, scientifically surveyed data might enjoy? The answer may well be a resounding "No"; however, this is all relative to the data in question. As an example, if we look at the History of Marine Animal Populations (HMAP) historical datasets, while every attempt has been made to provide detailed supporting documentation that accompanies the datasets, the data are nevertheless open to a level of interpretation and may possibly include incorrect values. In the case of HMAP Dataset 6 Newfoundland (1675–98) (Pope, 2003), the highly respected historian Peter Pope produced meticulously collated data based primarily on archived British Admiralty records and other state documents which detailed actual landings of various ports and staging areas around Newfoundland. There is little reason to believe that these records are false, but degrees of inaccuracy may exist based on human error through transcription and differences in reporting practice. Furthermore, there is little means of verifying these figures where no corroborating data exist. It is the work of historians to assess and clean the data to whatever extent they may and then to report the levels of accuracy that they find.

We may regard history as "a chronological record of significant events (such as those affecting a nation or institution) often including an explanation of their causes" and "a treatise presenting systematically related natural phenomena (as of geography, animals, or plants)" (Merriam-Webster, 2021, Online, History, https://www.merriam-webster.com/dictionary/history?utm_campaign=sd&utm_medium=serp&utm_source=jsonld). If we are discussing current survey data that have been verified scientifically, then it stands to reason that these data are both reliable and historical. History starts now; the sum of all past human endeavor, and arguably of all aspects of past natural occurrences that impact humanity, defines the study of history. Data collected currently are historical as soon as they are recorded. This implies that all data are historical, but in practical terms, we need to accept that there are recognizable departures between scientific survey data that are typically machine observed, mechanically obtained, computer enhanced, and humanly verified, and historical data that are extracted from analog reports and accounts (such as archived documents) by experts with focused and informed knowledge.

To discount historical data just because they are not able to be verified in the same manner as scientific data would lead to an obvious generated gap in knowledge and thereby invalidate all current data. By this, we mean that without every reasonable attempt being made to include historical information as the temporal development of what is currently regarded as good data, the information held presently is an isolated and unfounded point that has no relevance in human considerations. For example, a collection of data that reports a number of a specific species of marine animals in a particular location may be interesting from a biological perspective, but only if it is reported within a framework that bears comparisons from the past; failing the past connection, the data are simply an ungrounded statement that immediately becomes history in and of itself as time passes. The next set of comparable data that appears will also be meaningless without comparison to the first set and so on. Historical data provides the platform that enables patterns to be developed in four-dimensional space-time and enables the ability to model predictions for the future. As an example, the NorFish Project has provided several data series that are openly available, but could be regarded as being of limited value, or even meaningless without supporting documentation that qualifies the data (Holm et al., 2021).

Historical sources

Obtained from sources such as archival documents, primary texts, and even oral reporting, historical data are prone to misrepresentation, misunderstanding, and miscommunication to name a few of the myriad of misprefixed similes that abound. The inherent difficulties of accessing, collating, compiling, and presenting historical data are the reasons why capable historians are deployed in the process; historians have the knowledge and expertise to explore the sources, interpret the raw data that are forthcoming, and collate these into viable data series, where possible. A typical historical archive document may comprise numbers, words, charts, maps, tables, and/or images that portray a meaningful grouping of data that the historian can interpret, sometimes translate, corroborate, investigate, and collate. During the collation and dissemination, exacting standards of accuracy, voracity, and robustness are being applied. When historical data are made available as digitally recorded matter, these are provided in various formats and presented in a variety of ways that enable researchers to interrogate and extract relevant information from them, to address the questions they are seeking answers for. In *Historical Research*, Danto (2008) compares and contrasts the types of sources used in historical research and emphasizes the use of corroborative materials; while primary literature is vital, secondary sources provide an underpinning role that bolsters the veracity of historical research.

Historical data sources are found in a large variety of disparate formats and present a large cross-section of human activity over time. Resources may be gleaned from institutional libraries, publications old and new, expedition logbooks, project reports, gray literature sources, museums and natural history collections, oral interviews, transcripts, letters of communication, and so on. In the modern age, the internet has enabled exemplars of information to be made available through highly relevant online resources such as Ocean Biodiversity Information System (OBIS, https://www.obis.org/), Flanders Marine Institute (VLIZ, http://www.vliz.be/en/), Biodiversity Heritage Library (BHL, https://www.biodiversitylibrary.org/), World Register of Marine Species (WoRMS, http://www.marinespecies.org/), and a plethora of similarly respected data repositories. For archival materials, serious pitfalls are present in the online world where relevance, accuracy, and reliability must always be questioned; while websites like Wikipedia or even the resource web pages of many institutes, research bodies, and universities (e.g., British History Online, Institute of Historical Research) can be very useful tools, due to their open-ended submission nature, every statement should be carefully considered and where necessary independently verified before acceptance. More detailed insights into the possible pitfalls of archival and online research materials may be explored in the work of Duff et al. (2004).

Reliability of historical data

A very real problem for historical data series is the gaps in information that are, inevitably, a reality of the process of obtaining viable information; recorded information may be absent altogether, it may be contradictory where several sources present different data, or it may be qualitative and not scientifically verifiable. In general terms, historical data become less scientifically verifiable depending on its age, cultural background, and the veracity of the source providers.

Several factors may contribute to making working with historical data a serious challenge. Historical publications may lack standardization among volumes of the same series/expedition, or even within the same publication (Faulwetter et al., 2016; Mavraki et al., 2016). How results are presented can vary greatly; for example, human observations can often be misleading—a sighting by an individual of a 1000 fish may be seen by another as merely a 100 fish. Similarly, species lists for documenting sampling dates and depths may vary from other lists that include records of individual species counts. The information in many historical publications is often unstructured, hidden in free text, not annotated with metadata, and thus not easily retrievable (Mavraki et al., 2016). This should not exclude these works from valid research; they were often created before modern standards of measures, modeling and recording were first introduced. In some cases, information from tables may be repeated in the text with small differences, making it difficult to understand which part of the given information is actually correct. Different volumes of the same series/expedition can also be published in different languages, depending on the taxonomic expert who analyzed the material and authored the publication. In many cases, coordinates may be missing or are inaccurate, depending on the quality and accuracy of the instrumentation deployed and the abilities of the human observer using the equipment. Sometimes only a location name gives an indication of where samples have been collected or reported data originate. When coordinates are available, one can still stumble upon the problem of missing or inaccurate georeferencing, for example, when sample locations of marine species appear on land, such as in the center of an island. In some historical publications, coordinates may refer to the "Paris" and not the "Greenwich" meridian line without a proper indication of the intended reference. Historical publications may refer to uncommon and nonstandard systems of measurements that require conversion to a standardized system (e.g., SI – International System of Units-metric). These publications and archival materials may reference old toponyms and political boundaries or findings from other events or expeditions for which limited information is available. They often include taxonomic inconsistencies such as outdated synonyms, misspellings, and old classification schemes; these require updating, correction, and revision to remove ambiguities and misinterpretation (Mavraki et al., 2016). For example, English whalers of the 18th century often noted whales as "fish" in logbooks, but within the contextual framework of the reports, the true identity may be gleaned (Decou, 2018); information is often reported and interpreted in different ways by different people.

Some common procedures exist in the management of historical data in order to alleviate many of the issues and problems identified (Faulwetter et al., 2016; Mavraki et al., 2016; Tsikopoulou et al., 2016):

1. The assignment of coordinates where possible; when the latitude and longitude are not documented, it may still be possible to pinpoint these (within a specified level of uncertainty) by using a Geographic Information System (GIS). If maps are available, these can be imported in GIS, and coordinates may be assigned. Where specific places are identified, the Marine Regions Gazetteer (https://marineregions.org/) may be consulted to retrieve coordinates.
2. Where necessary, new "artificial" sampling events may be created for samples (each sample is assumed to represent a unique combination of

position, date, time, gear, length of wire, and haul duration) when inconsistencies between different expedition volumes and ambiguities do not allow for a clear assignment for well-described sampling events.

3. The accurate identification of species is vital to ensure taxonomic precision. Common names for species may be recorded alongside the correct scientific names and metadata associated with those species. WoRMS (WoRMS Editorial Board, 2020), AlgaeBase (Guiry and Guiry, 2020), and FishBase (Froese and Pauly, 2019) are online resources that provide invaluable assistance in this endeavor.

4. A detailed review of conversion factors is vital to ensure that nonstandard or noncurrent measurement systems are correctly interpreted based on modern systems (e.g., metric). A measurement of a "last" of herring could indicate a conversion rate of 1.9764 metric tonnes, while a "fathom" may represent 1.8 m of ocean depth.

5. Documenting missing data is vital in providing awareness of gaps that need to be recognized in recorded data series. For example, a sudden drop in the number of fish caught in a specific period may be due to natural or human factors (weather, war, etc.), but it may also be due to information not being recorded. These gaps need to be understood and reported.

Unwritten, historical data

Having first addressed the written historical data and how it can be used to provide background ecosystem information, it is necessary that we focus on the unwritten history that has been gathered by indigenous people, passed along through generations and fine-tuned through these communities' learned experiences (Convention on Biological Diversity, 2011; Nicholas, 2018). IK or Traditional Knowledge (TK) has long been overlooked by the scientific community; it is unwritten, historical data, and tells the tale of centuries of tradition that have survived to modern times.

Indigenous languages are sometimes referred to as "the historical record of what has been there" (https://www.insidescience.org/news/conserve-species-traditional-local-knowledge-and-science-can-complement-each-other, 2019-12-24) and more and more scientists, as well as the wider scientific community and policymakers, are realizing that an enormous treasure of data and information is hidden in IK, which has mostly not been captured in the modern scientific way so far. In the past, IK was occasionally valued, when it supported scientific results and sometimes seen as a myth, when it challenged scientific findings (Nicholas, 2018).

Over the past few decades, the scientific community has slowly started to change its approach toward IK. By sharing the Western scientific approach and engaging with local (indigenous) communities, scientists are becoming more and more aware of how IK can add value to their research. It provides useful information that would not be easily collected otherwise, and it can bring benefits to both Western science and local communities, by sharing their research observations and results. Numerous studies have found that the population wildlife estimates acquired via TK are in accordance with those obtained via Western methods (https://www.insidescience.org/news/conserve-species-traditional-local-knowledge-and-science-can-complement-each-other, 2019), which not only gives scientists more confidence in coordinating their research with indigenous communities but also provides these communities with more information about their own observations. Alongside these examples is the archaeological evidence of mariculture by indigenous people in coastal British Columbia that determined that this practice predated the European settlements (Nicholas, 2018).

Following the inclusion of IK and indigenous peoples and communities in international conventions, several initiatives across the world have been set up by the scientific community and indigenous and/or local communities. The Convention on Biological Diversity (CBD, https://www.cbd.int/), that entered into force in 1993, explicitly recognizes the importance of indigenous people in its text (Article 8j) and any program of work under this convention needs to take into account the IK of indigenous peoples and local communities (https://www.cbd.int/convention/wg8j.shtml, 2020). Furthermore, the CBD has developed a Traditional Knowledge Information Portal, an electronic tool for TK research, which focuses on making information relevant to and about TK available (https://www.cbd.int/tk/about.shtml, 2015). A number of initiatives aim not only at conservation or restoration efforts but also monitoring, as is the case of the Local Environmental Observer (LEO) network, implemented in 2012 as a network of local observers and topic experts who share knowledge about unusual animal, environment, and weather events (https://www.leonetwork.org/en/docs/about/about) (LEO Network, 2017). A 2019 Intergovernmental Science-Policy Platform on Biodiversity and Ecosystem Services' (IPBES) report highlighted that "Regional and global scenarios currently lack and would benefit from an explicit consideration of the views, perspectives, and rights of Indigenous Peoples and Local Communities, their knowledge and understanding of large regions and ecosystems, and their desired future development pathways" (https://ipbes.net/global-assessment, 2019) (Intergovernmental Science-Policy Platform on Biodiversity and Ecosystem Services IPBES, 2019).

Even though both the Indigenous and Western Knowledge share the same attributes, verification through repetition, inference and prediction, empirical observations, and recognition of patterns, IK is more than just numbers, categories, and hierarchies, as it is often collected with an associated context (Nicholas, 2018). Structuring IK in a Western scientific fashion is not a trivial task and doing so would mean stripping the information from its context and presenting it only as facts, decreasing its value. Scientific biological databases and data standards used in marine biology are not yet prepared to deal with the complexity of these data and work needs to be developed, so that the wealth of IK is captured not only for future generations but also to assist policy makers and researchers with their assessments. In 2017, the World Intellectual Property Organization (WIPO) published a Toolkit with several resources and checklists to assist researchers documenting TK and ensure that all issues, from intellectual property to confidentiality or documenting via a database or register, are addressed (WIPO, 2017).

In summary

History is literally the sum of all human knowledge. However, for a maritime historian with a focus on marine life, the field is considerably narrowed. There are three primary research avenues that must be understood and explored in the search for a dataset of an historical nature: what is known, what is unknown, and what is unknowable. In each case, a thorough investigation must be undertaken in order to arrive at a process of data extraction that identifies what happened in time, and to what extent the information is accurate.

Where legacy data are concerned, their integration into a harmonized, comparable format and their quality control is a challenge. Data may be inaccurate, inconsistent, repeated, missing (temporarily or permanently), or incompatible. This may occur across different volumes of the same series/expedition, in conflicting or incongruent archive sources or even within the same publication. Obvious errors are often identified during digitization, translation, transcription, or observation. Where the original authors can no longer be consulted, some of these problems will remain unresolved indefinitely, and there is currently no standard strategy or "best practice" on how to deal with such issues. In a best-case scenario, researchers may focus on specific standards (like the OBIS adopted Darwin Core [DwC]) and thereby share both the outputs they have created and the levels of knowledge of interpretive methodologies that enable historical data extraction.

The capture of historical information can be a tedious and time-consuming effort. Nevertheless, biodiversity legacy literature contains a wealth of information on the biosphere and can provide valuable insights into the past state of the world's ecosystems. Consequently, loss of data equals loss of unique resources and ultimately to the loss of our natural wealth. The use of backward projections used by historians including extrapolation and interpolation that historical sources provide may enable researchers to contextualize the current state of the oceans with a view to forecasting future scenarios with greater accuracy and relevance (Nicholls et al., 2021).

Even though IK has existed since human existence, it is only fairly recently that efforts began to incorporate it in Western scientific studies. Slowly, mentalities are changing and several international initiatives have been and continue to be established to promote the collaboration between Western science and IK. Similar to what happens with historical data, capturing IK is not a trivial task, essentially because it cannot be treated in the same way as other data. In IK, context is of the uttermost importance and removing the data from its context devalues it. The wealth of information provided by IK is essential to the understanding of our living resources, but, all the same, the connection with Western scientific data can assist the indigenous communities in better understanding their rights and the world around them.

Research Data Life Cycle

T. S. Eliot - The Rock (Eliot, 1934)

[…]

The endless cycle of idea and action.

[…]

Where is the wisdom we have lost in knowledge?

Where is the knowledge we have lost in information?

[…]

The Research Data Life Cycle (RDLC) enables us to visualize and map the different phases research data go through, and associate them with essential short-term and, crucially, long-term data management requirements. Although the concept of an RDLC is relatively new, many services and guiding principles needed to support scientists and data users are already in place. Fully implementing the steps and principles of an RDLC is becoming easier as modern research practice and technology enables easier connection between and traceability of digital objects, and allows the development of specialized data services. Through this, scientists and data users can find the necessary support at every level of the cycle.

The scientific method is based on reproducibility and, in order to accomplish this, sharing the data is not enough. It should become common practice for scientists to also share scripts, codes, and workflows that were employed to analyze the data used in the research, which can be done via several online development platforms and repositories such as GitHub (https://github.com/) and Zenodo (https://zenodo.org/), to name but a couple.

Several initiatives came into existence to assist scientists in how to share their data, build the communities to discuss and develop guidelines for good data management, and make sure they engage with their user community. An example of such an initiative is the ESIP (www.esipfed.org), founded in 1998 and focusing on supporting the work that falls at the intersection of science, data, and users.

The RDA (https://www.rd-alliance.org/), launched in 2013, a community-driven initiative similar to ESIP, focuses on the development and adoption of infrastructures that promote data sharing and data-driven research, and aim to accelerate the growth of a cohesive data community. Under RDA's umbrella, the CoreTrustSeal organization was set up as a replacement of the earlier Data Seal of Approval certification and the WDS Regular Members certification. Data repositories with the CoreTrust-Seal reflect the core characteristics of trustworthy data repositories. This certification is seen as a first step toward a global framework for repository certification (https://www.coretrustseal.org/). Ideally, all marine scientific data find their way to certified repositories, where their continuity is ensured.

There are several schemes and interpretations of an RDLC, and they all share the same general flow, which can be summarized as follows: we collect data to gather information, this information brings us knowledge, and from this knowledge, we get wisdom. This is also known as the Data—Information—Knowledge—Wisdom model or DIKW pyramid (Rowley, 2007).

A successful RDLC requires that some of the key early stages of the life cycle be documented prior to undertaking the research. For this reason, DMPs have become a mandatory section of grant applications by many funding agencies. In short, it specifies how research data will be handled both during and after the end of a project. The first step in the research cycle is writing a DMP and ideally is completed at the project proposal writing stage or minimally before the project is started (e.g., Michener, 2015).

> By failing to prepare, you are preparing to fail.
>
> **Benjamin Franklin.**

Nowadays, to avoid loss of data and information, funding agencies require project proposals and evaluation processes to contain a detailed DMP. The major components include information about the data and the data format, the metadata content and format, policies that will be applied to access, share, and reuse the collected data, how the long-term storage and data management will be ensured, and the budget required to perform all these tasks.

Although DMPs have become standard practice in the last 15 years, and more journals are requesting that the data behind published analyses are archived and made available, a large amount of data are still at the brink of being lost or have already disappeared (Vines et al., 2013, 2014).

Estimates about the number of bytes of data that are produced each day vary. One study indicates we produce about 2.5 quintillion of bytes of data (in the broadest sense) each day, and that 90% of all available data have been produced in the last 3—5 years (PricewaterhouseCoopers LLP, 2019), whereas a recently published report by the European Commission indicates the produced volume of data in the world will rapidly grow from 33 zettabytes in 2018 to an expected 175 zettabytes in 2025 (European Commission, 2020). Although scientific data have generally been produced at rapid speed in the last 20 years and DMPs are becoming the norm, there is still no solid (waterproof) method to safeguard all data. A 2013 study revealed that about 80% of worldwide publicly funded scientific data were lost within 20 years of its publication, and therefore long-term preservation of research data should not be managed by individual researchers but through public data archiving and sharing (Vines et al., 2013). This seems to illustrate that the RDLC and its corresponding

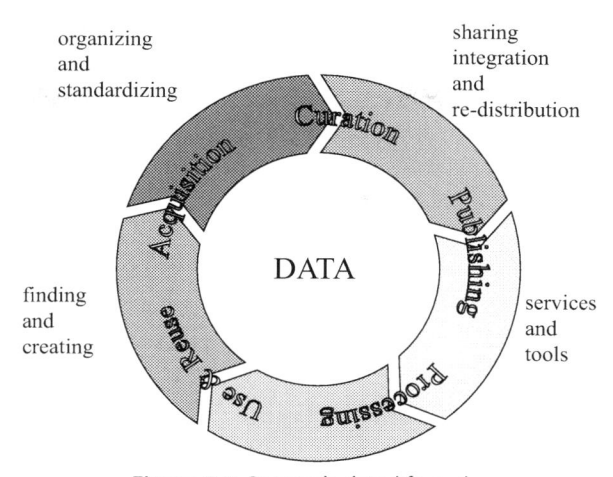

organizing
and
standardizing

sharing
integration
and
re-distribution

finding
and
creating

services
and
tools

Figure 2.2 Research data life cycle.

DMPs are not yet fully integrated into the common knowledge/habits of scientists worldwide. If scientists changed their mindset from "my data" to "our data" the marine biological data community could see improvements in retention of data, information, and wisdom that might otherwise be lost. Further, providing scientists with incentives to work alongside data managers could also lead to improvements in those areas. Finally, placing emphasis on data citation and attribution as comparable to a scientific publication provides another avenue for increasing scientists' involvement in the RDLC. To help address the fear that data could be misinterpreted or misused, the authors encourage collaborating with data stewards and IT developers to make sure all relevant information is stored together with the data in a readily, easily accessible format.

Various books have been written on the topic of RDLCs and DMPs already. With the risk of repeating some common information and statements, we will highlight the existing guidelines and focus on data services linked to the five steps of the actual life cycle (Fig. 2.2), with a focus on marine biology.

Phase 1—data acquisition: finding and creating data

Hypothesis testing is the primary reason scientists collect data. From a researcher's point of view, the RDLC consists of (1) gathering data, (2) organizing data, (3) analyzing data, (4) publishing results and findings based on the data, and (5) presenting these findings to peers at conferences and communicating main conclusions and key messages to the wider audience (e.g., citizens, policymakers, and stakeholders) (e.g., Griffins et al., 2018 and references therein, Nieva de la Hidalga et al., 2020). Communication to the wider audience is the most important piece of communication efforts because data, results, and findings can become part of the common good, and can be brought to use in global initiatives such as the UN Decade of Ocean Science for Sustainable Development (2021−30) and the UN Global Sustainable Development Goals (https://sdgs.un.org/goals), the post 2020 targets of the CBD (https://www.cbd.int/conferences/post2020) and the IPBES assessments (https://ipbes.net/).

In the field of marine science, ease of data collection varies widely. There can be differences in equipment, time, organization, skills, human resources, and funding required to evaluate marine biodiversity in different locations, ranging from coastal areas to e.g., deep sea or the poles. A basic coastal biodiversity research may require supplies such as boots, a bucket, and a shovel, as well as a basic knowledge of the tides, whereas deep-sea research or research in remote locations may demand access to a research vessel and ship time, different levels of logistics depending on the sampling location, high-tech equipment (e.g., Remotely Operated Vehicles), and excellent knowledge of several oceanographic disciplines, e.g., seafloor topography, currents, ice dynamics, hydroacoustics, etc.

Although miles apart, the above examples illustrate an important topic within marine biological science. Specifically, marine life cannot be studied independently from the study of other marine disciplines such as physics, chemistry, and bathymetry. Therefore, marine biologists collect nonbiological information such as water temperature, salinity, sediment characteristics, tidal, and general weather data that are closely associated with their biological samples. As these biotic and abiotic data go hand in hand, it is important to keep them together, specifically in the next steps of the RDLC.

Phase 2—data curation: organizing and standardizing data and its associated metadata

The term data curation has several definitions, but they all center around maintaining, preserving, and adding value to digital research data throughout its life cycle (e.g., https://www.dataversity.net/what-is-data-curation/). Generally, data curation focuses on metadata maintenance and management, rather than the data themselves, but one key aspect is that data without appropriate metadata can be completely worthless. The relevance and importance of good metadata are often underestimated.

In data management (and many other fields, e.g., journalism and teaching) training, documenting basic metadata are often described via the 5 Ws: Who, What, Where, When, and Why (e.g., Flint, 2018). Within the science field, a sixth question is often added: How. By answering these six questions, which is impossible by a simple yes/no, a wealth of important information about the data collected is documented and will assist users with understanding important aspects of the data. The origin of the 5Ws and How dates back to Aristotle, where they were used as elements of circumstances (Sloan, 2010). Although the classical origin of these questions to the ethics discipline has been largely lost over time, they remain a standard way of formulating or analyzing rhetorical questions (e.g., Copeland, 1991; Robertson Jr., 1946). The need for good metadata was identified in a study related to the COVID-19 pandemic, which emphasized that good metadata maximize data reusability that often holds the key for data-driven research discoveries (Schriml et al., 2020). This statement can be easily applied to all fields of science.

As with data, standards and formats are in place for metadata, such as those defined by the International Organization for Standardization (ISO, https://www.iso.org/home.html) and Ecological Metadata Language. Well-structured and standardized metadata are critical for proper data management but may be overlooked by data providers. Myriad reasons for this exist (Schriml et al., 2020), but the consequences remain the same: when metadata are incomplete, it becomes harder to fully use the data in integrated data exercises, which is a major requirement when looking at scientific questions on a global scale. Data providers are encouraged to prepare complete metadata, as metadata are as valuable as the data and are an integral part of a DMP.

Long after the data have served their original goal, they can be reused indefinitely, as long as they are discoverable and the associated metadata are complete for the specific purpose.

The scientific community could benefit from moving toward a vision where contributions, e.g., data, code, protocols, etc., are equally relevant and important compared to publications and can potentially provide multiple and long-lasting scientific rewards.

Metadata repositories are proliferating and provide users with searchable catalogs to metadata records. For a (nonextensive) list of widely used metadata repositories, we refer to Box 2.1. Notably, putting metadata in a repository does not always mean the data are stored there as well. There can be several reasons not to share data right away (moratorium, time needed to publish, sensitive data, etc.), but, by making metadata available, there is an

Box 2.1 Nonextensive list of widely used and well-known online metadata repositories

Nonextensive list of widely used and well-known online metadata repositories containing or linking to marine biological data

- Arctic Data Portal (ADP, https://arcticdata.io/)
- Australian Ocean Data Network (AODN, https://portal.aodn.org.au/)
- Common Metadata Repository (NASA — CMR, https://cmr.earthdata.nasa.gov/search)
- Conservation of Arctic Flora and Fauna (CAFF, https://www.caff.is/)
- DataONE (https://www.dataone.org/)
- European Data Portal (https://data.europa.eu/en)
- European Directory of Marine Environmental Data (EDMED, https://www.seadatanet.org/Metadata/EDMED-Datasets)
- Global Change Master Directory (NASA — GCMD, https://earthdata.nasa.gov/earth-observation-data/find-data/idn)
- Integrated Marine Information System (IMIS, https://www.vliz.be/en/integrated-marine-information-system)
- International Long Term Ecological Research Network (iLTER, https://www.ilter.network/)
- Marlin Metadata System (CSIRO, https://marlin.csiro.au/)
- National Centers for Environmental Information (NOAA — NCEI, https://www.ncei.noaa.gov/)
- Natural Environment Research Council Data Catalog (NERC, NERC, https://datasearch.nerc.ac.uk/geonetwork/srv/eng/catalog.search#/home)
- Ocean Data and Information Network for Africa (ODINAFRICA, https://www.odinafrica.org/)
- Ocean Data and Information System Catalog of Sources (IOC — ODIS, https://catalogue.odis.org/)
- USGS Science Data Catalog (https://data.usgs.gov/datacatalog/)

increase in the visibility of the research and associated institute(s). Moreover, it may alert other researchers about the dataset, and they can contact the data provider for access to the data or to clarify any details or questions they might have.

The curation part of the Life Cycle also includes data governance and management, aimed at facilitating sharing and redistribution. Similar to the standards that are in place for metadata, there are also standards for data (Box 2.2). Equally important in the context of the RDLC is the fact

Box 2.2 Common guiding principles, data standards, and vocabularies in marine biology

(Most) common guiding principles, data standards, and vocabularies in marine biology.

Guiding principles

- CARE: Principles for Indigenous Data Governance
  - o Published in 2020 (Carroll et al., 2020)
  - o Applies to (Indigenous) people, purpose oriented
- FAIR: Findable, Accessible, Interoperable, and Reusable
  - o https://www.go-fair.org/fair-principles/
  - o Published in 2016 (Wilkinson et al., 2016)
  - o Applies to (meta)data
- INSPIRE: Infrastructure for spatial information in Europe
  - o European Directive—https://inspire.ec.europa.eu/about-inspire/563
  - o Published in 2007, full implementation required by 2021
  - o Applies to spatial (geographic) information
- TRUST: Transparency—Responsibility—User community—Sustainability—Technology
  - o https://www.rd-alliance.org/trust-principles-trustworthy-data-repositories-%E2%80%93-update
  - o Launched in 2019
  - o Applies to data repositories

(Meta) data standards

- ISO: International Organization for Standardization
  - o https://www.iso.org/standards.html
  - o Applies to different aspects of (meta)data (e.g., ISO 19115, 19139)
- EML: Ecological Metadata Language
  - o https://eml.ecoinformatics.org/
  - o Released in 1997
  - o Applies to metadata
- Audubon Core
  - o https://www.tdwg.org/standards/ac/
  - o Ratified in 2013
  - o Applies to metadata for biodiversity multimedia resources
- MIxS: Minimum Information about any (x) Sequence
  - o https://gensc.org/mixs/
  - o Released in 2005
  - o Genomic Standards Consortium family of minimum information standards

Continued

Box 2.2 Common guiding principles, data standards, and vocabularies in marine biology (*cont'd*)

   o Applies to metadata
- Darwin Core
  - o https://dwc.tdwg.org/
  - o Released in 2009
  - o Applies to biological data
- Access to Biological Collection Data (ABCD)
  - o https://abcd.tdwg.org/
  - o Released in 2007
  - o Applies to specimens and observations

Vocabularies

- WoRMS: World Register of Marine Species
  - o https://www.marinespecies.org
  - o Released in 2000
  - o By use of LSID
  - o Applies to taxon names, as part of biological data
- NERC Vocabulary Server and SeaDataNet Common Vocabularies
  - o http://vocab.nerc.ac.uk/ and https://www.seadatanet.org/Standards/ Common-Vocabularies
  - o Released in 2005
  - o By use of URI
  - o Applies to a broad range of concepts relevant to marine data and related domains including labels for observed variables, reported units, measurement platforms, sensors, and sampling gears
- Marine Regions
  - o www.marineregions.org
  - o Released in 2012
  - o By use of MRGID
  - o Applies to geographical (marine) place names

that users of publicly accessible data should be able to understand how the data were collected, in order to assess its fitness for use. Akin to the data itself, data collection protocols should be documented and shared as part of the metadata. Systems such as the Ocean Best Practices system (https://www. oceanbestpractices.org/) and Protocols.io (https://www.protocols.io/)

allow researchers to document and cite their methods for data collection leading to reuse of methods already created and better understanding of datasets for future reuse of the data (Box 2.2).

There are several standards for biological data and they are not always interchangeable. From the ones mentioned in Box 2.2, we highlight DwC, used by the OBIS, the Global Biodiversity Information Facility (GBIF), and the Atlas of Living Australia (ALA). DwC is maintained by the Biodiversity Information Standards association (also known as TDWG) which maintains several other biological data standards including the ABCD and Audubon Core.

The use of data standards like DwC facilitates the assimilation of multiple biological datasets into integrated data systems like OBIS or GBIF. Biological data are complex and tend to be documented in ways that are unique to each research project. By applying a standard, data from multiple projects can be integrated together and subsetting is then facilitated, allowing for a quicker Data to Wisdom process. To help improve the data usability and interoperability, several extensions have been introduced, including the Extended Measurements or Facts (De Pooter et al., 2017), MIxS Sample (Minimum Information about any (x) sequence, https://gensc.org/mixs/), and several Global Genome Biodiversity Network (https://www.ggbn.org/ggbn_portal/) extensions.

Just the mere fact that multiple formats and standards are in use stresses the need for interoperability in integration exercises, allowing all these different formats to easily communicate and intertwine with each other, while giving users a seamless experience when accessing and querying the data. Syntactic and semantic interoperability provides support for not only cross-system interoperability but also across different domains and disciplines. Ultimately, information from all marine scientific disciplines is required if we want to understand marine ecosystem functions and manage its resources in a sustainable way, as is envisioned in the UN Decade of Ocean Science for Sustainable Development.

Scientists need to keep up with a continuously evolving world of technology and discovery, leading to new ways of processing digital samples such as video and imagery, or the (semi)automatic processing of vast amounts of samples of, e.g., zooplankton by making use of a ZooSCAN (Gorsky et al., 2010). These innovations trigger the development of new guidelines on how to collect, curate, and manage the data, in parallel with directions and priorities these new branches of research will take in the future.

Phase 3—data publishing: data sharing, integration, and redistribution

In recent years, the importance of making data publicly accessible has been addressed by governments and international initiatives. Publishing should not merely be seen as the publication of findings in scientific literature but also as sharing (raw) data with the world, allowing everyone to access, reuse, and cite them.

In 2013, in the United States, three directives were published that required government agencies to make data and research results openly accessible, in machine readable ways, using open formats, standards, and including metadata. President Obama issued Executive Order 13462— Making Open and Machine Readable the New Default for Government Information. The Office of Science and Technology Policy issued a Memorandum entitled "Increasing Access to the Results of Federally Funded Scientific Research" and the Office of Management and Budget issued memorandum M-13-13 entitled "Open Data Policy-Managing Information as an Asset." The combination of these three policies defined a new way of managing data and set the path for increased access to government and government-funded data.

In Europe, in 2018, the international initiative cOALition S was launched, with the goal to make full and immediate Open Access to research publications a reality. cOALition S is built around Plan S, whereby research funders mandate that access to scholarly publications generated through research grants they allocate must be fully and immediately open and cannot be monetized (www.coalition-s.org). Although this has not yet been fully achieved, the scientific community can act by making the (raw) data, upon which results and conclusions are based, publicly available through the vast network of existing initiatives and systems.

Even with governments pushing for open access to publicly funded data, it helps when data providers are properly credited for making their data available. One way to achieve this is with Digital Object Identifiers (DOIs) which are widely used by peer-reviewed journals to increase the visibility and traceability of an article and has simplified the maintenance of citation indices currently used to assign academic credit. When it comes to data traceability, dataset DOIs work in an analogous way to publication DOIs, in that whenever and wherever a statement relies on data, the corresponding data should be cited and, through its DOI, the version used in a publication can be retrieved. A dataset for which a DOI has been assigned is fixed, cannot be changed with time, and should stay accessible in a persistent manner. This DOI will also always carry a minimum of metadata attached to it. Similar to what happens for publication DOIs, a dataset DOI can be retraced to its origin through https://www.doi.org. Initiatives such as Data Citation Index (Thomson Reuters) will link the indexed datasets to publications that cite them, and via the use of an ORCID (Open Researcher and Contributor ID), researchers can claim authorship of the datasets they have created (e.g., Paskin, 2005 and references therein; Mayernik and Maull, 2017 and references therein).

Phase 4—data processing: services and tools

In order to help scientists with data transformation, collation, and analysis, services and tools are of the essence. The setup of Research Infrastructures (RIs) helps scientists to deal with large amounts of data, from a variety of sources. Such an infrastructure "… is the foundation for an ocean observing system, providing the platforms and services to deliver environmental data, information, and knowledge. Essential components include both the hardware and core resources including people, institutions, data, and e-infrastructure that maintain and sustain operations" (European Marine Board, 2013). RIs are thus facilities, resources, and services used by scientists to carry out research and develop innovation. A number of RIs (e.g., LifeWatch ERIC, ELIXIR, EMBRC ERIC, Euro BioImaging ERIC, DiSSCo, iDigBio, and NCEAS) have been developing various Virtual Research Environments, which include many virtual Laboratories offering one stop-over access points, high computational capacity, and collaborative research platforms that support the needs of digital biodiversity science (e.g., Matsunaga et al., 2013; Arvanitidis et al., 2016, 2019, Dañiobeitia et al., 2020).

Although we—as marine (biological) data managers—are all working on the marine environment and offering our knowledge and expertise to a wide audience, from students to policymakers and the broader public, there will never be a single system that will cater to every end user's needs. The future clearly lies in an intersystem communication and interoperability and identifying and understanding the infrastructures that are already in place, including their interconnections; this is key in being able to address the wide range of user's needs, so that when a new service request arises, open communication between infrastructures facilitates cross-discipline interaction and collaboration. In addition to providing access to collated and available data, an equal share of tools and services need to be in place to assist scientists in bringing their data to the right level of content, quality, and standardization, so they can more easily be submitted to the relevant repositories and infrastructures. A number of services, e.g., those related with quality control, are available through different portals, although not all are well known to scientists.

As species biodiversity is a key element of almost all marine biology—related research, an important quality control step is to cross-check the taxon list with controlled and standardized taxonomic vocabularies, to ensure that the currently accepted and correctly spelled taxon names are being used. Several data systems have already developed such services; for example, WoRMS and LifeWatch (see Box 2.2). Similar services are being developed/already exist to match (raw) data files with the recommended guidelines, formats, standards, and vocabularies, thereby greatly augmenting the management steps that can be undertaken by the scientists themselves and lessening the data managers' workload, e.g., LifeWatch E-services or the LifeWatch and EMODnet Biology QC tool. In an ideal world, the effort required to get a dataset ready for integration and reuse is equally divided between the originators and data managers, where each can make use of a spectrum of available tools and services to enhance and speed up the process. Arguably second only to taxonomy, location is the primary consideration for developing accurate, useable datasets. The "R" package accommodates formatted data series and is highly adaptable in its approach to geographic locators such as global coordinates. Using "R" libraries such as "robis," "ggplot," and "maps," a detailed, visual representation can be produced that clearly indicates the mapped locations of data points on stylized maps suited to the data being verified. The relevant codes may be shared and/or reused for future iterations of the data or for new datasets.

Phase 5—data use and reuse

Integrating your data in larger systems and initiatives automatically gives them more visibility and (re)usability. Allowing your data to become part of initiatives such as OBIS (through its regional and thematic nodes) and GBIF indirectly allows them to be used in large-scale integrated analyses that address long-standing biological and ecological questions that are relevant not only from the scientific viewpoint but also to, e.g., policymakers. Too many of these questions remain (partially) unanswered, demonstrating that we are not yet completely there when it comes down to sharing, redistributing, and accessing scientific data to address global questions (e.g., Sutherland et al., 2006, 2013, Benson et al., 2018).

Following data integration into larger systems and initiatives, it is important that the users have an idea which data are fit for their purpose. Several approaches can accommodate for this, e.g., implementing data filters on a portal or quality flags at record or dataset level. This type of measures can be found in several initiatives that make biological biodiversity data available, e.g., EMODnet Biology, (Eur)OBIS, and GBIF. The indication of completeness and quality of a record or dataset does not only help a user get a better idea of the content and usability, from their point of view/research question, but it also helps the originator to track possible oversights which can be improved, updated, and subsequently integrated in a future data ingestion cycle for these initiatives. Also important in this aspect are the metadata and provenance information that helps a user better understand the data and its fitness for reuse.

Even though several initiatives address data quality as "good," "bad," or "poor," it can be beneficial to replace these subjective and potentially misleading labels with references to the fitness for purpose of the data. This is a crucial aspect whether it refers to datasets or records and it is dependent on the (scientific) question being addressed. When a scientist plans to perform seasonal analyses, each record should have at least a year and month information. Records with a date at only year level will not be suited for this particular objective, but will have a value in more general biodiversity-related analyses. Similar parallels are applicable to presence versus abundance information or species level versus higher-level taxon identifications. Data will have a higher grade of reusability when the mapping with controlled vocabularies is done, increasing the integration and usability options when users combine data from diverse datasets.

The Research Data Life Cycle in summary

For a scientist, the RDLC does not end when findings are published; it progresses to the next phase when the data are shared and comply with the FAIR principles. Data may never really be "finished" or "complete"; in the shared arena, data remain open to continual interpretation, reuse, and interrogation. From this point of view, scientists should not be afraid of change, as this is in fact the only certainty in data management. While facts do not change, interpretations do, and how data are presented and represented are vital to their validation.

The use of several formats, standards, and vocabularies throughout the RDLC would benefit from open discussions and consensus across the marine scientific community. Improvements in interoperability can be tackled within a specific domain as well as across domains addressing long-standing marine ecological questions that have so far remained unanswered and are urgent to address in the light of the UN Decade of Ocean Science for Sustainable Development.

> Theories desert us, while data defend us. They are our true resources, our real estate, and our best pedigree. In the eternal shifting of things, only they will save us from the ravages of time and from the forgetfulness or injustice of men.
> **Santiago Ramón y Cajal (1906 Nobel Prize winner) (Ramon y Cajal, 1999).**

Essential variables: their relevance for policies and conventions

Policymakers, and especially the scientists advising our policymakers, are major users of phase 4 and 5 of the RDLC, stressing again the importance of completing the full cycle to ensure their data to have a long lasting and highly relevant life, possibly surpassing the original goals for which it was collected.

> A scholar's positive contribution is measured by the sum of the original data that he contributes. Hypotheses come and go but data remain.
> **Santiago Ramón y Cajal (1906 Nobel Prize winner) (Ramon y Cajal 1999).**

To advise communities, managers, and policymakers of changes occurring in the ocean in response to human use, several standardized, regular, and long-term observations could be taken. Since all variables cannot be monitored everywhere at any given time (nor need to be), a framework of "Essential Variables" (EVs) has been implemented by the scientific community. Currently, there are three types of EVs: climate, ocean, and biodiversity variables. The Essential Climate Variables (ECVs) were established by the Global Climate Observing System of the WMO (WMO, 2016) and can be either a physical, chemical, or biological variable or a group of linked variables that critically contribute to characterize the Earth's climate. The ECVs are key to informing negotiations under the United Nations Framework Convention on Climate Changes and the Intergovernmental Panel on Climate Change.

The EOVs were established by the UNESCO's (United Nations Educational, Scientific and Cultural Organization) IOC under the Framework for Ocean Observing to guide the Global Ocean Observing System (GOOS) program (Lindstrom et al., 2012; Tanhua et al., 2019). Like the ECVs, the EOVs of GOOS also include a set of physical, biogeochemical, and biological variables which were defined by three GOOS expert panels: panel on physics and climate (built on the Ocean Observations Physics and Climate Panel or OOPC), panel on biogeochemistry (built on the International Ocean Carbon Coordination Project or IOCCP), and the panel on biology and ecosystems.

The GOOS panels on physics, climate, and biogeochemistry defined their EOVs based on specific scientific and societal requirements driven mostly by climate change and the need for weather forecasts. The GOOS biology and ecosystems panel defined the biological EOVs by considering societal needs (societal drivers and pressures) according to more than 20 international conventions and/or multilateral agreements relevant to marine life (see Table S2 of Miloslavich et al., 2018a). It used a Driver—Pressure—State—Impact—Response model to identify biological and ecological EOVs that are relevant for science, inform society, and are technologically feasible. For this, the GOOS BioEco Panel (1) examined relevant international agreements to identify societal drivers and pressures on marine resources and ecosystems, (2) evaluated the temporal and spatial scales of biological variables measured by more than 100 ocean observing programs, and (3) analyzed the impact and scalability of these variables and how they contribute to address societal and scientific issues. Societal drivers identified in (1) include sustainable use of biodiversity, biodiversity conservation, biodiversity knowledge, environmental quality and threat prevention and mitigation, capacity building, sustainable economic growth, ecosystem-based management, and food security. The anthropogenic pressures of concern of the agreements were mainly loss of habitat and biodiversity resources (including losses through overfishing) followed by climate change, pollution and eutrophication, coastal development, invasive species, solid waste, ocean acidification, extreme weather events, noise, and mining (Miloslavich et al., 2018a). The identified biological EOVs not only considered societal relevance to inform several international conventions and agreements and built upon a long history of exploration and observing from an engaged scientific community but also built on previously proposed technical and scientific frameworks (e.g., UNESCO/IOC, 2014; Grimes, 2014; Constable et al., 2016; NOC, 2016).

Finally, the Essential Biodiversity Variables (EBVs) were established under the framework of the Group on Earth Observation Biodiversity Observation Network and have been grouped into six classes: genetic composition, species populations, species traits, community composition, ecosystem structure, and ecosystem function (Pereira et al., 2013, Fig. 2.3). However, this EBV concept was rooted mostly on terrestrial communities.

In the past 5 years, the observing community has made significant progress in harmonizing all these different types of EVs both across and within disciplines. Specifically, for life in the ocean, biological EOVs and EBVs have converged, with EBVs measuring specific aspects within the biological EOVs. The EBVs will evaluate taxa across scales of spatial and temporal diversity within species (allelic diversity, species distribution, population abundance, population structure by age/size class, and phenology), across species (taxonomic diversity), and of the ecological context (primary productivity, secondary productivity, habitat structure, ecosystem extent/fragmentation, and ecosystem composition/functional type) (Muller-Karger et al., 2018a).

To implement biological EOVs, the biology and ecosystems panel of GOOS has adopted a strategy of systematically building or strengthening a community of practice for each of these. This has led to the revitalization of the Global Coral Reef Monitoring Network (Obura et al., 2019), the unification of several seagrass networks (Duffy et al., 2019), and the

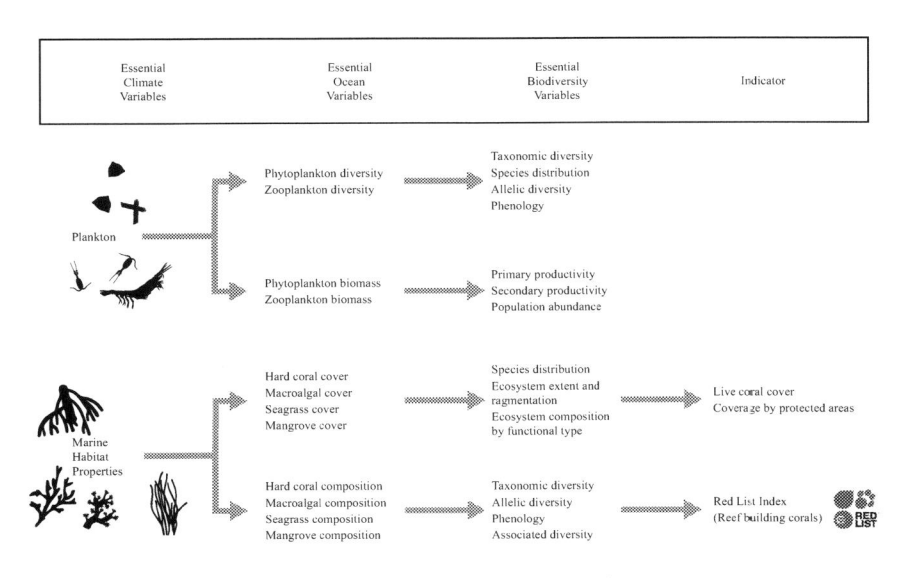

Figure 2.3 Essential variables and indicators.

establishment of the Global Ocean Macroalgal Observing Network and initiated the first steps toward the drafting of implementation plans for several of these biological EOVs (Miloslavich et al., 2018b; Miloslavich et al., 2019a,b). In this way, the biology and ecosystem panel has adopted a collaborative international approach stressing the importance of implementing standardized best practices, developing new technologies, strengthening data sharing and interoperability, and enhancing capacity building and technology transfer. These developments are needed to increase the availability and impact of scientific advice to decision-makers.

For global implementation, several steps need to be taken within each EOV network. These include the following: (1) identifying existing datasets at all geographical scales, (2) reviewing the technological monitoring approaches and defining Standard Operating Procedures, (3) recommending approaches to consolidate existing data and associated metadata in a data system under FAIR principles, and (4) planning the implementation of an international, standardized, innovative, and cost-effective system for a fit-for-purpose EOV monitoring (Miloslavich et al., 2019a). Over the next decade, technological developments facilitating automated measurements, along with the needed improvements in metadata and data architecture and making data FAIR, will be critical. Equally important will be to agree on common metadata standards facilitating scalability and sustainability (Miloslavich et al., 2019a).

Currently, the metadata for the physical EOVs are centralized by a Joint WMO-IOC Center for in situ Ocean and Marine Meteorological Observing Program Support (now called OceanOPS https://www.ocean-ops.org/board). OceanOPS is under the guidance of an Observations Coordination Group, which has the role of engaging networks to address new requirements, develop metrics to assess the performance of the observing system, facilitate international exchange of data and metadata, and encourage best practices on data management (Tanhua et al., 2019).

OBIS is identified as one of the main global data systems to integrate the data for biological EOVs and is recognized by the IOC member states as a long-term repository for oceanographic data and associated metadata (clause 5 of the IOC Oceanographic Data Exchange Policy (https://iode.org/policy)). Besides integrating and quality controlling (raw) data, OBIS is currently developing an online metadata interface and portal so that biological observing networks and their attributes can (1) be displayed and queried interactively, (2) are publicly accessible, and (3) can be regularly updated as observing programs change. This new feature will also promote groups to monitor and develop EOV time series data on OBIS.

In order to improve data availability issues, targeted efforts by the international community have concentrated on the development of the EOVs to promote cost-effective, interoperable observations across multiple platform types and on EBVs to provide information on the traits, health, and diversity of organisms and communities. Efforts to collate existing historical data should complement this approach by providing, wherever possible, data-based descriptions of past conditions that are consistent with these selected variables and monitoring strategies. It is thus key that the past and the present are brought together.

Use cases and stories

Throughout the section, several projects and initiatives have been mentioned although none have been described in detail. Elaborating on all these projects and initiatives, their link with the RDLC and why they have been so important in the evolution of marine biological sciences could be a book on its own; for this reason, we constrained the details to a few highlights, divided into four major (decadal) timespans: (1) before 1990, (2) 1990—99, (3) 2000—09, and (4) 2010—20.

Before 1990: looking back into history ...

Before scientists turned to the digital era, data were already collected and organized in oral transfers and paper formats (see Section Historical data). Several projects have been dealing with the digitization of these valuable resources to avoid their loss and to allow the data to be part of hind- and forecasts in relation to the life in oceans and seas.

That we sometimes underestimate the knowledge of what was collected in early centuries and the insights we have acquired in how oceans and seas were exploited in those early days is obvious, as Callum Roberts demonstrates in his book *The Unnatural History of the Sea* (Robert, 2007), or the many efforts that are now being done to catch up with IK, to be captured from oral tradition, where language is the actual historical record of what has been there (https://www.insidescience.org/news/conserve-species-traditional-local-knowledge-and-science-can-complement-each-other; Nicholas, 2018). The knowledge systems developed by indigenous communities are the result of countless generations building on the learned experiences (either at individual or collective level), which have been verified by the elders and conveyed through guided experiential learning, oral traditions, or other means of record keeping (Nicholas, 2018).

In 1975, an important milestone was reached in sharing information on published sources, by the establishment of the International Association of Aquatic and Marine Science Libraries and Information Centers (IAMSLIC). IAMSLIC was specifically set up to share information across different institutions and scientists on marine and aquatic realms (https://iamslic.wildapricot.org/).

From the pre-1990s, we highlight two historical study cases, giving more detailed insights in the process of digitizing paper-based resources and making them available to the scientific community.

Newfoundland English cod fishery 1698–1832

The HMAP project highlighted how human interactions have directly impacted fish stocks and fishing grounds over time but required intensive methodological approaches to ensure the veracity of the data it supplied. HMAP has made it possible to digitize and make publicly available a dataset on the Newfoundland English cod fisheries from 1698 to 1832.

The Newfoundland fishery of the early modern period has been documented and explored by maritime historians at some length. The case study pertains to work carried out by Haines (2004) and Pope (2003) that relate to archived Admiralty and Colonial records held at the British National Archives, Kew, London; original data were also obtained from archived correspondence and Board of Trade documents pertaining to the period.

The process of compiling the dataset involved some considerable effort in reading what were often almost unintelligible faded scripts on degraded paper dating back to the 17th and 18th centuries. The work of the historians furthermore required them to mine for data that were not always in formats that could be readily accessed. Numerical values for cod fish catches were typically recorded in antiquated measurements ranging from quintals to barrels and tons, to actual numbers of specimens. Values were typically reported as dried salted fish, which implied that they had been gutted and headed, then dried using salt and cold, dry, fresh air for several days. Conversion factors had to be carefully considered and even the actual numbers of vessels had to be carefully checked to avoid over or underreporting. Locations of the animals reported had to be determined based on their port of departure and a specific awareness of the fishing grounds of the Grand Banks and the Newfoundland coast. Quantifying number of vessels, catch weights, and coordinates was a painstaking task that required skill and perseverance.

Once compiled, the data had to be interrogated and tested against the original archival documents. In some instances, values were incorrect or even false in the case of a reporter's possible bias or exaggeration. For example, where a record adopted values given by a merchant in a letter to a tax official, it is quite possible that the reported number of fish caught may have been deliberately underestimated in order to pay less tax. In cases where reports of complaints about other fishermen are concerned, values may have been overstated in order to maximize the impact of an argument. These considerations had to be factored into the process of determining accurate values.

The next phase required the data to be processed into a uniform and accessible format that could be shared and reused by future users. The DwC metadata system was adopted and the data were channeled to accommodate the required and recommended fields of this format. Once a coherent and reliable format was compiled, the data were able to be published and disseminated through OBIS and become elements in the big data arena, shared among many other datasets of similar or aligned purpose. However, the data in and of itself held little value outside of the data mining research community until they were made available as a single and coherent dataset which included accompanying documentation that explained the basis of each record and the meaning of each adopted field. Conversion factors were explained and values which had to be revised, calculated, or assumed were clearly identified and explained as well.

The final output was a dataset, together with accompanying explanatory documentation, made available in a permanent, secure online repository; the (raw) data that had been submitted to OBIS were processed and submitted in the DwC metadata format to ensure data and metadata compatibility. The secure data were made available in the public domain using a Creative Commons licensing system that enabled researchers to accurately reference the resource and the original data providers in an academically robust format.

The reusability of these data is evidenced in the reliance of current researchers on the information it provides. For example, the NorFish project (an Environmental History of the North Atlantic 1400–1700) bases part of its developed theme of understanding the "Fish Revolution" in the early modern North Atlantic on these data; the data have been reviewed, interpreted, and incorporated with other relevant data to build an overview of overall fishery activities during the period (Holm et al., 2019).

Rescuing legacy data from historical oceanographic expeditions

Recognizing the importance of recording marine biodiversity became widespread from as early as the beginning of the 20th century. Many expeditions were organized to explore and investigate the oceans, including researching marine biodiversity around the world. Collected information, including samples and observations, was recorded in a plethora of scientific volumes; in general, the aims, sampling station data, and metadata (such as coordinates, depth, gear type, etc.) were published in introductory volumes. Specimens were often preserved and sent for expert evaluation and taxonomic identification. The outcome was a host of publications in varying formats, languages, and styles.

A typical example of this process was the Danish Oceanographic Expeditions (commenced 1903), under the leadership of Johannes Schmidt, which used a variety of research vessels in the Atlantic Ocean, the Mediterranean Sea, and adjacent waters of the Sea of Marmara and the Black Sea. One of these research vessels, the Thor, investigated marine biodiversity in the Mediterranean between 1908 and 1910. The aim of the expedition was to (1) understand the hydrography of specific regions including their merging zones, (2) record the occurrence (or absence) of the most common taxa within the existing environmental conditions at varying depths, and (3) to increase understanding of the ecology of marine species. To achieve this, sampling stations were allocated in a manner that enabled the comparison of species composition in areas with different environmental profiles, with both biological and environmental data being recorded at each station (Mavraki et al., 2016).

Within the activities of the LifeWatchGreece RI, four legacy datasets of the Thor expeditions, published in separate volumes, were digitized, quality controlled, and integrated into a harmonized dataset according to the DwC schema by Mavraki et al. (2016) with the aim to prevent potential loss of valuable legacy data on the Mediterranean marine biodiversity. The datasets covered 2043 samples collected from 567 stations between 1904 and 1930 in the Mediterranean and adjacent seas. The overall initiative resulted in 1588 occurrence records of calcareous algae (Rhodophyta), pelagic polychaetes, and fish (families Engraulidae and Clupeidae). Basic environmental data (e.g., sea surface temperature and salinity) as well as meteorological conditions were also recorded for most sampling events.

Typical difficulties encountered during the overall digitization process included inconsistency of data structure among and within the different volumes; identical codes for sampling stations in different locations/expeditions; taxonomic inconsistencies and updates; absence of unique sample codes; lack of precision in coordinates (e.g., coordinates for marine stations falling on land) or provision of toponyms instead of coordinates; and difficulties in the interpretation of ambiguous symbols in tables (e.g., asterisks vs. ditto marks vs. no data). The digitization process, along with the primary obstacles encountered and strategies to overcome such problems, was described in step-by-step detail in a publication (Mavraki et al., 2016) in a special collection of articles on LifeWatchGreece RI (Arvanitidis et al., 2016).

Recognizing the importance for mobilizing historical marine data, EMODnet Biology has developed a long-term data archaeology and rescue strategy. The overall objective of this initiative is to fill spatial and temporal gaps in aquatic species occurrences and make the rescued historical data available freely through the EMODnet portal and global biogeographic and biodiversity information aggregators (e.g., OBIS and its regional nodes, WoRMS and its subregisters, LifeWatch Species Information Backbone, GBIF, etc.). The experience gained has revealed several challenges at all stages of the data "life cycle," from dataset identification and metadata extraction to the digitization, standardization, and quality control of historical datasets. Nevertheless, facing these challenges is of paramount importance since loss of such data equals loss of unique resources required to understand global changes and ultimately to the loss of our natural wealth.

The digitizing of the Thor expeditions provided hitherto unexplored awareness of the biodiversity of the Mediterranean Sea, but also highlighted the difficulties inherent in attempting to retrospectively standardize a series of scientific evidence into a data stream.

The 1990s (1990—99)

The 1990s were marked by a growing realization that long-term comparative investigations on marine biodiversity were needed, and that these ought to span more ground than the backyard of any institution. Regionally and globally focused institutions and initiatives were already in place, e.g., IODE, ICES, MBA, and CPR, but the means to easily share, exchange, and combine data in different formats and standards had not been implemented, although ideas and intentions existed.

On a regional scale, the North Sea Benthos Project (NSBP) was initiated in 1999. This initiative—under the ICES Benthos Ecology Working

Group—aimed at bringing together national efforts around the North Sea, to resample the stations where benthic samples were collected in 1986 as part of the North Sea Benthos Survey (http://www.vliz.be/vmdcdata/nsbs/about.php), when several institutes and vessels from five countries (Germany, Scotland, the Netherlands, France, and Belgium) engaged in the sampling in the southern part of the North Sea. The sampling had not only focused on the macrobenthic infaunal communities but it also generated data on the physical—chemical state of the associated sediment and on the meiofauna and epifauna. The new survey was aimed at evaluating possible changes in the assemblages, in relation to natural or human influences. Where national dedicated efforts were not possible, contributions from ongoing national research and monitoring were sought, allowing for a comparable holistic assessment (Rees et al., 2007).

The NSBP was a successful endeavor (Rees et al., 2007). The report itself contained a clear statement on how the data should be made available after the project, either after an embargo period or after publication of the results, whichever came first. All data within the NSBP database became freely available during the subsequent decade and is still being used in studies related to the biodiversity of the benthos in the North Sea.

In 1996, efforts were put toward collating species lists and their distributions from a number of documents by developing the Marine Species Database for Eastern Africa (MASDEA), as part of the Kenya—Belgium Project (http://www.vliz.be/vmdcdata/masdea/). MASDEA has since been upgraded to the African Register of Marine Species (AfReMaS), which was developed to encompass species from the whole African coastal states. In addition to this initiative, data sharing was championed by the project "Regional Cooperation in Scientific Information Exchange in the West Indian Ocean (RECOSCIX-WIO)" that took place between 1989 and 1996 by the IOC with the Kenya Marine and Fisheries Research Institute as a Regional Dispatch Center through a network of marine libraries in Eastern Africa.

The 2000s (2000—09)

With the new millennium, new opportunities toward collaborative approaches and data sharing and integration came to life. This decade also gave birth to social media platforms such as Facebook (2004) and Twitter (2006), which have helped tremendously in spreading knowledge on biodiversity based on the available data. Major institutions, initiatives, and projects still use these and other social media platforms to reach out not only to the scientific community but also to policymakers and a broader audience, thereby increasing the general knowledge and awareness of their activities.

A wide suite of projects and initiatives were launched during this decade. We will highlight two of them, which are considered to have had, and continue to have, a major impact on the way the marine biological science discipline has evolved.

Census of Marine Life

The CoML project was initiated in 2000 as a 10-year project to improve our knowledge on marine biodiversity (www.coml.org; Alexander et al., 2011).

As CoML aimed to make all collected data available online, there was a need to develop a system to make this possible, which became the Ocean Biogeographic Information System (OBIS) which, recently, in 2020, was renamed as Ocean Biodiversity Information System (https://www.obis.org).

Among the hundreds of publications the Census produced, it also created a collection with a synthesis of the state of knowledge of marine biodiversity worldwide (Marine Biodiversity and Biogeography — Regional Comparisons of Global Issues, 2010 PLoS Collections: https://doi.org/10.1371/issue.pcol.v02.i09), which represents an unprecedented effort of compilation and analysis of global literature and data on marine biodiversity and provides a valuable insight for science and for policymakers (Costello et al., 2010; O'Dor et al., 2010). This collection was largely contributed by the 13 National and Regional Committees of the Census. These were established through early "Known, Unknown, Unknowable" workshops to diagnose the state of knowledge in the countries and large regions such as the sub-Saharan Africa, the Caribbean, South America, and Europe (Alexander et al., 2011).

One of the CoML program's projects was NaGISA or "Natural Geography in Shore Areas." NaGISA's goals were both to create a global quantitative baseline of coastal biodiversity including species ranges and habitat distribution, and identifying hotspots of marine coastal biodiversity, and to provide information to educate stakeholders (e.g., policymakers, students, and the public). In the South American region, between 2005 and 2010, NaGISA collected data using a standardized protocol in Venezuela, Trinidad and Tobago, Uruguay, and Argentina (Rigby et al., 2007). The protocol was later adopted and expanded by the South American Research Group in Coastal Ecosystems (SARCE) (SARCE, 2012).

Marine biodiversity ecosystem functioning

In parallel with the start of CoML, preparations were being made in Europe by Carlo Heip to initiate a major, long-term NoE, aiming at building long-lasting alliances between European marine researchers, their institutes, and universities and to instigate the development of central data systems at the service of the marine scientific community. Carlo Heip's plans and vision came together in the MarBEF EU Network of Excellence that ran from 2004 to 2009.

As a central part of the data management work within MarBEF, EurOBIS was established. MarBEF also gave birth, in 2004, to the first online register of marine species names, through the European Register of Marine Species (ERMS), which later evolved to the WoRMS. ERMS was originally available in book format and as MS Excel files, as opposed to the UNESCO-IOC Register of Marine Organisms (URMO), led by Jacob van der Land in the 1990s, which was the first attempt to compile an electronic list of all marine species then published on a 3.5″ floppy disk (van der Land, 1994, 2008).

Transferring ERMS and URMO's holdings to an online editing platform opened up a wide scale of possibilities: through a simple internet connection, these registers could be freely consulted from anywhere in the world, services could be developed to allow scientists to easily access taxon names, and also involve experts to easily update the taxon records. ERMS also virtually brought together a vast network of taxonomists, who dedicated time to keeping the register up to date thus ensuring high levels of quality of its content. Although ERMS was not the first online available species register—FishBase (Froese and Pauly, 2019), AlgaeBase (Guiry and Guiry, 2020), among others preceded it—ERMS and URMO were unique in the sense that they did not focus on a single taxon group but grouped an entire environment and engaged a large community of experts in its maintenance. In parallel with these initiatives, similar endeavors were undertaken in the late 1990s and early 2000s with the goal of organizing species names into registers and thereby triggering the start of biodiversity informatics (e.g., Catalogue of Life, Integrated Taxonomic Information System [ITIS], etc.) (Costello et al., 2013).

The ITIS (https://www.itis.gov/), originally formed in 1996 and currently supported by U.S. Department of the Interior, U.S. Geological Survey, and the Smithsonian Institution, was designed to provide consistent and reliable information on species taxonomy. The original ITIS database design was used as a model to create Aphia, the database behind the European and later the World Register of Marine Species (ERMS—WoRMS) (Vandepitte et al., 2015).

Within MarBEF, several of the Responsive Mode Programs aimed at bringing together biological data from several backgrounds/original goals, institutes, and countries together, to be able to address European/global issues, which could previously not be addressed due to the lack of integrated databases. It is fair to say that several of these projects within MarBEF have opened up possibilities to address more global biodiversity questions, instead of focusing on specific regions. The MacroBen (Vanden Berghe et al., 2009), MANUELA (Vandepitte et al., 2009), and LargeNet (Vandepitte et al., 2010) databases all captured both the biotic data (presence/abundance/biomass) and the associated abiotic data (sediment characteristics, levels of disturbance, physicochemical data, etc.). At the end of MarBEF, in 2009, the biological biodiversity data had largely found their way to several large-scale systems such as (Eur)OBIS and GBIF, but the associated abiotic data remained "hidden" in the offline integrated databases. These data were made available upon request, but could not be stored in (Eur) OBIS or GBIF together with their biological data.

The legacy of CoML and MarBEF

The initial onset of ERMS (2004) was carried through to the WoRMS, and still today, the network of taxonomists and thematic and regional experts involved in the Register, on particular habitats or marine regions, for example, is ever growing (Vandepitte et al., 2018; WoRMS Editorial Board, 2020). In the years following the start of WoRMS, the Register became adopted as the authoritative reference of marine species names by marine data centers.

In 2004, EurOBIS became one of the regional nodes for OBIS. Through this network, all biogeographic data collected in EurOBIS were shared with OBIS, contributing to the general goal of CoML. Both EurOBIS and OBIS are still operating and improving their systems and tools to make sure they can serve the goals and needs of the biological scientific community in the best possible way.

Building on the CoML NaGISA efforts, the SARCE was established in 2009 to continue to assess and fill knowledge gaps on marine biodiversity and biomass along the Pacific, Atlantic, and Caribbean coasts of South America. SARCE was able to demonstrate not only that species diversity of South American rocky shores followed a latitudinal gradient (temperature related) but also that regional and local environmental processes were important in understanding the increase in regional diversity toward the tropics (Cruz-Motta et al., 2020).

MarBEF also initiated the World Conference on Marine Biodiversity (WCMB), whose first edition took place at the end of the project, in November 2008, and has been organized ever since, with a 3/4-year cycle. The WCMB has become the major focal assembly to share research outcomes, management and policy issues, and discussions on the role of biodiversity in sustaining ocean ecosystems.

The 2010s (2010—20)

Successful initiatives that began in the previous decade continue to exist and thrive, such as OBIS and its well-established regional and thematic nodes, WoRMS, and many other biogeographic and taxonomic databases (e.g., FishBase, SeaLifeBase, ITIS, Catalogue of Life, the GBIF, etc.). All these systems continuously evolve with the technical developments ensuring interlinkage and availability of a wide array of tools and services that assist users in obtaining access and make the best use of the systems in place.

In 2015, the OBIS community took the initiative to ensure that datasets containing biotic and associated nonbiotic data were kept together, by developing an extended DwC format, named OBIS-ENV-DATA (De Pooter et al., 2017). This format allows the joint management and exchange of biodiversity observation data and environmental measurements, enhancing the value and benefits extracted from the data for marine sciences, biological analysis, and modeling. Two years after its implementation, the OBIS-ENV format was published and accepted under TDWG/DwC as an official extension and has since been widely used by the biological community (https://tools.gbif.org/dwca-validator/extension.do?id=http://rs. iobis.org/obis/terms/ExtendedMeasurementOrFact).

It is undisputed that the data within all these systems are being used, on their own or combined across systems and disciplines, as can be ascertained from the many publications that cite them and the purpose they have for policy and biodiversity management. Data retrieved from several of these systems have led to global analysis linked to, e.g., climate change effects on global marine biodiversity (Jones and Cheung, 2014). The outcomes of this particular study then found their way into two different policy documents: FAO report on impacts of climate change on fisheries and aquaculture (Barange et al., 2018) and GEO-6: Global Environment Outlook: Regional Assessment for Africa (UNEP, 2016).

The first World Ocean Assessment (WOA I), in 2015, provided an important scientific basis for the inclusion of ocean issues by governments, intergovernmental processes, policymakers, and other stakeholders involved in ocean affairs (The Group of Experts of the Regular Process, 2017). The assessment reinforced the science policy interface and established the basis for future assessments. In early 2020, the UN members received an invitation from the cochairs of the "ad-hoc working group of the Whole on the Regular Process" to proceed with a review of the first assessment that resulted in a second World Ocean Assessment (WOA II, 2021a,b). The second assessment brought together a group of experts brought together knowledge from across oceanographic disciplines to assess the current global situation (https://www.un.org/regularprocess/)(ISBN 978-92-1-130422-0. xxiii, 543 pp.; ISBN 978-92-1-130422-0. xvii, 500 pp.).

The combination of the WOA I and WOA II, alongside related initiatives, will assist with the implementation of the 2030 Agenda for Sustainable Development, specifically for its ocean-related goals. The publication of the WOA II will coincide with the onset of the UN Decade of Ocean Science for Sustainable Development (2021−30) and will thus be able to provide a good basis for the state of the art on our knowledge on this vast and majorly important ecosystem.

Other initiatives which used the most current and available scientific, technical information, and expertise included the establishment of the Ecologically and Biologically Significant Areas through the CBD, after the 2008 CBD Conference of Parties (COP) 9, some of which have evolved to be designated as Important Marine Mammal Areas globally (https://www.marinemammalhabitat.org/immas/).

Throughout the decade, we have not only seen the continuation of existing initiatives but also the revival of previously discontinued initiatives, leading to new endeavors and collaborations. The SARCE network, for example, saw a reactivation in 2016, with the establishment of the Pole to Pole project of the Marine Biodiversity Observation Network (MBON P2P). The MBON P2P was conceived as an international network of collaborating research institutions, marine laboratories, parks, and reserves seeking to address common problems related to sustaining ecosystem services through conservation ecology. MBON P2P is currently being developed as a regional collaboration throughout the Americas to collect biological data in coastal habitats (rocky shores and sandy beaches) using common methodologies and sharing best practices (Canonico et al., 2019). Data collected by the MBON P2P project will cross its own borders: the data will be uploaded to OBIS following the "From the Sea to the Cloud" strategy, making sure that the data are readily available and also contribute to the GOOS of the IOC of UNESCO. GOOS leads the implementation and coordination of sustained marine observing needs and practices around the world under the framework for ocean observing (Tanhua et al., 2019) by making available data on biological EOVs (Miloslavich et al., 2018b).

At the turn of the decade (2009–10), the European Marine Observation and Data Network (EMODnet) was established, a network of organizations supported by the EU's integrated maritime policy. These organizations work together to observe the sea, process the data according to international standards, and make that information freely available as interoperable data layers and data products. EMODnet provides access to European marine data, layers, and products across seven discipline-based themes (bathymetry, biology, chemistry, geology, human activities, physics, and seabed habitats), with the ultimate goal of providing cross-disciplinary access for scientists, policymakers, and other stakeholders described in the quadruple helix. Within the biological theme, the major focus is on making available data products, taking into account the identified challenges (Lear et al., 2020).

Operational Oceanographic Products and Services (OOPS), sometimes also referred to as monitoring indices, are being developed. Such indices are a simplification of the reality of the oceanographic state and of the processes that have an effect on ecosystem functioning; however, they support the reporting and description of the ecosystem. OOPS combine various variables for a variety of oceanographic disciplines and are dynamic, in a sense that they are regularly updated and aim to provide an almost real-time view on the status of the ecosystem. These types of products are highly relevant and needed, e.g., in the framework of EVs, but achieving them was unthinkable roughly 10–25 years ago (https://www.ices.dk/news-and-events/news-archive/news/pages/zoom-in-on-zooplankton-data.aspx).

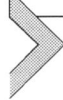

Toward the next decade: what are the challenges we are facing?

Progress on different marine biology challenges is being made. We are addressing not only scientific challenges but also those related to data accessibility and interpretation among others. Many of these challenges are being met because as a community, ocean scientists are generating data that are being used by various stakeholders, including academia, policy, industry, and the public at large.

Each challenge we face is highly dependent on the perspective. From a data provider point of view, a distributed infrastructure can lead to confusion when it is not understood where the data originate from, whereas those more aware of data systems may not face this challenge. In parallel, one can distinguish between the user and the provider viewpoints: one can argue that the end user may not need to know where the data come from exactly, but the provider does need to understand this, as that might define how they will provide their data, in terms of formats, standards, and vocabularies as well as submission possibilities. While the marine biological community is making data freely available, the existence of different systems, infrastructures, standards, networks, or applications might lead to users' confusion. Therefore, an awareness of all the connections must be demonstrated and clearly explained to address this constraint and allow users to make the best use of the tools and services available.

Although data systems are being used extensively, there is a need for RIs to link together, allowing an even wider geographical and temporal coverage of available data and a more user-friendly way to integrate different types of data (e.g., hydrography, biology, chemistry, etc.) for different user groups. The 2010s have given rise to several RIs, all with the potential to help us progress into the next decade and address the needs and demands of the scientific community, policymakers, industry, and the wider public. Within these networks and infrastructures, it will remain important that each clearly displays how they interlink and collaborate with other initiatives, where there might be collaboration possibilities and how data and services can be used either independently or by linking with other systems. If users cannot understand the many interconnected relationships, it will create more difficulties in working effectively and efficiently in the virtual world of marine (biological) science (Box 2.3).

From the authors' experience, it can be said that the marine biological data community is sitting at the cross-roads between the past and the future: we not only need to manage the large amounts of data that are being made available continuously and instantaneously (almost real-time) through digital channels (e.g., ZooSCAN, satellite images, underwater video, etc.), but we also need to manage the vast amount of valuable biological data that are still buried in journal articles, cruise reports, or in other nondigital, offline formats. Although it would take a tremendous push to liberate these data and make them accessible, their value is priceless, as they represent a baseline for understanding changes in biodiversity over time. As a variety of (meta) data standards, vocabularies and ontologies exist for biological data, choosing

Box 2.3 The acronym soup

From the opening speech of the GBIF Governing Board Chair at the 24th GBIF Governing Board meeting (Peter Schalk, 2018).

Acronym soup …

Imagine me [*the GBIF Governing Board chair*] visiting our Ministry of Science, Culture, and Education, and reporting to our State Secretary …

About my visit to the **COP13**, organized by the **CBD**. Where advice was given by **SBSTTA**. Important statements were made by **TDWG**. Representatives of **IUBS**. Follow up **GTI** was discussed. Representatives of the **CoL** and **EoL** flanked by **BHL** and **BOLD** agreed on collaborating on **COL-PLUS**. This was welcomed in the meeting by the **FAO** and **UNEP** representatives, who said funding may be made available through **UNDP** and **GEF**. In Europe, **CETAF** members got organized into **DISSCO**, now underpinned by the **ICEDIG** and **SYNTHESYS** projects that were funded under **H2020**. **LifeWatch** was absent. In the United States of America, **iDiGBio** flanks this initiative. **ALA** will join **CoL-PLUS**.

those that best fit the data, methodology, and management goals can be confusing and time-consuming process which might not always result in a clear decision. The marine biological community could benefit from an army of trainers to inform or educate scientists and nonscientists on how to efficiently and correctly digitize paper-based data; an army of data managers or data scientists to conduct all the work necessary to standardize and make accessible all the data previously and currently being collected; as well as institutional support in acknowledging the importance of data management and participation in the development of best practices. To overcome and avoid some of these pitfalls, it is essential that we continue to train future data managers and data scientists.

The marine biological community, in contrast with other disciplines, faces a significant lag in time between when data are collected and when they are made available for reuse (Muller-Karger et al., 2018b). Processing biological samples is and will remain time consuming: identifying species with the aim of getting an insight of the biodiversity at a location requires not only good identification skills, but also knowledge of the field of taxonomy This discipline required to describe newly discovered and undescribed species on the many shelves at labs and universities is under pressure: fewer and fewer young scientists go into the field of taxonomy, as describing a species is not (no longer) seen as a "sexy" part of the marine biology discipline, although it is very highly needed (Boxshall, 2020; Orr et al., 2020;

Costello, 2020). More and more focus is going toward the field of genetic identification, demonstrated by the many research papers in this field (e.g., Palumbi, 1996; Holland, 2000; Rodrigues da Silva et al., 2010; Jorger and Schrodl, 2013). And although genetic analysis can greatly speed up the identification of species in the marine environment, it should not be seen as the discipline that replaces taxonomy. On the contrary, both disciplines—taxonomy and genetics—can collaborate and make each other stronger. The general situation of a decreasing number of taxonomists is made worse when the taxonomic keys they developed are not updated or lose value if they are not synchronized or used alongside the new science of molecular biology and eDNA. For decision-makers to be able to respond appropriately to current changes, this time lag and lack of taxonomic expertise can be troublesome: they need current and correct data to feed into their policy processes. The whole community needs to move into a state of "as close to real time sharing and integration of data as possible" while tackling the challenges this brings with it.

> Taxonomy is described sometimes as a science and sometimes as an art, but really it's a battleground.
> **Bill Bryson, A Short History of Nearly Everything (Bryson, 2003).**

From the authors' perspective, it is clear that, besides the lack of skilled taxonomists, taxonomic name resolution also remains a challenging area, with data managers, many of whom are not taxonomists, struggling as the intermediary link between data providers with particular taxonomic concepts and authoritative catalogs that do not always align in their classification. Taxonomic services, such as taxon matching tools, can return different results depending on the service used and also use varying Application Program Interface methods, sometimes with inconsistent levels of documentation and requiring new skills or an understanding of the new formats in use. Additionally, the rise of the "omics" approach creates an ambiguity in how to best represent a taxon and how to integrate the concepts together. Although there is interaction between the fields of taxonomy and genomics, their linkage is not yet optimal, possibly leading to confusion on the actual species diversity and the interpretation of the data of these disciplines when they are being combined. It is thus key to document in detail which services exist, what they can and cannot do, and how ambiguities and doubts can be addressed.

Challenges do not just present themselves in the first phases of the Data Life Cycle; when data are acquired and curated, they are virtually present in all phases of the cycle, on different levels and in a variety of size categories. Processing, using, and reusing data are not always straightforward; as an example, let us look at the macroalgae-related EOV where challenges were evident when bringing the data together. Metadata compiled from about 80 networks revealed a huge heterogeneity of spatial and temporal sampling, methodological variability, data availability, and format and sustainability of the data collection (Miloslavich et al., 2019b).

To integrate existing and future data, a data management architecture that allows standards-based data and metadata, from multiple sources, to be harvested in a centralized global data repository linked with OBIS through web services is suggested. For this, data standards and formats for each of the monitoring methods for the different EOVs, along with vocabularies for each variable recorded, nontaxonomic categories, methods, instruments, and units, will need to be agreed upon. A practical example using the OBIS-ENV approach in DwC as part of the OBIS manual was developed for macroalgae cover: https://obis.org/manual/dataformat/ #example (Miloslavich et al., 2019b).

A key challenge in the early phases of the RDLC is for the actors of the data creation process (manufacturers of scientific sensors and platforms, research scientists, and technical engineers) to work closely with data and information specialists in their field of research, to ensure that the data they collect can be reused in multiple ways. In order to do this, they need to apply or refer to data standards in machine readable formats at the early stages of the data life cycle.

Currently, the large-scale integration process of linking multidisciplinary data is somewhat inefficient and prone to errors because the mapping of the data granules to standards is done retrospectively by data managers working in data repositories. These data managers often need extra information and background research in order to ensure that the data are correctly identified against standardized vocabularies. Occasionally, this is simply impossible, as information is not available, and the data will have to be labeled with a "more generic" machine-readable term, hence decreasing its reusable value in a cloud-based data ecosystem. Can this process be facilitated? Yes, given clear, global, and interdisciplinary communication. Scientists are sometimes unaware of the work done by information specialists in their own scientific domain; "reinventing the wheel" should be avoided at all costs and one should build on existing operating systems. When these are found not to match the requirements, active dialogue within the scientific community working in the domain can facilitate their improvement.

Funding is crucial and not always easy to obtain for cross-domain activities. Although funders do not always commit to long-term monitoring programs, these are very important for research and continuity in measuring marine biodiversity, especially since they provide not only the baseline for future policies but also allow for continuous monitoring once those policies are implemented.

However, with the rise of the "digital environment," new opportunities will come. We just need to make sure that funds are used to strengthen, develop, and connect existing services, rather than develop new ones in isolation. If new services are needed, they should build upon existing systems and services making use of the acquired expertise and further funding can support the development within existing infrastructures.

The availability of data is and will remain key in marine sciences. Availability of data always needs to be seen in its broadest sense: as data cannot always be widely shared, they are of equal importance that their existence and contact details are known. Several reasons can exist to not have data publicly accessible, varying from data not being digitized to highly sensitive or under a moratorium until published. These situations can be compared to libraries, where very old, rare, or otherwise vulnerable books can only be consulted at the library itself, sometimes only under a preserved atmosphere and/or under supervision, to avoid damage and in which a controlled release can be a challenge on its own. In essence, these books are available, when you know where to go and who to contact, so they are known and findable to the public, although perhaps not through digital means. Data availability also links with the ease to connect to the internet, which might be very straightforward in developed countries, but still proves to be very problematic in developing or least developed countries. The next decade should bridge the online access gap as well: a vast amount of marine scientific data exist in those regions, but there are few possibilities to make these easily accessible or to assist the local scientists in managing these data and bringing them to global standards, usability, interoperability, and to join the open science movement (e.g., Mwelwa et al., 2020). This situation has an extra dimension, a downside: not being able to share their data is one thing, but not being able to easily access all existing infrastructures and services is equally negative: the knowledge and expertise in those countries cannot be brought to practice, unless they are able to access and join initiatives in more developed countries, where online access is not a constraint.

A common factor for the scientific community is the challenge of finding sufficient resources to keep their work running. This is not different for the marine biological community; we know that we can add value to the existing data by fully documenting them using the standards developed. A number of biological data are partially documented and partially standardized and could become more valuable with additional documentation, but bringing all available data to the highest level possible will require a serious investment and commitment from both scientists and funders. Standards are being developed to improve data and their interoperability, but they are not yet fully implemented in existing systems and on (recent) data; adding to this, the data backlog also needs to be taken into account.

Pursuing increased (inter)operability is at the top of the list for many infrastructures and initiatives, without ignoring the legacy behind us. Looking toward the future, efficient workflows can be encouraged at the very early stage of data conception and collection through the adoption of, e.g., Personal Identifiable Data (PID), well-governed and versioned vocabularies and formats. The use of global vocabularies and registries for referencing instruments, observations or observing platforms, and other data details would ideally become second nature to scientists, in addition to any commonly adopted preferred labels ("pet names"). The more points of connectivity there are within any given dataset, the more opportunities for interoperability and increased scientific and decision-making value.

It's clear that there are no problems in identifying the many challenges that need to be tackled, although one can say that an equal number of past challenges have already been addressed throughout the past decade. Some challenges dating from almost a decade ago (Heip and McDonough, 2012) still stand, although progress is being made: we still struggle, for example, with fully understanding the importance of marine biodiversity to human well-being and understanding in detail how populations adapt to changing environmental conditions or changing species interactions. Although major progress has been made in these fields, the need to handle them is highlighted in the UN Decade of Ocean Science for Sustainable Development.

In conclusion

Within the scientific world, marine biology might still be lagging behind its sister disciplines, even though it emerged around the same time. The progress made in the past 2 decades has been exponential on several fronts: from gaining access to data, to connecting the biological data with other disciplines, but the top of this exponential development seems to not have been reached yet. Many valuable guidelines, standards, formats, systems, and infrastructures are already in place to lead marine biological data into the next decade, and the community can focus on fine-tuning the (un)identified challenges.

Regardless of the knowledge one is pursuing, the speed of acquisition and processing should never overrule quality. Without good quality and interoperable data, one cannot perform sound science nor can one deal with the long-standing ecological questions, related to our oceans and seas that need urgent answers. As a community in a data-driven world, we have the task to assist the scientific community in how to do things and how to help them connect with colleagues inside and outside their fields of expertise. Cross-discipline collaboration and cross-system interoperability will be key words in the next decade, both needed to bring data, knowledge, and expertise together. It is highly unlikely that the world will evolve toward a single system that can serve all the needs of the user communities or answer all questions; the new decade is the time to interlink everything and to ensure long-term, continuous availability and interoperability of data and systems.

> Data that is loved tends to survive.
>
> **Kurt Bollacker, computer scientist.**

Acknowledgments

The authors are grateful to Ward Appeltans (Project manager OBIS, UNESCO/IOC Project Office for IODE) and Pieter Provoost (Data manager OBIS, UNESCO/IOC Project Office for IODE) for their advice and feedback at the very first drafting stages of the chapter. A special thank you to Prof. Koen Martens (Royal Belgian Institute of Natural Sciences—RBINS) for revealing interesting insights in the transfer of publishing research papers "through paper" to utilizing online tools for manuscript submissions. Any use of trade, firm, or product names is for descriptive purposes only and does not imply endorsement by the U.S. Government.

References

Alexander, V., Miloslavich, P., Yarincik, K., 2011. The Census of marine life—evolution of worldwide marine biodiversity research. Mar. Biodivers. 41, 545—554. https://doi.org/10.1007/s12526-011-0084-1.

Anderson, K., Rozwadowski, H.M., 2016. Soundings and Crossings: Doing Science at Sea, 1800—1970. Watson Publishing International, LLC, pp. 145—178, 9780881351446.

Arvanitidis, C., Chatzinikolaou, E., Gerovasileiou, V., Panteri, E., Bailly, N., Minadakis, N., Hardisty, A., Los, W., 2016. LifeWatchGreece: construction and operation of the national research infrastructure (ESFRI). Biodivers. Data J. 4, e10791. https://doi.org/ 10.3897/BDJ.4.e10791.

Arvanitidis, C., Warwick, R., Somerfield, P., Pavloudi, C., Pafilis, E., Oulas, A., Chatzigeorgiou, G., Gerovasileiou, V., Patkos, T., Bailly, N., Hernandez, F., Vanhoorne, B., Vandepitte, L., Appeltans, W., Keklikoglou, K., Chatzinikolaou, E., Michalakis, N., Filiopoulou, I., Panteri, E., Gougousis, A., Bravakos, P., Christakis, C., Kassapidis, P., Kotoulas, G., Magoulas, A., 2019. The collaborative potential of research infrastructures in addressing global scientific questions. Biodiv. Inf. Sci. Stand 3, e37289. https://doi.org/10.3897/biss.3.37289.

Barange, M., Bahri, T., Beveridge, M., Funge-Smith, S., Gudmundsson, A., Kalikoski, D., Poulain, F., Vannuccini, S., Wabbes, S., 2018. Impacts of Climate Change on Fisheries and Aquaculture- Synthesis of Current Knowledge, Adaptation and Mitigation Options, 627th ed. Fisheries and Aquaculture Technical Paper, Rome.

Benson, A., Brooks, C.M., Canonico, G., Duffy, E., Müller-Karger, F.E., Sosik, H.M., Miloslavich, P., Klein, E., 2018. Integrated observations and informatics improve understanding of changing marine ecosystems. Front. Mar. Sci. 5, 1—8. https://hdl.handle. net/10.3389/fmars.2018.00428.

Borgman, C.L., Wallis, J.C., Mayernik, M.S., 2012. Who's got the data? Interdependencies in science and technology collaborations. J. Comput. Support. Cooperat. Work 21 (6), 485—523. https://doi.org/10.1007/s10606-012-9169-z.

Boxshall, G., 2020. Self-help for taxonomists: three things we must do for taxonomy to survive. Megataxa 001 (1), 039—042. https://doi.org/10.11646/megataxa.1.1.7.

Breyne, M., Seys, J., Lescrauwaet, A.-K., Debergh, H., Haspeslagh, J., Lust, H., Mees, J., August 2010. The World's very first marine research station in Ostend (Belgium). Down Earth 18 (4).

Bristol, C.L., 1926. The oceanographic station at Salammbo, North Africa. Science 63 (1635), pp. 448—448.

Bryson, B., 2003. A Short History of Nearly Everything. Abridged. Random House, Inc., New York.

Calisher, C.H., 2007. Taxonomy: what's in a name? Doesn't a rose by any other name smell as sweet? Croat. Med. J. 48 (2), 268—270. PMC 2080517. PMID 17436393.

Canonico, G., Buttigieg, P.L., Montes, E., Muller-Karger, F.E., Stepien, C., Wright, D., et al., 2019. Global observational needs and resources for marine biodiversity. Front. Mar. Sci. 6. https://doi.org/10.3389/fmars.2019.00367.

Carroll, S.R., Garba, I., Figueroa-Rodríguez, O.L., Holbrook, J., Lovett, R., Materechera, S., Parsons, M., Raseroka, K., Rodriguez-Lonebear, D., Rowe, R., Sara, R., Walker, J.D., Anderson, J., Hudson, M., 2020. The CARE principles for indigenous data governance. Data Sci. J. 19 (1), 43. https://doi.org/10.5334/dsj-2020-043 (In this issue).

Constable, A.J., Costa, D.P., Schofield, O., et al., 2016. Developing priority variables ("ecosystem Essential Ocean Variables" — eEOVs) for observing dynamics and change in southern ocean ecosystems. J. Mar. Syst. 161, 26—41.

Convention on Biological Diversity, 2011. Traditional Knowledge and the Convention on Biological Diversity (Accessed 4 September 2020). https://www.cbd.int/traditional/ intro.shtml.

Copeland, R., 1991. Rhetoric, Hermeneutics, and Translation in the Middle Ages- Academic Traditions and Vernacular Texts. Cambridge University Press, 9780511597534.

Costello, M.J., 2020. Taxonomy as the key to life. Megataxa 001 (2), 105—113. https:// doi.org/10.11646/megataxa.1.2.1.

Costello, M.J., Coll, M., Danovaro, R., Halpin, P., Ojaveer, H., Miloslavich, P., 2010. A census of marine biodiversity knowledge, resources, and future challenges. PloS One 5. https://doi.org/10.1371/journal.pone.0012110.

Costello, M.J., Bouchet, P., Boxshall, G., Fauchald, K., Gordon, D., Hoeksema, B.W., Poore, G.C.B., Van Soest, R.W.M., Stöhr, S., Walter, T.C., Vanhoorne, B., Decock, W., Appeltans, W., 2013. Global coordination and standardisation in marine biodiversity through the World Register of Marine Species (WoRMS) and related databases. PloS One 8 (1), 20. https://hdl.handle.net/10.1371/journal.pone.0051629.

Cruz-Motta, J.J., Miloslavich, P., Guerra-Castro, E., Hernández-Agreda, A., Herrera, C., Barros, F., Navarrete, S.A., Sepúlveda, R.D., Glasby, T.M., Bigatti, G., Cardenas-Calle, M., Carneiro, P.B.M., Carranza, A., Flores, A.A.V., Gil-Kodaka, P., Gobin, J., Gutiérrez, J.L., Klein, E., Krull, M., et al., 2020. Latitudinal patterns of species diversity on South American rocky shores: local processes lead to contrasting trends in regional and local species diversity. J. Biogeogr. 47 (9), 1—14. https://doi.org/10.1111/jbi.13869.

Darwin, C., 1838—43. The Zoology of the Voyage of H.M.S. Beagle (5 Volumes). Smith, Elder and Co., London.

Dañobeitia, J.J., Pouliquen, S., Johannessen, T., Basset, A., Cannat, M., Pfeil, B.G., Fredella, M.I., Materia, P., Gourcuff, C., Magnifico, G., Delory, E., del Rio, F.J., Rodero, I., Beranzoli, L., Nardello, I., Iudicone, D., Carval, T., Gonzalez Aranda, J.M., Petihakis, G., Blandin, J., Kutsch, W.L., Rintala, J.-M.. Gates, A.R., Favali, P., March 31, 2020. Toward a comprehensive and integrated strategy of the European marine research infrastructures for ocean observations. Front. Mar. Sci. 7. https://doi.org/10.3389/fmars.2020.00180.

Danto, E.A., 2008. Historical Data Sources. Historical Research. Oxford Scholarship Online. https://doi.org/10.1093/acprof:oso/9780195333060.001.0001, 2009, (Accessed 21 January 2021).

DeCou, C., 2018. When whales were fish. Lateral Magazine - Philosophy & History, p. 29. http://www.lateralmag.com/articles/issue-29/when-whales-were-fish, (Accessed 2 September 2020).

de la Hidalga, A.N., Hardisty, A., Martin, P., Magagna, B., 2020. The ENVRI Reference Model. Towards Interoperable Research Infrastructures for Environmental and Earth Sciences: A Reference Model Guided Approach for Common Challenges. Springer International Publishing, pp. 61—81, 978-3-030-52829-4.

De Pooter, D., Appeltans, W., Bailly, N., Bristol, S., Deneudt, K., Eliezer, M., Fujioka, E., Giorgetti, A., Goldstein, P., Lewis, M., Lipizer, M., Mackay, K., Marin, M., Moncoiffe, G., Nikolopoulou, S., Provoost, P., Rauch, S., Roubicek, A., Torres, C., Van de Putte, A., Vandepitte, L., Vanhoorne, B., Vinci, M., Wambiji, N., Watts, D., Salas, E.K., Hernandez, F., 2017. Toward a new data standard for combined marine biological and environmental datasets - expanding OBIS beyond species occurrences. Biodivers. Data J. 5, e10989 hdl.handle. net/10.3897/BDJ.5.e10989.

Dolan, J., 2007. On Kofoid's trail: marine biological laboratories in Europe and their life histories. Limnol. Oceanogr. Bull. Am. Soc. Limnol. Oceanogr. 16, 73—76. Hal-00670780.

Duff, W., Craig, B., Cherry, J., 2004. Historians' use of archival sources: promises and pitfalls of the digital age. Publ. Historian 26 (2), 7—22. https://doi.org/10.1525/tph.2004.26.2.7.

Duffy, J.E., Benedetti-Cecchi, L., Trinanes, J., Muller-Karger, F.E., Ambo-Rappe, R., Boström, C., et al., 2019. Toward a coordinated global observing system for seagrasses and marine macroalgae. Front. Mar. Sci. 6. https://doi.org/10.3389/fmars.2019.00317.

Eliot, T.S., 1934. The Rock. Faber & Faber, London.

Engesæter, S., 2002. The importance of ICES in the establishment of NEAFC. ICES Mar. Sci. Symp. 215, 572—581.

European Commission, 2020. Communication from the Commission to the European Parliament, the Council, the European Economic and Social Committee and the Committee of the Regions: A European Strategy for Data. Available at: https://ec.europa.eu/info/sites/info/files/communication-european-strategy-data-19feb2020_e n.pdf.

European Marine Board, 2013. Navigating the Future IV. Position Paper 20 of the European Marine Board, Ostend, Belgium, 9789082093100.

Faulwetter, S., Pafilis, E., Fanini, L., Bailly, N., Agosti, D., Arvanitidis, C., Boicenco, L., Catapano, T., Claus, S., Dekeyzer, S., Georgiev, T., Legaki, A., Mavrak, i D., Oulas, A., Papastefanou, G., Penev, L., Sautter, G., Schigel, D., Senderov, V., Teaca, A.m, Tsompanou, M., 2016. EMODnet workshop on mechanisms and guidelines to mobilise historical data into biogeographic databases. Res. Ideas Outcomes 2, e10445. https://doi.org/10.3897/rio.2.e10445.

Flint, L.N., 2018-08-07. Newspaper Writing in High Schools, Containing an Outline for the Use of Teachers. Forgotten Books.

Froese, R., Pauly, D. (Eds.), 2019. FishBase. World Wide Web Electronic Publication Version 12/2019 www.fishbase.org.

Gorsky, G., Ohman, M.D., Picheral, M., Gasparini, S., Stemmann, L., Romagnan, J.B., Cawood, A., Pesant, S., García-Comas, C., Prejger, F., 2010. Digital zooplankton image analysis using the ZooScan integrated system. J. Plankton Res. 32 (3), 285–303. https://doi.org/10.1093/plankt/fbp124.

Grenier, L., 1998. Working with Indigenous Knowledge: A Guide for Researchers. The International Development Research Centre (IDRC), 0889368473, 100pp.

Griffin, P.C., Khadake, J., LeMay, K.S., et al., 2018. Best practice data life cycle approaches for the life sciences [version 2; peer review: 2 approved]. F1000Research 6, 1618. https://doi.org/10.12688/f1000research.12344.2.

Grimes, S., 2014. D.6.2 Report on the Essential Ocean Ecosystem Variables and on the Adequacy of Existing Observing System Elements to Monitor Them. GEOWOW EC Grant Agreement No. 282915. GEOWOW, Brussels, p. 130.

Guiry, M.D., Guiry, G.M., 2020. AlgaeBase. World-wide electronic publication, National University of Ireland, Galway (Accessed 8 September 2020). https://www.algaebase.org.

Haines, M., 2004. HMAP Dataset 5: Newfoundland, 1675–1698. Last Modified 2002. https://hydra.hull.ac.uk/resources/hull:2156 (Accessed 15 May 2019).

Hector, A., 2009. Charles Darwin and the importance of biodiversity for ecosystem functioning. Vierteljahrsschr. der Naturforschenden Ges. Zürich 154 (3/4), 69–73.

Heip, C., McDonough, N., 2012. Marine Biodiversity: A Science Roadmap for Europe. Marine Board Future Science Brief 1, European Marine Board, Ostend, Belgium, 978-2-918428-75-6.

Heip, C., Hummel, H., van Avesaath, P., Appeltans, W., Arvanitidis, C., Aspden, R., Austen, M., Boero, F., Bouma, T.J., Boxshall, G., Buchholz, F., Crowe, T., Delaney, A., Deprez, T., Emblow, C., Feral, J.P., Gasol, J.M., Gooday, A., Harder, J., Ianora, A., Kraberg, A., Mackenzie, B., Ojaveer, H., Paterson, D., Rumohr, H., Schiedek, D., Sokolowski, A., Somerfield, P., Sousa Pinto, I., Vincx, M., Weslawski, J.M., Nash, R., 2009-01-01. Marine Biodiversity and Ecosystem Functioning. Printbase, Dublin, Ireland.

Hofmann, E.E., Gross, E., 2010. IOC contributions to science synthesis. Oceanography 23 (3), 152–159. https://doi.org/10.5670/oceanog.2010.30.

Holland, B.S., 2000. Genetics of marine bioinvasions. Hydrobiologia 420, 63–71. https://doi.org/10.1023/A:1003929519809.

Holm, P., Ludlow, F., Scherer, C., Travis, C., Allaire, B., Brito, C., Hayes, P.W., Matthews, A., Rankin, K.J., Breen, R.J., Legg, R., Lougheed, K., Nicholls, J., 2019. The North Atlantic fish revolution, c. AD 1500. Quat. Res. 1–15. https://doi.org/10.1017/qua.2018.153.

Holm, P., Nicholls, J., Hayes, P.W., Ivinson, J., Allaire, B., 2021. Accelerated extractions of North Atlantic cod and herring, 1520—1790. Fish and Fisheries 1—19. https://doi.org/10.1111/faf.12598.

Inaba, K., 2015. Japanese marine biological stations: preface to the special issue. Reg. Stud. Mar. Sci. 2, 154—157.

Inglis, J., 1993. Traditional Ecological Knowledge: Concepts and Cases. International Program on Traditional Ecological Knowledge and International Development Research Centre (Canada), 1552502570, 150pp.

Intergovernmental Science-Policy Platform on Biodiversity and Ecosystem Services, IPBES, 2019. Summary for Policymakers of the Global Assessment Report on Biodiversity and Ecosystem Services (Version Summary for Policymakers). Zenodo. https://doi.org/10.5281/zenodo.3553579.

IOC, July 2020. United Nations Decade of Ocean Science for Sustainable Development 2021 - 2030. Implementation Plan. Version 2.0.

Jones, M.C., Cheung, W.W.L., 2014. Multi-model ensemble projections of climate change effects on global marine biodiversity. ICES J. Mar. Sci. 72 (3), 741—752. https://doi.org/10.1093/icesjms/fsu172.

Jones, S., Pergl, R., Hooft, R., Miksa, T., Samors, R., Ungvari, J., Davis, R.I., Lee, T., 2020. Data management planning: how requirements and solutions are beginning to converge. Data Intell. 2 (1—2), 208—219. https://doi.org/10.1162/dint_a_00043.

Jörger, K.M., Schrödl, M., 2013. How to describe a cryptic species? Practical challenges of molecular taxonomy. Front. Zool. 10 (59). https://doi.org/10.1186/1742-9994-10-59.

Lear, D., Herman, P., Van Hoey, G., Schepers, L., Tonné, N., Lipizer, M., Muller-Karger, F.E., Appeltans, W., Kissling, W.D., Holdsworth, N., Edwards, M., Pecceu, E., Nygård, H., Canonico, G., Birchenough, S., Graham, G., Deneudt, K., Claus, S., Oset, P., 2020. Supporting the essential — recommendations for the development of accessible and interoperable marine biological data products. Mar. Pol. 117, 103958. https://doi.org/10.1016/j.marpol.2020.103958.

LEO Network, 2017. About LEO Network (Accessed 2 September 2020). https://www.leonetwork.org/en/docs/about/about.

LifeWatch Belgium, n.d. Aristotle and Marine Biodiversity. http://www.lifewatch.be/en/2016-news-aristotle. (Accessed 31 August 2020).

Lindstrom, E., Gunn, J., Fischer, A., McCurdy, A., Glover, L.K., 2012. A Framework for Ocean Observing. Task Team for an Integrated Framework for Sustained Ocean Observing. UNESCO.

Marion, A.-F., 1897. La Station zoologique d'Endoume-Marseille. In: Revue générale internationale scientifique, littéraire et artistique, vol. 14, pp. 179—198.

Matsunaga, A., Figueiredo, R., Thompson, A., Traub, G., Beaman, R., Fortes, J.A.B., 2013. Digitized biocollections (iDigBio) cyberinfrastructure status and futures. In: TDWG 2013 Annual Conference - Abstract. https://mbgocs.mobot.org/index.php/tdwg/2013/paper/view/412/0.

Mavraki, D., Fanini, L., Tsompanou, M., Gerovasileiou, V., Nikolopoulou, S., Chatzinikolaou, E., Plaitis, W., Faulwetter, S., 2016. Rescuing biogeographic legacy data: the "Thor" expedition, a historical oceanographic expedition to the Mediterranean Sea. Biodivers. Data J. 4, e11054. https://doi.org/10.3897/BDJ.4.e11054.

Mayernik, M.S., Maull, K.E., 2017. Assessing the uptake of persistent identifiers by research infrastructure users. PloS One 12 (4), e0175418. https://doi.org/10.1371/journal.pone.0175418.

McClenachan, L., Ferretti, F., Baum, J.K., 2012. From archives to conservation: why historical data are needed to set baselines for marine animals and ecosystems. Conser. Lett. 5, 349—359. https://doi.org/10.1111/j.1755-263X.2012.00253.x.

Merriam-Webster, 2021 (Accessed 21 June 2021). https://www.merriam-webster.com/dictionary/history.

Michener, M.K., 2015-10-22. Ten simple rules for creating a good data management plan. PLoS Comput. Biol. 11 (10). https://doi.org/10.1371/journal.pcbi.1004525.

Miloslavich, P., Bax, N.J., Simmons, S.E., Klein, E., Appeltans, W., Aburto-Oropeza, O., et al., 2018a. Essential ocean variables for global sustained observations of biodiversity and ecosystem changes. Glob. Chang. Biol. 24, 2416−2433. https://doi.org/10.1111/gcb.14108.

Miloslavich, P., Pearlman, J., Kudela, R., 2018b. Sustainable observations of plankton, the sea's food foundation. Eos 99. https://doi.org/10.1029/2018EO108685.

Miloslavich, P., Bax, N., Satterthwaite, E., 2019a. Designing the global observing system for marine life. Eos 100. https://doi.org/10.1029/2019EO127053.

Miloslavich, P., Johnson, C., Benedetti-Cecchi, L., 2019b. Keeping a watch on seaweeds: the forests of the world's coasts. Eos 100. https://doi.org/10.1029/2019EO113401.

Muller-Karger, F.E., Miloslavich, P., Bax, N.J., Simmons, S., Costello, M.J., Pinto, I.S., et al., 2018a. Advancing marine biological observations and data requirements of the complementary Essential Ocean Variables (EOVs) and Essential Biodiversity Variables (EBVs) frameworks. Front. Mar. Sci. 5. https://doi.org/10.3389/fmars.2018.00211.

Muller-Karger, F.E., Hestir, E., Ade, C., Turpie, K., Roberts, D.A., Siegel, D., Miller, R.J., Humm, D., Izenberg, N., Keller, M., et al., 2018b. Satellite sensor requirements for monitoring essential biodiversity variables of coastal ecosystems. Ecol. Appl. 28, 749−760. https://doi.org/10.1002/eap.1682.

Mwelwa, J., Boulton, G., Wafula, J.M., Loucoubar, C., 2020. Developing open science in Africa: barriers, solutions and opportunities. Data Sci. J. 19 (31), 1−17. https://doi.org/10.5334/dsj-2020-031.

National Ocean Council (NOC), 2016. Biological and Ecosystem Observations within United States Waters II: A Workshop Report to Inform Priorities for the United States. Report by the Interagency Ocean Observation Committee on Biological Integration and Observation BIO Task Team (IOOC Bio Task Team).

Nicholas, G., 2018. It's Taken Thousands of Years, but Western Science Is Finally Catching up to Traditional Knowledge. The Conversation (Accessed 15 September 2020). https://theconversation.com/its-taken-thousands-of-years-but-western-science-is-finally-catching-up-to-traditional-knowledge-90291.

Nicholls, J., Allaire, B., Holm, P., 2021. The Capacity trend method: a new approach for enumerating the Newfoundland cod fisheries (1675−1790). Hist. Methods 54 (2), 80−93. https://doi.org/10.1080/01615440.2020.1853643.

Obura, D.O., Aeby, G., Amornthammarong, N., Appeltans, W., Bax, N., Bishop, J., et al., 2019. Coral reef monitoring, reef assessment technologies, and ecosystem-based management. Front. Mar. Sci. 6. https://doi.org/10.3389/fmars.2019.00580.

Odishaw, H., 1959. International geophysical year. Science 129 (3340), 14−25. https://doi.org/10.1126/science.129.3340.14. ISSN 0036-8075. JSTOR. 1755204. PMID 17794348.

Orr, M.C.C., Ascher, J.S., Bai, M., Chesters, D., Zhu, C.-D., 2020. Three questions: how can taxonomists survive and thrive worldwide? Megataxa 001 (1), 019−027. https://doi.org/10.11646/megataxa.1.1.4.

O'Dor, R., Miloslavich, P., Yarincik, K., 2010. Marine biodiversity and biogeography − regional comparisons of global issues, an introduction. PloS One 5. https://doi.org/10.1371/journal.pone.0011871.

Palumbi, S.R., 1996. What can molecular genetics contribute to marine biogeography? An urchin's tale. J. Exp. Mar. Biol. Ecol. 203 (1). https://doi.org/10.1016/0022-0981(96)02571-3.

Paskin, N., 2005. Digital object identifiers for scientific data. Data Sci. J. 4, 12—20. https://doi.org/10.2481/dsj.4.12.

Pearlman, J., et al., 2019. Evolving and sustaining ocean best practices and standards for the next decade. Front. Mar. Sci. https://doi.org/10.3389/fmars.2019.00277.

Pereira, H.M., Ferrier, S., Walters, M., Geller, G.N., Jongman, R.H.G., Scholes, R.J., et al., 2013. Essential biodiversity variables. Science 339, 277—278.

Pope, P., 2003. HMAP Dataset 6: Newfoundland, 1698—1833 (Accessed 15 May 2019). https://hydra.hull.ac.uk/resources/hull:2157.

PricewaterhouseCoopers LLP, 2019. Putting a Value on Data. www.pwc.co.uk/data-analytics/documents/putting-value-on-data.pdf.

Ramón y Cajal, S., 1999. Advice for a Young Investigator (Translated by Neely Swanson & Larry W. Swanson). MIT Press.

Structure and dynamics of the North Sea Benthos. In: Rees, H.L., Eggleton, J.D., Rachor, E., Vanden Berghe, E. (Eds.), 2007. ICES Cooperative Research Report No, vol. 288, p. 258. https://doi.org/10.17895/ices.pub.5451.

Rigby, P.R., Iken, K., Shirayama, Y., 2007. Sampling Diversity in Coastal Communities: NaGISA Protocols for Seagrass and Macroalgal Habitats. Kyoto University Press, Japan, p. 145.

Roberts, C., 2007. The Unnatural History of the Sea, vol. XVII. Island Press, Washington D.C, 978-1-59726- 102-9, pp. 436.

Robertson Jr., D.W., 1946. A note on the classical origin of "circumstances" in the medieval confessional. Stud. Philol. 43 (1), 6—14.

Rodrigues da Silva, N.R., da Silva, M.C., Genevois, V.F., Esteves, A.M., de Ley, P., Decraemer, W., Rieger, T.T., Correia, M.T.S., 2010. Marine nematode taxonomy in the age of DNA: the present and future of molecular tools to assess their biodiversity. Nematology 12 (5). https://doi.org/10.1163/138855410X500073.

Rowley, J., 2007. The wisdom hierarchy: representations of the DIKW hierarchy. J. Inf. Commun. Sci. 33 (2), 163—180. https://doi.org/10.1177/0165551506070706.

Schriml, L.M., Chuvochina, M., Davies, N., Emiley Eloe-Fadrosh, A., Finn, R.D., Hugenholtz, P., Hunter, C.I., Hurwitz, B.L., Kyrpides, N.C., Meyer, F., Mizrachi, I.K., Sansone, S.-A., Sutton, G., Tighe, S., Walls, R., 2020. COVID-19 pandemic reveals the peril of ignoring metadata standards. Sci. Data 7, 188. https://doi.org/10.1038/s41597-020-0524-5.

Sloan, M.C., 2010-07-03. Aristotle's nicomachean ethics as the original locus for the septem circumstantiae. Classical Philol. 105 (3). https://doi.org/10.1086/656196.

South American Research Group in Coastal Ecosystems (SARCE), 2012. Protocol and Sampling Design for Marine Diversity Assessments for the South American Research Group on Coastal Ecosystems. South American Research Group on Coastal Ecosystems (SARCE), Caracas, Venezuela. https://doi.org/10.25607/OBP-5.

Steinhart, G., Chen, E., Arguillas, F., Dietrich, D., Kramer, S., 2012. Prepared to plan? A snapshot of researcher readiness to address data management planning requirements. J. Sci. Librariansh. 1 (2). https://doi.org/10.7191/jeslib.2012.1008.

Sutherland, W.J., Armstrong-Brown, S., Armsworth, P.R., Brereton, T., Brickland, J., Campbell, C.D., Chamberlain, D.E., Cooke, A.I., Dulvy, N.K., Dusic, N.R., Fitton, M.G., Freckleton, R.P., Godfray, H.C.J., Grout, N., Harvey, H.J., Hedley, C., Hopkins, J.J., Kift, N.B., Kirby, J., Kunin, W.E., Macdonald, D.W., Marker, B., Naura, M., Neale, A.R., Oliver, T., Osborn, D., Pullin, A.S., Shardlow, M.E.A., Showler, D.A., Smith, P.L., Smithers, R.J., Solandt, J.-L., Spencer, J., Spray, C.J., Thomas, C.D., Thompson, J., Webb, S.E., Yalden, D.W., Watkinson, A.R., 2006. The identification of 100 ecological questions of high policy relevance in the UK. J. Appl. Ecol. 43 (4), 617—627.

Sutherland, W.J., Freckleton, R.P., Godfray, H.C.J., Beissinger, S.R., Benton, T., Cameron, D.D., Carmel, Y., Coomes, D., Coulson, T., Emmerson, M.C., Hails, R.S., Hays, G.C., Hodgson, D.J., Hutchings, M.J., Johnson, D., Jones, J.P.G., Keeling, M.J., Kokko, H., Kunin, W.E., Lambin, X., Lewis, O.T., Malhi, Y., Mieszkowska, N., Milner-Gulland, E.J., Norris, K., Phillimore, A.B., Purves, D.W., Reid, J.M., Reuman, D.C., Thompson, K., Travis, J.M.J., Turnbull, L.A., Wardle, D.A., Wiegand, T., 2013. Identification of 100 fundamental ecological questions. J. Ecol. 101 (1), 58−67 hdl.handle.net/10.1111/1365-2745.12025.

Tanhua, T., McCurdy, A., Fischer, A., Appeltans, W., Bax, N., Currie, K., et al., 2019. What we have learned from the framework for ocean observing: evolution of the global ocean observing system. Front. Mar. Sci. 6. https://doi.org/10.3389/fmars.2019.00471.

The Group of Experts of the Regular Process, 2017. In: The First Global Integrated Marine Assessment: World Ocean Assessment I. Cambridge University Press/United Nations, Cambridge, ISBN 978-1-316-51001- 8, 973 pp.

WOA II. The Second World Ocean Assessment, Volume 1, 2021a. United Nations, New York, ISBN 978-92-1-130422-0 xxiii, 543 pp.

WOA II. The Second World Ocean Assessment, Volume II, 2021b. United Nations, New York, ISBN 978-92-1-130422-0 xvii, 500 pp.

Treloar, A., 2014. In: The Research Data Alliance: Globally Co-ordinated Action against Barriers to Data Publishing and Sharing, vol. 27. Learned Publishing, pp. S9−S13. https://doi.org/10.1087/20140503.

Tsikopoulou, I., Legaki, A., Dimitriou, P., Avramidou, E., Bailly, N., Nikolopoulou, S., 2016. Digging for historical data on the occurrence of benthic macrofaunal species in the southeastern Mediterranean. BDJ.4.e10071 Biodivers. Data J. 4, e10071. https://doi.org/10.3897/BDJ.4.e10071.

UNEP, 2016. GEO-6: Regional Assessment for Africa. United Nations Environment Programme.

UNESCO-IOC, 2014. Report of the First Workshop of Technical Experts for the Global Ocean Observing System (GOOS) Biology and Ecosystems Panel: Identifying Ecosystem Essential Ocean Variables (EOVs). UNESCO-IOC, Paris.

UNESCO-IOC register of marine organisms. A common base for biodiversity inventories. In: van der Land, J. (Ed.), 1994. Families and Bibliography of Keyworks. DOS-Formatted Floppy Disk. National Museum of Natural History (Naturalis), Leiden.

United Nations, 1992. 1992 Rio Declaration on Environment and Development. UN Doc. A/CONF.151/26 (vol. I), 31 ILM 874.

van der Land, J. (Ed.), 2008. UNESCO-IOC Register of Marine Organisms (URMO). Available at: http://www.marinespecies.org/urmo (Accessed 15 September 2020).

Van Wyhe, J., 2007. Mind the gap: did Darwin avoid publishing his theory for many years? Notes Record Roy. Soc. Lond. 61, 177−205. https://doi.org/10.1098/rsnr.2006.0171.

Vanden Berghe, E., Claus, S., Appeltans, W., Faulwetter, S., Arvanitidis, C., Somerfield, P.J., Aleffi, I.F., Amouroux, J.M., Anisimova, N., Bachelet, G., Cochrane, S.J., Costello, M.J., Craeymeersch, J.A., Dahle, S., Degraer, S., Denisenko, S., Deprez, T., Dounas, C., Duineveld, G., Emblow, C., Escaravage, V., Fabri, M.-C., Fleischer, D., Grémare, A., Herrmann, M., Hummel, H., Karakassis, I., Kedra, M., Kendall, M.A., Kingston, P., Kotwicki, L., Labrune, C., Laudien, J., Nevrova, E.L., Occhipinti-Ambrogi, A., Olsgard, F., Palerud, R., Petrov, A., Rachor, E., Revkov, N.K., Rumohr, H., Sardá, R., Sistermans, W.C.H., Speybroeck, J., Janas, U., Van Hoey, G., Vincx, M., Whomersley, P., Willems, W., Wlodarska-Kowalczuk, M., Zenetos, A., Zettler, M.L., Heip, C.H.R., 2009. MacroBen integrated database on benthic invertebrates of European continental shelves: a tool for large-scale analysis across Europe. Mar. Ecol. Prog. Ser. 382, 225−238. https://doi.org/10.3354/meps07826.

Vandepitte, L., Vanaverbeke, J., Vanhoorne, B., Hernandez, F., Bezerra, T.N., Mees, J., Vanden Berghe, E., 2009. The MANUELA database: an integrated database on meiobenthos from European marine waters. Meiofauna Mar. 17, 35—60.

Vandepitte, L., Vanhoorne, B., Kraberg, A., et al., 2010. Data integration for European marine biodiversity research: creating a database on benthos and plankton to study large-scale patterns and long-term changes. Hydrobiologia 644, 1—13. https://doi.org/10.1007/s10750-010-0108-z.

Vandepitte, L., Vanhoorne, B., Decock, W., Dekeyzer, S., Trias Verbeeck, A., Bovit, L., Hernandez, F., Mees, J., 2015. How Aphia—the platform behind several online and taxonomically oriented databases—can serve both the taxonomic community and the field of biodiversity informatics. J. Mar. Sci. Eng. 3, 1448—1473.

Vandepitte, L., Vanhoorne, B., Decock, W., Lanssens, T., Dekeyzer, S., Verfaille, K., Horton, T., Kroh, A., 2018. 10 years of the World Register of Marine Species (WoRMS): where do we stand and where are we heading? Peer J. Preprints 6, e26682v1. https://doi.org/10.7287/peerj.preprints.26682v1.

Vines, T.H., Andrew, R.L., Bock, D.G., Franklin, M.T., Gilbert, K.J., Kane, N.C., Moore, J., Moyers, B.T., Renaut, S., Rennison, D.J., Veen, T., Yeaman, S., 2013. Mandated data archiving greatly improves access to research data. FASEB. J. 27 (4), 1304—1308. https://doi.org/10.1096/fj.12-218164.

Vines, T.H., Albert, A.Y.K., Andrew, R.L., Débarre, F., Bock, D.G., Franklin, M.T., Gilbert, K.J., Moore, J.S., Renaut, S., Rennison, D.J., 2014-01-06. The availability of research data declines rapidly with article age. Curr. Biol. 24 (1), 94—97. https://doi.org/10.1016/j.cub.2013.11.014.

Voultsiadou, E., Gerovasileiou, V., Vandepitte, L., Ganias, K., Arvanitidis, C., 2017. Aristotle's scientific contributions to the taxonomic classification, nomenclature and distribution of marine organisms. Mediterr. Mar. Sci. 18, 468—478.

Went, A.E.J., 1972. Seventy years a growing. A history of the international council for the exploration of the sea. Rapp. P.-V. Reun. CIEM 165, 249.

Wilkinson, M.D., Dumontier, M., Aalbersberg, I.J., Appleton, G., Axton, M., Baak, A., Blomberg, N., Boiten, J.-W., Silva Santos, L.B., Bourne, P.E., Bouwman, J., Brookes, A.J., Clark, T., Crosas, M., Dillo, I., Dumon, O., Edmunds, S., Evelo, C.T., Finkers, R., Gonzalez-Beltran, A., Gray, A.J.G., Goth, P., Goble, C., grethe, J.S., Heringa, J., 't Hoen, A.C., Hooft, R., Kuhn, T., Kok, R., Kok, J., Lusher, S.J., Martone, M.E., Mons, B., Packer, A.L., Persson, B., Rocca-Serra, P., Roos, M., van Schaik, E., Sansone, S.A., Schultes, E., Sengstag, T., Slater, T., Strawn, G., Swertz, M.A., Thompson, M., van der Lei, J., van Mulligen, E., Velterop, J., Waagmeester, A., Wittenburg, P., Wolstencroft, K., Zhao, J., Mons, B., 2016. The FAIR guiding principles for scientific data management and stewardship. Sci. Data 3 (1). https://doi.org/10.1038/sdata.2016.18.

Wolff, T., 2010. The Birth and First Years of the Scientific Committee on Oceanic Research (SCOR). Scientific Committee on Oceanic Research, Baltimore, Maryland, USA.

World Meteorological Organization (WMO), 2016. The Global Observing System for Climate: Implementation Needs. World Meteorological Organization Available at: https://library.wmo.int/opac/doc_num.php?explnum_id= 3417.

World Intellectual Property Organization (WIPO), 2017. Documenting Traditional Knowledge — A Toolkit. WIPO, Geneva. www.wipo.int/tk/en/resources/tkdocumentation.html, (Accessed 2 September 2020).

WoRMS Editorial Board, 2020. World Register of Marine Species. VLIZ. https://doi.org/10.14284/170 (Accessed 8 September 2020). www.marinespecies.org. http://www.marinespecies.org.

Further reading

Kissling, W.D., Ahumada, J.A., Bowser, A., Fernandez, M., Fernández, N., García, E.A., Guralnick, R.P., Isaac, N.J.B., Kelling, S., Los, W., McRae, L., Mihoub, J.-B., Obst, M., Santamaria, M., Skidmore, A.K., Williams, K.J., Agosti, D., Amariles, D., Arvanitidis, C., Bastin, L., De Leo, F., Egloff, W., Elith, J., Hobern, D., Martin, D., Pereira, H.M., Pesole, G., Peterseil, J., Saarenmaa, H., Schigel, D., Schmeller, D.S., Segata, N., Turak, E., Uhlir, P.F., Wee, B., Hardisty, A.R., 2018a. Building essential biodiversity variables (EBVs) of species distribution and abundance at a global scale. Biol. Rev. 93, 600−625.

Kissling, W.D., Walls, R., Bowser, A., Jones, M.O., Kattge, J., Agosti, D., Amengual, J., Basset, A., van Bodegom, P.M., Cornelissen, J.H.C., Denny, E.G., Deudero, S., Egloff, W., Elmendorf, S.C., Alonso García, E., Jones, K.D., Jones, O.R., Lavorel, S., Lear, D., Navarro, L.M., Pawar, S., Pirzl, R., Rüger, N., Sal, S., Salguero-Gómez, R., Schigel, D., Schulz, K.-S., Skidmore, A., Guralnick, R.P., 2018b. Towards global data products of essential biodiversity variables on species traits. Nat. Ecol. Evol. 2, 1531−1540.

CHAPTER THREE

Data management infrastructures and their practices in Europe

Dick M.A. Schaap[1], Antonio Novellino[2], Michele Fichaut[3], Giuseppe M.R. Manzella[4,5]

[1]Mariene Informatie Service MARIS B.V., Nootdorp, the Netherlands
[2]ETT SpA, Genova, Italy
[3]IFREMER/SISMER, Brest, France
[4]The Historical Oceanography Society, La Spezia, Italy
[5]OceanHis SrL, Torino, Italy

Introduction

The oceans, seas, and coastal environments are home to diverse marine ecosystems. They provide a wealth of resources as well as influence the climate and offer many economic opportunities. However, marine ecosystems are sensitive to long-term global change and combinations of multiple local pressures, which can affect ecosystems in unpredictable ways. That includes understanding the dynamic spatial and temporal interplay between ocean physics, chemistry, and biology. Stressors, including climate change, pollution, and overfishing, affect the ocean. We need a better understanding when predicting their interactions and identifying tipping points to decide on management priorities, thus requiring smarter observations when assessing the state of the ocean and predicting how it may change in the future.

Ocean observation data are essential for addressing global challenges: to ensure food security and to contribute to the UN 2030 Agenda for Sustainable Development (SD), notably UNSD Goals 2 (zero hunger), 13 (climate), 14 (life below water), and 17 (partnership), monitoring their targets for 2030.

Measurements of "sustainability" have been discussed in many fora (e.g., OECD, 2017; Eurostat, 2018). The United States has established indicators on Sustainable Development Goals (https://sdg.data.gov/), notably 10 indicators for Goal 14 (Conserve and sustainably use the oceans, seas and marine resources for sustainable development) and 8 indicators for Goal 13 (Take urgent action to combat climate change and its impacts). The Eurostat of the European Union set indicators on Sustainable Development in 2018

Ocean Science Data
ISBN: 978-0-12-823427-3
https://doi.org/10.1016/B978-0-12-823427-3.00007-4

© 2022 Elsevier Inc.
All rights reserved.

and targets and their rationale have been listed in a paper by Miola and Schlitz (2019). In Australia, CSIRO is working to support the capacity to conserve and sustainably use marine resources and to engage in national, regional, and global scientific programs. CSIRO's work on modeling and monitoring the marine environment for the UN Environment Program specifically addressed knowledge gaps identified to be of global significance. Other initiatives have been established worldwide supporting sustainable development goals. In 2014, the European Commission began the Copernicus Marine Environment Monitoring Service (CMEMS) to provide data and information about the state and dynamics of the World Ocean and European Regional Seas. A foremost CMEMS objective is to respond to user needs, European directives, and policies and sectors of the blue economy in general.

Marine and ocean data are thus instrumental for research and monitoring, predicting and managing the marine environment, but also for assessing fish stocks and biodiversity, supporting offshore engineering, hazard and disaster management, the tourist industry, and many other socioeconomic activities at sea and along the coasts. One of today's biggest challenges is the need to maximize the full potential of observation networks and use them in a vast range of services supporting "blue growth" (e.g., Eikeset et al., 2018). The main goals of initiatives are the production of objective, reliable, and comparable information for those concerned with framing, implementing, and further developing environmental policies, and for the use by a wider public.

A general framework for the design and implementation of sustained ocean observation systems has been provided by the Global Ocean Observing System (GOOS) initiative established in response to the recommendations of the Second World Climate Conference held in Geneva in 1990 (Moltman et al., 2019). Coordination at the international level of observations around the world ocean on themes such as climate, operational services, and ecosystem health has opened up opportunities for new collaborations in science, technology, and society.

Increasingly, marine and ocean data and information are also derived from Earth Observation (EO). Nowadays, space technologies, infrastructures, services, and data also provide important input for addressing societal challenges and large global concerns. A worldwide network of government, academic and research institutions, public and private data providers, businesses and experts, in general, have the ambition to produce information management systems "wherein decisions and actions are informed by coordinated, comprehensive, and sustained Earth observations" (Group on Earth Observations - http://earthobservations.org/index.php). Copernicus, the European Union's Earth observation and monitoring program (https://www.copernicus.eu/en), started operations in 2014, producing a wealth of data and information regarding Earth subsystems (land, atmosphere, oceans, and inland waters) and cross-cutting processes (climate change, disaster management, and security), most of which are made available on a free, open, and full basis. Copernicus is the European contribution to the GEO's Global Earth Observation System of Systems (GEOSS — http://www.earthobservations.org/index.php). GEOSS facilitates the sharing of environmental data and information collected from the large array of observing systems contributed by countries and organizations within GEO. Worldwide, nations stimulate activities with a focus on contributing to and drawing benefit from GEOSS.

Recently, the G7 science ministers recognized, for the "Future of the Seas and Oceans" (https://www8.cao.go.jp/cstp/kokusaiteki/g7_2017/20170928annex1.pdf), that gaining more knowledge about the status of the ocean and its changes is the key priority. That requires more science as well as demanding more availability and improved access to all marine and ocean observation data, not only from European originators but also from complementary international sources as these become available.

Given the range of threats and the relative lack of knowledge about marine ecosystems, the United Nations (UN) has declared 2021—30 the "Decade for Ocean Science for Sustainable Development" (https://www.oceandecade.org/) with the aim of creating "improved conditions for sustainable development of the Ocean."

There are, therefore, many globally coordinated initiatives that intend to fight against a decline in ocean health and, at the same time, bring together public and private actors within a single framework.

This chapter will present the most significant programs that have made it possible to reach a level of coordination of initiatives relating to the science for sustainable development of the ocean, examining, in particular, those initiatives that allow access to data and information. Practical cases of implementation of observation data organization and management will be provided by using major European programs as examples. These programs, and equivalent ones conducted in other areas, are providing opportunities to start a new roadmap in the integration of data management infrastructures at the international level.

The importance of marine data

From the 1970s onwards, most international agreements on marine environment protection have included provisions which oblige countries, both individually and jointly, to keep the condition of sea areas they agree to protect under surveillance.

Using the principles defined during the 1972 United Nations Conference on the Human Environment (https://www.un.org/en/conferences/environment/stockholm1972), nations throughout the world have, over the years, put into place several policies and tools to respond to this challenge (e.g., Wright et al., 2016). Initially, monitoring of marine areas was rather narrow in scope with much of the routinely collected, and most readily available, data focusing on a few areas of environmental importance or concern. During the last decades of the 20th century, it was clearly stated that marine areas had environmental problems associated with human activities, and the scale of the problems was not fully quantified or understood. It was recognized that there was a general need to improve the scope, quality, and availability of marine environmental data.

A holistic functional approach to ecosystems is at the base of worldwide, integrated ocean—coastal zone management. In Europe, this approach was translated into regulations, the most important one being the Marine Strategy Framework Directive (MSFD - https://ec.europa.eu/environment/marine/eu-coast-and-marine-policy/marine-strategy-framework-directive/index_en.htm). The European Union is now stepping up its ambition with the European Green Deal (European Commission, 2019). The European Green Deal sets the path for Europe to become the first climate-neutral continent by 2050 and provides a roadmap with actions to boost the efficient use of resources by moving to a clean, circular economy; reverting biodiversity loss and cutting pollution. In this context, inspired by the Apollo 11 mission that put a man on the moon, "moon-shot" missions seek to enable large-scale transformations in key areas through bold, concrete, game-changing solutions, which provide public good to Europe's citizens. This mission focuses on developing systemic and transformative solutions for "healthy oceans, seas, coastal and inland waters."

The Mission "Healthy oceans, seas, coastal and inland waters" aims to know, restore, and protect our ocean and waters by 2030, by reducing human pressures on marine and freshwater environments, restoring degraded ecosystems, and sustainably harnessing the essential goods and services they provide.

Marine environmental monitoring services

Various application domains are related to possibilities of accessing heterogeneous data and related metadata, in order to understand and use them. Heterogeneity in networking infrastructures is resolved through the use of a common reference model within and between organizations.

In the early 1990s, the International Organization for Standards developed a Reference Model of Open Distributed Processing supporting interworking, interoperability, and portability (Raymond, 1995).

Data from multiple sources, multiple platforms, and sensors are today combined in reanalysis and forecast models, and support services and the definition of paths toward sustainable uses of the oceans. In situ and satellite marine environmental monitoring systems provide observations from the seafloor to the overlying ocean and atmosphere in multiple environmental matrices (air, water, biota, seafloor).

Data are standardized and collated into comparable and searchable datasets. Data are analyzed and transformed into value-added products including maps, anomalies, and other statistical information, enabling investigation of changes such as harmful algal blooms. They can be used by end-users for a wide range of applications in a variety of areas related to climate change impacts including monitoring and reporting for sustainable development, nature conservation, regional and local marine spatial planning, fisheries, shipping, and tourism. These data stretch back for decades and are complemented by physical and biogeochemical variables such as temperature, sea level, chlorophyll concentration, primary productivity, and nutrient concentrations.

Insights from these observations need to be regularly communicated to the marine climate modeling community, since data assimilated from both physical and biogeochemical observations have the potential to improve model projections. When predictions from these models are combined with data on marine ecosystem responses to multiple environmental and human pressures, they can provide policymakers and ecosystem managers with more relevant information on which to base key decisions, e.g., best-case fisheries management scenarios that promote sustainable exploitation under given future climate projections.

Earth Observation technologies are generating an unprecedented high volume of data. Traditional data management systems and data-processing software cannot handle "big data" that requires technologies and methodologies incorporating new forms of data integration from datasets that are diverse and massive.

Climate models are becoming increasingly complex and coping with the rapidly expanding diversity, volume, and complexity of ocean–model outputs is a key challenge. This leads to subsequent challenges of analysis using conventional methods. New analytical workflows and the adoption of a "zero download" paradigm whereby scientists process and analyze their data in the cloud, combined with machine learning techniques, could address this emerging issue. Understanding the complex spatial and temporal variability of the physical and biological dynamics of the ocean depends on numerous region-specific factors. Progress can be made by applying a combination of unsupervised machine learning with differing sources of modeling and observations to identify emergent patterns with spatial and temporal commonalities. Within multimodel ensembles, machine learning can also be used to identify systematic biases in model projections, which are linearly linked to observable model bias in a concept known as *emergent constraint*. This allows reliable models to be identified, unreliable models to be rejected, and eventually reduces climate projection uncertainties. Additionally, machine learning can be applied to a suite of marine climate and ecosystem data to address complex societal relevant issues, e.g., to predict extreme or harmful marine climate events such as marine heatwaves.

A global ocean observing system framework

Sustainable development and ecosystem services have required the development of an integrated, sustained, and global observing and data delivery framework, including coastal and open oceans (Malone et al., 2014). The Global Ocean Observing System (GOOS) is a worldwide infrastructure established to ensure the sustained flow of standardized, interoperable ocean measurements from in situ networks, satellite observations, governments, UN agencies, and individual scientists. GOOS was first set up in 1991 by IOC-UNESCO, with other sponsors, as a platform for global cooperation and information exchange, ensuring current and future sustainability of the global oceans.

It is a coordinated but highly decentralized and federated system with its 13 GOOS Regional Alliances (GRAs), and it collaborates with partner infrastructure such as JCOMM, IODE, International Council for Science, World Data Centers for the planning and implementation of observations for the world's oceans, aimed ultimately at delivering data, related services, and information products in support of research and applications.

JCOMM, in concert with these networks, has also established initial capabilities to monitor the status of the global observing system (https://www.ocean-ops.org/board).

The main types of in-situ observing systems comprise:

- Drifting Argo Floats for the measurement of temperature and salinity profiles to ∼2000 m and, by tracking them, mean subsurface currents. In addition, the latest model BGC–Argo allows measurements of biogeochemical properties of the oceans.

- Research vessels which deliver complete suites of multidisciplinary parameters from the surface to the ocean floor, but with very sparse and intermittent spatial coverage and at very high operational cost.

- XBTs by research vessels and ships of opportunity underway for the measurement of temperature and salinity profiles to ∼450−750 m depth.

- Surface Moorings, particularly capable of measuring subsurface temperature profiles continuously over long periods. Currents are often monitored, with meteorological measurements usually being made too. Biofouling restricts the range of measurements that can be made from long deployments in the photic zone but surface salinity and biogeochemical measurements are attempted.

- Ferry-Box and other regional ship of opportunity measurement programs for surface transects which may include temperature, salinity, turbidity, chlorophyll, nutrients, oxygen, pH and algal types.

- The network of tide gauges, which provides long-term reference and validation of sea level data.

- Gliders, which complement floats and moorings and can perform transects of physical and biogeochemical parameters from the surface to 1000 m at a lower cost than ships.

- Surface drifters are cheap and light-weight platforms that passively follow the horizontal flow at the surface via a drogue/sail. They complement satellites for sea surface temperature and surface current measurements.

- Permanent long-range (up to 200 km) HF-radar monitoring systems in specific regions of national/international interests and importance, typically as part of an observatory.

- Sea mammals can be fitted with noninvasive miniaturized ocean sensors that can help with collecting measurements in remote and extremely cold places such as polar areas.

Tables 3.1 and 3.2 summarize the applicable best practice/standards. Other important technical documents to consider are International

Table 3.1 Platform/instrument-dependent best practices and QC.

Platform	Network program	Best practices and QC manuals
ARGO	ARGO program (http://www.argo.net/)	Argo float data and metadata from Global Data Assembly Center (Argo GDAC) (ARGO, 2021; https://www.seanoe.org/data/00311/42182/)
XBT	eXpendable BathyThermographs (XBTs) (https://www.aoml.noaa.gov/phod/goos/xbtscience/data_management.php)	SOOPIP Quality Control Cookbook for XBT data (http://woce.nodc.noaa.gov/woce_v3/wocedata_1/woce-uot/document/qcmans/csiro/csiro.htm) Procedures used at AOML to quality control real-time XBT data collected in the Atlantic Ocean (http://woce.nodc.noaa.gov/woce_v3/wocedata_1/woce-uot/document/qcmans/aoml/aoml_1.htm)
Glider	OceanGliders (https://www.oceangliders.org/) Everyone's Gliding Observatory (https://www.ego-network.org/)	SeaDataNet data management protocols for glider data (SeaDataCloud deliverable 9.14) Manual for quality control of temperature and salinity data observations from gliders (https://ioos.noaa.gov/wp-content/uploads/2015/10/Manual-for-QC-of-Glider-Data_05_09_16.pdf)
Air-sea interaction profiler		SCOR WG 142: Recommendation for oxygen measurements from Argo floats, implementation of in-air measurement routine to assure highest long-term accuracy (http://doi.org/10.13155/45917)
CTD		The acquisition, calibration, and analysis of CTD data (http://ioc-unesco.org/components/com_oe/oe.php?task=download&id=7111&version=1.0&lang=1&format=1)

Table 3.1 Platform/instrument-dependent best practices and QC.—cont'd

Platform	Network program	Best practices and QC manuals
HFR	European HFR node of the Global HFR network	Common procedures for HFR QC management (http://www.jerico-ri.eu/download/jerico-next-deliverables/JERICO-NEXT-Deliverable_5.14_V1.pdf)
Sea mammals	AniBOS	Data processing and validation http://www.meop.net/database/data-processing-and-validat.html

Table 3.2 Data quality standard reference.

Parameters	EuroGOOS DATAMEQ	QARTOD
Temperature in the water column	https://archimer.ifremer.fr/doc/00251/36230/	https://www.ioos.noaa.gov/ioos-in-action/temperature-salinity/
Salinity in the water column	https://archimer.ifremer.fr/doc/00251/36230/	https://www.ioos.noaa.gov/ioos-in-action/temperature-salinity/
Oxygen in the water column	https://www.atlantos-h2020.eu/download/7.2-QC-Report.pdf	https://www.ioos.noaa.gov/ioos-in-action/manual-real-time-quality-control-dissolved-oxygen-observations/
Currents in the water column	https://archimer.ifremer.fr/doc/00251/36230/ https://www.atlantos-h2020.eu/download/7.2-QC-Report.pdf	https://www.ioos.noaa.gov/ioos-in-action/currents/
Light attenuation in water		https://www.ioos.noaa.gov/ioos-in-action/oceanic-optics/
Fluorescence in the water column		https://www.ioos.noaa.gov/ioos-in-action/oceanic-optics/
Wind speed and direction		https://www.ioos.noaa.gov/ioos-in-action/wind-data/
pCO_2	https://www.atlantos-h2020.eu/download/7.2-QC-Report.pdf	

Council for Science (ICSU) publications, produced by the Scientific Committee on Oceanic Research (SCOR) in the context of the GO-SHIP program. Tables 3.3 and 3.4 list some relevant SCOR and GO-SHIP documents.

Table 3.3 SCOR relevant documents on best practices.

Contribution	Methodologies, studies
Joint panel on oceanographic tables and standards	• Algorithms for calculations of fundamental properties of seawater (http://ioc-unesco.org/components/com_oe/oe.php?task=download&id=7111&version=1.0&lang=1&format=1), UNESCO technical papers in marine science 44 (1983) • Progress on oceanographic table and standards 1983 −86, (http://unesdoc.unesco.org/images/0007/000756/075627eb.pdf) UNESCO technical papers in marine science 50 (1986)
Continuous current velocity measurements	• An intercomparison of some current meters, UNESCO Technical Papers in Marine Science 11.
Evaluation of CTD data	• The acquisition, calibration, and analysis of CTD data (http://ioc-unesco.org/components/com_oe/oe.php?task=download&id=7111&version=1.0&lang=1&format=1), UNESCO technical papers in marine science 54 (1988)
Methodology for oceanic CO_2 measurements	• Methodology for oceanic CO_2 measurements (http://unesdoc.unesco.org/images/0012/001241/124165Eo.pdf), UNESCO technical papers in marine science 65 (1992)
Thermodynamics and equation of state of seawater	• The international thermodynamic equation of seawater − 2010: Calculation and use of thermodynamic properties (http://www.teos-10.org/pubs/TEOS-10_Manual.pdf). IOC manuals and guides 56 • An historical perspective on the development of the thermodynamic equation of seawater (http://www.ocean-sci.net/8/161/2012/os-8-161-2012.pdf) - 2010 Special issue of *Ocean Science*
Deep ocean exchanges with the shelf	• Deep ocean exchanges with the shelf (http://www.ocean-sci.net/special_issue18.html) - special issue of ocean science

Table 3.4 GO-SHIP relevant documents on best practices.

Contribution	Methodology application	Authors and link to the document
Data acquisition overview	Reference quality water sample data: Notes on data acquisition	Swift (2010)
Methods for water sampling and analysis	Method for salinity (conductivity ratio) measurement	Kawano (2010)
	Recommendations for the determination of nutrients in seawater to high levels of precision and inter-comparability using continuous flow analysers	Hydes et al. (2010).
	Determination of dissolved oxygen in seawater by Winkler titration using the amperometric technique	Langdon (2010)
	Guide to best practices for ocean CO_2 measurement	Dickson et al. (2007)
	Sampling and measurement of chlorofluorocarbon and sulfur hexafluoride in seawater	Bullister and Tanhua (2010)
	Collection and measurement of carbon isotopes in seawater DIC	McNichol et al. (2010)
	Sampling and Measuring helium isotopes and tritium in Seawater	Jenkins et al. (2010)
CTD Methods	Notes on CTD/O_2 data acquisition and processing using seabird hardware and software	McTaggart et al. (2010)
	CTD oxygen sensor calibration procedures	Uchida et al. (2010)
	Calculation of the thermophysical properties of Seawater (2010)	IOC, SCOR, and IAPSO (2010)
	A manual for acquiring lowered doppler current profiler data	Thurnherr et al. (2010)
Underway Measurements	Ship-mounted acoustic doppler current profilers	Firing and Hummon (2010)
	A guide to making climate quality meteorological and flux measurements at sea (2006)	Bradley and Fairal (2007)

GOOS regional alliances—the European monitoring framework

The Global Ocean Observing System utilizes the Framework for Ocean Observing to guide its implementation of an integrated and sustained ocean observing system (Fig. 3.1). This systems approach, designed to be flexible and to adapt to evolving scientific, technological, and societal needs, helps deliver an ocean observing system with a maximized user base and societal impact. The GOOS Regional Alliances (GRAs) are nations and/or institutions teamed up to share GOOS principles and goals but apply them regionally. Thirteen GRAs represent different regions of the globe, emphasizing regional priorities, differing by need, resources, and culture. Some GRAs emphasize data sharing or regional capacity development, while others are building out extensive observation systems with dedicated marine service goals, such as oil spill response capabilities or typhoon forecasting.

The European Global Ocean Observing System (EuroGOOS) is the European component of GOOS, founded in 1994. EuroGOOS coordinates the development and operation of (European) regional operational systems (ROOSs) which encompass four systems at present for the Arctic (Arctic ROOS), the Baltic (BOOS), the North West Shelf (NOOS), and the Ireland–Biscay-Iberian area (IBI-ROOS) sea regions. EuroGOOS is strongly connected to the Mediterranean regional Alliance MONGOOS. These regional assemblies are instrumental in active cooperation in new developments, operations, and planning investments.

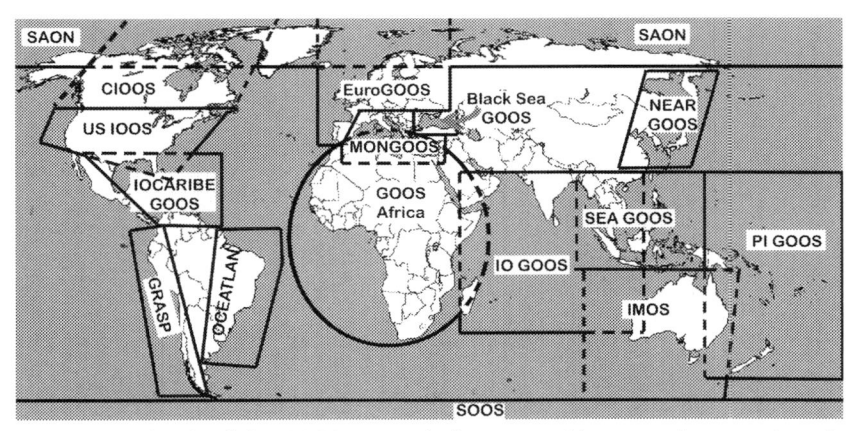

Figure 3.1 A sketch of the GOOS regional alliances and integrated ocean observing systems derived from https://goosocean.org/index.php?option=com_content&view=article&id=83&Itemid=121.

Recent EU marine data infrastructures and EU programs are widely based on EuroGOOS and ROOS achievements and goals. Projects address the following EuroGOOS community needs:

- Provision of easy access to data through standard generic tools, easy means of using the data without having to be concerned about data processing and who processes them, and that adequate metadata are available to describe how the data were processed.
- To combine in situ observation data with other information (e.g., satellite images or model outputs) in order to derive new products, build new services, or enable better-informed decision-making.

The ocean data management and exchange process within EuroGOOS are intended to reduce duplication of effort among agencies, improve quality, and reduce costs related to geographic information, thus making oceanographic data more accessible to the public and helping to establish key partnerships to increase data availability.

Besides, a EuroGOOS data management system will deliver a system that will meet European needs, in terms of standards and respecting the structures of the contributing organizations. The structure will include:

- Observation data providers, which can be operational agencies, marine research centers, universities, national oceanographic data centers, and satellite data centers.
- Integrators of marine data, such as the Copernicus in situ data thematic center (for access to near real-time data acquired by continuous, automatic, and permanent observation networks) or the SeaDataNet infrastructure (for quality controlled, long-term time series acquired by all ocean observation initiatives, missions, or experiments), ICES and EurOBIS for biodiversity observations, and the European Marine Observation and Data Network (EMODnet) thematic portals.
- The integrators that will support both data providers willing to share their observation data, and users who want to access oceanographic data from a range of providers encompassing multiple types of data from multiple regions. They also develop new services to facilitate data access and increase the use of both existing and new observational data.
- Links with international and cross-disciplinary initiatives such as GEOSS (Global Earth Observation System of Systems), both for technical solutions to improve harmonization as well as for dissemination of European data in an interdisciplinary global context.

The targeted integrated system deals with data management challenges that must be met to provide efficient and reliable data service to users.

These include:

- Common quality control for heterogeneous and near real-time data
- Standardization of mandatory metadata for efficient data exchange
- Interoperability of network and integrator data management systems

Data governance

Although important progress has been made, problems still exist in the data management framework; the integration and harmonization of data across the multiscale, multiplatform, multisensor ocean observing "system of systems" (Snowden et al., 2019). IODE published manuals and guides, starting in the 1960s that deal with operational procedures for data collection, quality assessment and quality control, standards and reference materials, data formats, etc.

(https://www.iode.org/index.php?option=com_oe&task=viewDoclist Record&doclistID=9).

The general vision behind the concepts on data and information services is that of "global ocean commons," i.e., a global public good (IOC, 2013) that must be used sustainably.

Data access and data interoperability

A building block for a distributed system in which component systems can exchange and understand information is the standardization of data formats, distribution protocols, and metadata.

Metadata is "data that provides information about other data." Metadata describes a broad range of information that allows observations to be understood and to become information and knowledge. It provides a context for research findings, ideally in a machine-readable format, and it is used to understand if data matches expectations. Metadata requires the use of standardized sets of terms to solve the problem of ambiguities. The collection and management of ocean data metadata is a complex stepwise process that includes identification of observing requirements/sensors, the configuration of sensors, the deployment at sea, and collection of observations that are regularly/periodically transmitted to a shore-side receiving station. The receiving station reformats the data messages, possibly applying some automated quality control, and then distributes the data over one of several communication pathways such as the Global Telecommunication System (GTS). Data Assembly Centers collect these data and develop added-value re-analysis/products. As an example, Fig. 3.2 presents an ideal

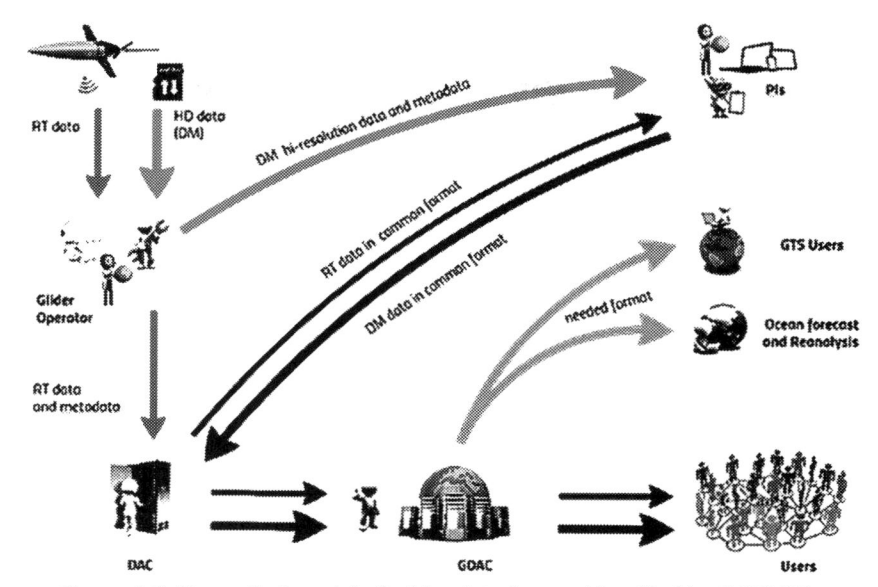

Figure 3.2 Theoretical model of glider data flow as identified by OGDMTT.

model for glider data as designed by the Ocean Glider Data Management Task Team (OGDMTT - https://www.oceangliders.org/taskteams/data-management/).

Each data collection step produces information that may be associated with a given observation, hence generating an immense volume of metadata information. Metadata provides a context for research findings, ideally in a machine-readable format, and it is used to understand if data matches expectations. Metadata requires the use of standardized sets of terms to solve the problem of ambiguities.

A best practice, or standard operating procedure, is the bottom-up methodology adopted by multiple organizations which has repeatedly produced (inter-intra-alia) superior results relative to other methodologies with the same objective.

Standards are generally top-down and may serve as benchmarks for evaluation, in addition to being processes. The International Organization for Standardization (ISO) defines standards as "documents of requirements, specifications, guidelines or characteristics that can be used consistently to ensure that materials, products, processes and services are fit for their purpose." Standards may become mandatory legislated standards, such as the European INSPIRE legislation (https://inspire.ec.europa.eu/inspire-legislation/26).

Snowden et al. (2019) proposed a fine simplified differentiation between three levels of metadata in the data collection chain:

Level 1—Instrument and Platform (I-P) documents the sensor and platform characteristics associated with an observation. The majority of this metadata is known at the time of deployment (or earlier) and does not change through during the finite life of the platform.

Level 2—Provenance and Lineage (P-L) describes the processing and history of observations, including information about their source, version, quality assessment and control, history, and accountability.

Level 3—Collection and Discovery (C-D) includes high-level information to use and interpret collected data.

According to the level, the data harmonization process may require the adoption of a best practice, or standards, or both. Within the context of EuroGOOS, the European component of the intergovernmental Global Ocean Observing System (GOOS), the Data Management, Exchange and Quality (DATAMEQ - http://eurogoos.eu/data-management-exchange-quality-working-group-data-meq/) working group help to improve harmonization and integration of European marine data. DATAMEQ WG works hand in hand with other international and European communities to develop an overall concept for the management of observation data, taking into consideration existing data management systems, as well as drafting and maintaining a minimum set of standards for data quality control related to observation data collection, processing, and exchange procedures. During recent years, these recommendations have been applied to various European projects (EuroSEA - https://eurosea.eu/, JERICOS3 - https://www.jerico-ri.eu/, etc.) which are, at the same time, contributing to improving the recommendations themselves.

If we generalize the concept, it is possible to identify, for each level, which best practice and which standard may be used to progress toward data harmonization. Some key considerations for data management and harmonization are:

Data versioning—for a typical deployment we can expect to see multiple versions of the dataset come online, ranging from preliminary datasets while the platform is in the water, through to highly refined versions having undergone expert scrutiny post-platform recovery. The resolution of the dataset may also vary between versions, both in terms of sampling frequency and the number of parameters reported. Formalized data management needs to account for these different dataset streams and implement suitable strategies for superseding and merging.

Data standards—the adoption of recognized standards facilitates the discoverability and exchange of observations and encourages greater interoperability. Use of controlled vocabularies to annotate metadata and data, together with the adoption of recognized data exchange formats, must underpin any data management strategy.

Quality Assurance (QA)/Control (QC)—both key requirements to add integrity and gain user confidence in datasets. Flexible, consistent approaches that are tailored to specific end-user requirements are strongly desirable. QA and QC approaches must be defined, harmonized, and documented in a robust infrastructure for glider data management.

Data delivery—timely, efficient data flow that meets all end-user requirements is imperative. As with data standards, there are recognized data delivery protocols and tools enabling appropriate data dissemination to end-users.

Operational monitoring—an important component of any data management system incorporating near real-time data is the capture of metrics to monitor data availability-accessibility and the health of the data management workflow between stakeholders.

Collaborative working—helps to pool knowledge and develop expertise. It also increases visibility, encourages commonality, and reduces duplication of effort. Ultimately it is an important prerequisite, enabling the building of shared infrastructures to pool and disseminate data to stakeholders.

Provenance and lineage

This metadata describes the processing and history of observations, including information about their source, version, quality assessment, and control. These policies will be applied to access, share, and reuse the collected data, and how long-term storage and data management will be ensured. A good practice is to attach this metadata to original data as soon as possible as close to the data originator as possible. This level also deals with consistent, well-designed file formats that are essential to facilitate standalone, interoperable information exchange.

The primary data format for distribution will be the OceanSites netCDF-4 classic model (http://www.coriolis.eu.org/Data-Services-Products/MyOcean-In-Situ-TAC/Documentation).

NetCDF (Network Common Data Form) is a set of software libraries and machine-independent data formats that is the international standard for common data and is the one adopted by all key European and international ocean-data management infrastructures (Global Data Assembly Centers, CMEMS, EMODnet, SeaDataNet, etc.). The recommended implementation of NetCDF is based on the community-supported Climate and Forecast Metadata Convention (CF), version 1.6, which provides a definitive description of the data in each variable, and the spatial and temporal properties of the data. Key European infrastructures are using the same core NetCDF structure with some specific infrastructure extensions based on former OceanSITES NetCDF implementation:

- OceanSITES data format manual
  http://www.oceansites.org/docs/oceansites_data_format_reference_manual.pdf
- CMEMS InsTAC format extension specification: https://doi.org/10.13155/59938
- SeaDataNet format extension specification: https://doi.org/10.13155/56547

Collection and discovery

This is the level of the catalog and interoperability and should be in line with FAIR principles. Harmonization of data collection and data discovery is implemented at pure metadata level and is used to understand if data matches expectations. Importantly, having this metadata discoverable and linked to a repository does not mean having the data in the same repository. Having this data available increases the visibility of research activities and stakeholders can learn about the existence of data and contact the data generator (i.e., the project).

At this level, it is important to follow ISO standards and it is recommended the adoption of:

- ISO 8601 Representation of date and time
- ISO 19108 Temporal characteristics of geographic information
- ISO 19113 Revised by 19157 standards for geographic information
- ISO 19115 Geographical information metadata
- ISO 19119 Taxonomy of services
- ISO 19139 Geographical information metadata implementation specification

ISO 19113 defines quality principles, which are applied in ISO 19115 (geographic metadata). The metadata records in the current GEOSS use the ISO 19115 data model and its companion XML encoding (ISO

19139). ISO 19115 Standard requires a basic minimum number of metadata elements that are essential for the data presentation (see also "data quality dimension" in Chapter 4):

- Dataset or dataset series on specific challenges ("what"),
- A geographic bounding box ("where"),
- Temporal extent ("when"),
- Contact point to learn more about or order the dataset ("who").

This minimum set of metadata can be extended with additional elements being used to increase interoperability and usability. Specifically, EuroGOOS DATAMEQ recommends the inclusion of a unique platform identifier that should be the WMO if available; a code assigned by the network platform steering group/regional data assembly center. In the case of ships, the unique ID would be the ICES code.

This concept is evolving: recent discussions within the glider network platform data management task team expressed the need for managing a higher granularity of the platform information, i.e., the mission. A glider can run several missions, and missions may have different scopes (i.e., different sensor settings), alternatively repeated tracks may see gliders with different equipment (upgrade of sensors with more recent/higher resolution devices).

Information about sensor settings is defined at level 1—Sensor and Platform. It is recommended to have a level 3 metadata item to link to mission-sensors information, i.e., a link to a service exposing sensor information. It could be a link to the SensorML.

- **Standard names for the parameters.** Concerning the parameter naming conventions, it is recommended to use the Climate and Forecast (CF) standard name convention and SeaDataNet—NVS.v.2. short names
- **Institution/data provider.** SeaDataNet developed and maintains the European Directive of Marine Organisations (EDMO). This catalog contains more than 3000 entries and is widely adopted as it has a unique ID identifying marine institutes.
- **Provenance and DOI.** Provenance is defined as level 2 metadata, and that level includes all the processing details and applied procedures. Level 3 provenance is mainly related to acknowledgment of sources and entities that were involved in data processing and validation. Whenever possible DOI (DOIs for ocean data, general principles and selected examples https://doi.org/10.13155/44515) should be included.

Although it does not concern dissemination metadata, an important element for implementing level 3 is the dissemination infrastructure. Data and metadata have to be open and freely accessible, and data server/service tools facilitating the implementation of data harmonization toward data users are (obviously) recommended.

Data harmonization through brokering approaches

The rapid developments of brokering approaches by data integrators obviate, to a large extent, the tedious and cumbersome implementation of rigid sets of standards. Major data integrators are working on developing brokers in order to improve cross-platform interoperability and lower entry barriers for both users and data providers. In a fully operational broker-approach, users, as well as data providers, will not be asked to comply with specific standard regulations or implement any specific interoperability technology. Data providers will ideally continue to use their tools and publish their resources according to their own standards, as long as common and internationally recognized formats are used (EuroGEOSS). In this way, the brokering approach loosens the necessity for implementing a common data model and exchange protocol, by providing the necessary mediation and transformation functionalities (Nativi et al., 2015).

Recently, GEOSS, in particular the EuroGEOSS project (http://www.eurogeoss-broker.eu/), developed a brokering framework to make heterogeneous resources from a variety of published data providers commonly found and accessible by the user community (Nativi et al., 2013). The brokering approach is now part of the GEOSS common infrastructure (GCI) called GEO Discovery and Access Broker (DAB) and builds on a broker for each main functionality: discovery, access, and semantic interoperability (Nativi et al., 2015).

Mature data integrators within AtlantOS, like SeaDataNet, are already making use of the GEOSS brokering services. For example, the XML encoding output of the SeaDataNet common data index service has been upgraded to the new SeaDataNet ISO 19139 Schema to comply with the EU INSPIRE Directive Implementing Rules. This conversion was amended using the GEO-DAB brokerage service that also plays an important function in the GEOSS portal.

ERDDAP (ERDDAP is a data server providing a simple, consistent way of downloading subsets of gridded and tabular scientific datasets in common file formats to make graphs and maps. ERRDAP implements FGDC Web

Accessible Folder (WAF) with FGDC-STD-001-1998 and ISO 19115 WAF with ISO 19115–2/19139) represents an emerging key technology for data dissemination, brokerage, and interoperability. Many international data integrators have already adopted this technology, and following the preliminary work done by NOAA and, at the European level, by EMOD-net Physics, other European infrastructures (CMEMS, SeaDataNet) are moving toward its adoption. More recently, GOOS started promoting ERDDAP as a capacity development tool aimed at lowering the technical barriers for supplying and accessing data, working toward the implementation of FAIR (Findable, Accessible, Interoperable, and Reusable) principles.

FAIRness of data and related services

The FAIR concept relates to "Data and services that should be Findable, Accessible, Interoperable, and Re-usable, both for machines and for people." The emphasis is on machine FAIRness. Technological advances provide innovative opportunities for new forms of science, which is one of the drivers behind, e.g., CODATA (https://codata.org/), the Committee on Data of the International Science Council (ISC), and the European Open Science Cloud (EOSC - https://www.eosc-portal.eu/). However, this demands well-described, accessible data that conform to community standards. The FAIR principles articulate the attributes data need to have to enable and enhance reuse, by humans and machines. There were a few predecessors before coming to the formulation of the FAIR Guiding Principles as now in use. An influential document was the OECD's 2007 "Principles and Guidelines for Access to Research Data from Public Funding" (https://www.oecd.org/sti/inno/38500813.pdf). The seminal Royal Society report of 2012, "Science as an Open Enterprise" (https://royalsociety.org/topics-policy/projects/science-public-enterprise/report/) argued that research data being open was not sufficient as data need to be accessible, assessable, interoperable, and usable. The 2013 G8 Science Ministers' statement (https://www.gov.uk/government/news/g8-science-ministers-statement) says: "Open scientific research data should be easily discoverable, accessible, assessable, intelligible, useable, and wherever possible interoperable to specific quality standards." These criteria were adopted in the initial data guidelines for the EU Horizon 2020 framework program later the same year. Echoing these criteria, the FAIR principles were conceived at the Lorentz conference in 2014 and published following consultation via the FORCE11

Group (https://www.force11.org/group/fairgroup/fairprinciples). The most influential document is the article by Wilkinson et al. (2016) in *Nature Scientific Data*, "The FAIR Guiding Principles for Scientific Data Management and Stewardship." This article introduced a table of FAIR Guiding Principles which can be used by Research Infrastructures (RIs) for checking their FAIRness and formulating measures for improving their FAIRness. The table published in the article is shown in Table 3.5. Most of the principles are related to metadata. Moreover, the emphasis is on machine FAIRness for data and their associated services.

Another recent and prominent publication about FAIR is "Turning FAIR into reality" (https://ec.europa.eu/info/sites/info/files/turning_fair_into_reality_1.pdf) by an EU committee of experts. The following has been extracted and summarized from their report.

Data are **Findable** when they are described by sufficiently rich metadata and registered or indexed in a searchable resource that is known and accessible to potential users. Additionally, a unique and persistent identifier

Table 3.5 FAIR Guiding Principles as defined in Wilkinson et al. (2016).
The FAIR Guiding Principles

To be Findable:
    F1. (meta)data are assigned a globally unique and persistent identifier
    F2. data are described with rich metadata (defined by R1 below)
    F3. metadata clearly and explicitly include the identifier of the data it describes
    F4. (meta)data are registered or indexed in a searchable resource
    To be Accessible:
    A1. (meta)data are retrievable by their identifier using a standardized
        communications protocol
    A1.1 the protocol is open, free, and universally implementable
    A1.2 the protocol allows for an authentication and authorization procedure,
    where necessary
    A2. metadata is accessible, even when the data are no longer available
    To be Interoperable:
    I1. (meta)data use a formal, accessible, shared, and broadly applicable language for
        knowledge representation.
    I2. (meta)data use vocabularies that follow FAIR principles
    I3. (meta)data include qualified references to other (meta)data
    To be Reusable:
    R1. meta(data) are richly described with a plurality of accurate and relevant
        attributes
    R1.1. (meta)data are released with a clear and accessible data usage license
    R1.2. (meta)data are associated with detailed provenance
    R1.3. (meta)data meet domain-relevant community standards

should be assigned such that the data can be unequivocally referenced and cited in research communications. The identifier enables persistent linkages to be established between the data, metadata, and other related materials in order to assist data discovery and reuse. Related materials may include the code or models necessary to use the data, research literature that provides further insights into the creation and interpretation of the data and other related information.

Accessible data objects can be obtained by humans and machines upon appropriate authorization and through a well-defined and universally implementable protocol. Anyone should be able to access at least the metadata. It is important to emphasize that Accessible in FAIR does not mean Open without constraint. Accessibility means that the human or machine is provided—through metadata—with the precise conditions by which the data are accessible and that the mechanisms and technical protocols for data access are implemented such that the data and/or metadata can be accessed and used at scale, by machines, across the web.

Interoperable data and metadata are described as community and/or domain standards for technical interoperability and vocabularies for semantic interoperability, and they include qualified references to other data or metadata. It is this that allows the data to be "machine-actionable." Interoperability is an essential feature in the value and usability of data. Legal interoperability of data has to be considered as well. In FAIR, legal interoperability falls under the principle that data should be "Reusable."

For data to be **Reusable**, the FAIR principles reassert the need for rich metadata and documentation that meet relevant community standards and provide information about provenance, reporting how data was created and information about consecutive data reduction or transformation processes to make data more useable, understandable, or "science-ready." The ability of humans and machines to assess and select data based on criteria relating to provenance information is essential to data reuse, especially at scale. Reusability also requires that the data be released with a "clear and accessible data usage license": in other words, the conditions under which the data can be used should be transparent to both humans and machines.

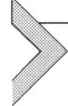

Ocean data standards for processing data and metadata

In the framework of the IODE initiatives on data management standards, there have been many discussions and solutions for processing data and metadata (e.g., UNESCO, 2006). The MEDI steering group recommended (UNESCO, 2006), inter alia, to identify and document metadata management best practices and develop relevant vocabularies to describe marine datasets. In 2008, a forum on oceanographic data management and exchange was held in Oostende (Belgium) (IOC, 2008) and participants agreed on some elements concerning metadata profiles:

- ISO 19115 or a community profile of ISO 19115 to be used for creating discovery metadata
- to adopt other ISO standards where appropriate, while recognizing some limitations

The participants to the forum were requested to:

- identify existing practices for determining data quality;
- identify where practices (QC tests) are common across standards;
- provide a recommendation for standard QC tests.

An Ocean Data Standards and Best Practices Project (ODSBPP) was established in 2013 within the framework of the WMO-IOC Joint Technical Commission for Oceanography and Marine Meteorology (JCOMM).

ODSBPP is recommending some standards such as (http://www.oceandatastandards.org/recommended-standards-main-menu-44):

- Technology for SeaDataNet Controlled Vocabularies for describing Marine and Oceanographic Datasets — A joint proposal by SeaDataNet and ODIP projects
- Recommendation for a Quality Flag Scheme for the Exchange of Oceanographic and Marine Meteorological Data

SeaDataNet ocean standards

SeaDataNet (https://www.seadatanet.org) is a major pan-European infrastructure for managing, indexing, and providing access to marine datasets and data products, acquired by European organizations from research cruises and other observational activities in European coastal marine waters, regional seas, and the global ocean. Founding partners are National Oceanographic Data Centers (NODCs), major marine research institutes, UNESCO-IOC, and ICES. The SeaDataNet network was initiated in the 1990s and over time its network of data centers and infrastructure with standards, tools, and services has expanded, during many EU projects, such as in the past decade by SeaDataNet, SeaDataNet2, Ocean Data Interoperability Platform with USA, Canada and Australia (ODIP 1 and 2), EMODnet thematic projects, and SeaDataCloud. SeaDataNet has close cooperation with various ocean observing communities such as EuroGOOS and Euro-Argo, and major marine data management initiatives and infrastructures, in particular with the European Marine Observation and Data network (EMODnet) and Copernicus Marine Environmental Monitoring Service (CMEMS).

SeaDataNet develops, governs, and promotes common standards, vocabularies, software tools, and services for marine data management, which are freely available from its portal and widely adopted and used.

SeaDataNet (SDN) **common standards** for the marine domain have been developed and are maintained, collaborating with European and international experts, adopting and adapting ISO and OGC standards, and achieving INSPIRE compliance, where possible. These standards are published at the SeaDataNet portal (http://www.seadatanet.org) and comprise:

- INSPIRE compliant marine **metadata** profiles of the ISO 19115—19139 standards for datasets (**CDI** = Common Data Index) and research cruises (**CSR** = Cruise Summary Reports);
- marine **metadata** formats for data collections (**EDMED** = European Directory of Marine Environmental Data), research projects (**EDMERP** = European Directory of Marine Environmental Research Projects), monitoring programs and networks (**EDIOS** = European Directory of Initial Ocean-observing Systems), and organizations (**EDMO** = European Directory of Marine Organizations);
- SDN controlled **vocabularies** for the marine domain, with international governance, user interfaces, and web services. These follow the W3C SKOS specification for encoding the data dictionaries and taxonomies served;
- standard **data exchange** formats (SDN ODV ASCII, SDN NetCDF with CF compliance) as applied for download services. These interact with other SDN standards such as the Vocabularies and Quality Flag Scale;
- standard **QA-QC procedures** for various data types, together with IOC-IODE and ICES.

SeaDataNet has focused, with success, on applying these standards for interconnecting data centers enabling the provision of **integrated online access** to comprehensive sets of multidisciplinary, in situ and remote sensing marine data, metadata, and products. For that purpose, a set of dedicated **SeaDataNet software tools and online services** has been developed, is maintained, and available on the SeaDataNet portal, to be used **by each data center** for sharing metadata and data resources through the

SeaDataNet infrastructure. It includes documentation and common software tools for converting and preparing metadata and data into SeaDataNet formats, for populating the SeaDataNet directory services, for statistical analysis and grid interpolation (DIVA), and a versatile software package (ODV) for data analysis, QA-QC, and presentation. The **SeaDataNet infrastructure** can support a wide variety of data types and serve several sector communities. The SeaDataNet infrastructure comprises a series of services, both for users and data providers.

European directory services

The SeaDataNet network of data centers maintains and publishes a series of European directory services which are widely used. These give a wealth of data and information, such as overviews of marine organizations in Europe, and their engagement in marine research projects, managing large datasets, and data acquisition by research vessels and monitoring programs for the European seas and global oceans:

- European Directory of Marine Organizations (EDMO) (https://edmo.seadatanet.org/search) (>4,250 entries)
- European Directory of Marine Environmental Data (EDMED) (https://edmed.seadatanet.org/) (>4,200 entries)
- European Directory of Marine Environmental Research Projects (EDMERP) (https://edmerp.seadatanet.org/search) (>3,200 entries)
- European Directory of Cruise Summary Reports (CSR) (https://csr.seadatanet.org/) (>58,300 entries)
- European Directory of the Ocean Observing Systems (EDIOS) (http://seadatanet.maris2.nl/v_edios_v2/search.asp) (>350 programs and >16,500 series entries)
- Common Data Index Data Discovery and Access service (CDI) (https://cdi.seadatanet.org/search) (>2.4 million entries)

The Common Data Index (CDI) data discovery and access service is a core service and provides unified online discovery and access to vast resources of datasets, managed by > 110 connected SeaDataNet data centers from 34 countries around European seas, from both research and monitoring organizations. Currently, it gives access to more than 2.4 million datasets, originating from more than 700 organizations in Europe, covering physical, geological, chemical, biological, and geophysical data, acquired in European waters and global oceans (Fig. 3.3).

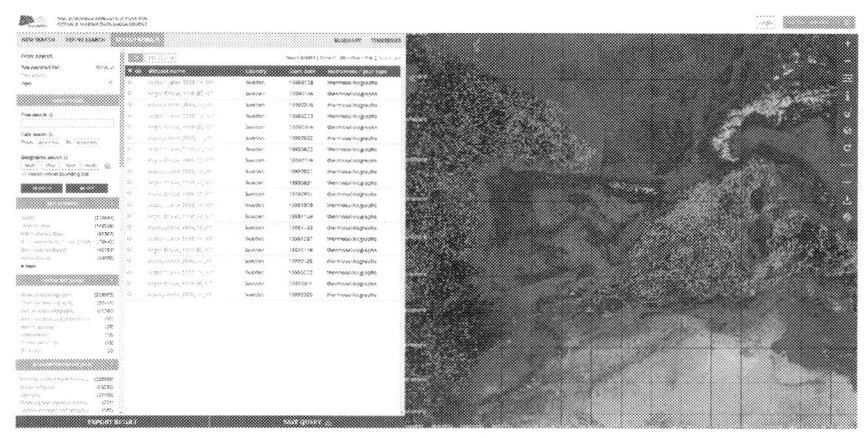

Figure 3.3 Dynamic user interface of the SeaDataNet CDI data discovery and access service.

The online CDI User Interface has recently been completely upgraded as part of the EU SeaDataCloud project, adopting cloud technology and cooperating with EUDAT (https://www.eudat.eu), a European network of academic computing centers. It gives users powerful search options and highly detailed insight into the availability and geographical spreading of marine datasets that are managed by the connected data centers. The User Interface moreover includes a simple shopping basket function for requesting access, to easily download datasets from a central data cache for unrestricted datasets and, if granted, from connected data centers for restricted datasets.

CDI metadata gives information on the what, where, when, how, and who of each dataset. It also gives standardized information on the applicable data access restrictions. The user can place selected datasets in a shopping basket. All users can freely query and browse in the CDI directory; however, submitting requests for data access via the shopping basket requires user registration in the SeaDataNet Marine-ID user register. Data requests can concern unrestricted and/or restricted datasets. Requests for unrestricted datasets are processed immediately after submission and requested datasets are made ready for download automatically from the SeaDataNet central unrestricted data cloud, whereas requests for restricted datasets are forwarded to connected data center managers for their consideration, with most cases dealing with data originators. Users receive confirmation e-mails of their dataset requests and subsequent processing, and can also check progress and undertake data downloading from their personal MySeaDataNet dashboard. Data centers also have their MySeaDataNet dashboards to follow all transactions for their datasets, and to handle requests for restricted data that require their attention.

The CDI metadata follows the ISO19115-19139 standard and is fully INSPIRE compliant. Moreover, for semantic mark-up it is supported by an extensive set of controlled vocabularies and European directories, such as for organizations (EDMO), projects (EDMERP), and Cruise Summary Reports (CSR). In this way, enriching metadata and optimizing machine-to-machine services contributes to optimizing FAIRness of the CDI service. That is further amplified by developing and providing SeaDataNet SPARQL endpoints for supporting machine-to-machine services, using the Linked Data principles, to CDI and other directories such as RDF resources. Moreover, datasets are harmonized for delivery to SeaDataNet ODV and NetCDF (CF) formats, where possible, adopting SeaDataNet controlled vocabularies internally, and complemented with the metadata of the CDI records. Use of common vocabularies in all metadatabases and data formats is an important prerequisite toward consistency, interoperability, and FAIRness.

There is the continuous maintenance of the SeaDataNet CDI service through supplementary population by the many data centers, which is further stimulated by the adoption of the CDI formats, tools, and service in many other projects, such as several EMODnet thematic projects. For this purpose, data centers make use of an online CDI Import and Validation dashboard for submitting and managing imports of new and updated CDI and data entries. The online dashboard also enables data centers to oversee and generate reports of all transactions directed to their data center.

NERC vocabulary services

In information science, controlled vocabularies are carefully selected lists of words and phrases, which are used to tag units of information (a document or work) so that they may be more easily retrieved in a search. Use of common vocabularies is an important prerequisite toward consistency and interoperability. SeaDataNet adopted this principle at an early stage and started building and using controlled vocabularies in order to mark-up metadata, data, and data products.

SeaDataNet common vocabularies (https://www.seadatanet.org/Standards/Common-Vocabularies) consist of lists of standardized terms that cover a broad spectrum of disciplines of relevance to the oceanographic and wider community. The vocabulary services are technically managed and hosted at the NERC Vocabulary Server (NVS2.0). NVS is a SKOS-vocabulary and is fit for Linked Data by having unique http URIs. The vocabularies are made available as web services for machines (SOAP API - http://vocab.nerc.ac.uk/vocab2.wsdl - and SPARQL endpoint - http://vocab.nerc.ac.uk/sparql/) and by means of client interfaces for end-users. The client interfaces provide end-users options for searching, browsing, and export of selected entries and can be found at the SeaDataNet portal (https://www.seadatanet.org/Standards/Common-Vocabularies).

Collections, concepts, and schemes are presented to the server as Uniform Resource Identifiers (URIs), in a consistent syntax. Each collection is synonymous with one of the controlled vocabularies or code lists. Each concept is synonymous with a unique vocabulary term and meaning. Each scheme can be viewed as an aggregation of one or more concepts. Semantic relationships (links) between those concepts may also be viewed as part of a concept scheme.

At present, NVS maintains and provides more than 290 different vocabularies of which more than 110 are relevant and governed as SeaDataNet vocabularies. The number of terms (concepts) in these vocabulary collections has steadily increased over time, in dialogue with research communities and under influence of many projects adopting the vocabularies for their data management. These included mapping activities for marking-up metadata and data, resulting in many requests for new terms to be added. At present NVS contains more than 150,000 terms divided among various vocabulary collections. For instance, the P01—Parameter Usage Vocabulary contains currently more than 43,000 terms. Next to vocabularies, the NVS also includes mappings between vocabularies, both internal (NVS to NVS) and external (NVS to well-established external vocabularies).

An illustrative example of internal mapping is the P08 (SDN Parameter Disciplines) => P03 (SDN Agreed Parameter Groups) => P02 (SDN Parameter Discovery Vocabulary) => P01 (SDN Parameter Usage Vocabulary) hierarchical mapping which is used for easing discovery services and classifying measurement parameters. In the CDI Data Discovery and Access, service P02 (and its P03 and P08 broader relations) is used in the CDI metadata, while P01 is used in the data. Also, many of the other SDN vocabularies are used in the metadata (depending on the SeaDataNet directories) and data, such as L05 for device categories, L06 for platform categories, and L22 for devices (measurement instruments). Good examples of external mapping are, for instance, mappings of NVS to the World Register of Marine Species (WoRMS) and Global Change Master Directory (GCMD).

The SeaDataNet list user interface has been built on top of the SOAP web service as provided by NVS. Each vocabulary has a user interface for querying, retrieving, browsing, and CSV export of terms. Several vocabularies are hierarchical and include a thesaurus button to browse hierarchically (Fig. 3.4).

Additional attention is given to exposing the P01 Parameter Usage Vocabulary, which is used for indicating, in the SeaDataNet data files (ODV and NetCDF (CF)), which parameters have been observed. The P01 is used intensively by data providers when mapping local datasets to the SeaDataNet target data formats. However, identifying the right P01s during mapping is also quite a challenge as P01 at present counts more than 43,000 terms and each P01 concept is built up of several elements following a semantic model. In particular, when mapping complex terms as in the case of chemistry, it takes considerable effort to locate the right terms or to identify missing terms that should be added. Example of P01 term (MMUSDTBT): "Concentration of tributyltin cation {tributylstannyl TBT + CAS 36,643-28-4} per unit dry weight of biota {*Mytilus galloprovincialis* (ITIS: 79,456: WoRMS 140,481) [Subcomponent: flesh]}." This truncation consists of components for a measurement property, chemical substance, measurement matrix relationship, and matrix. These components also have parts.

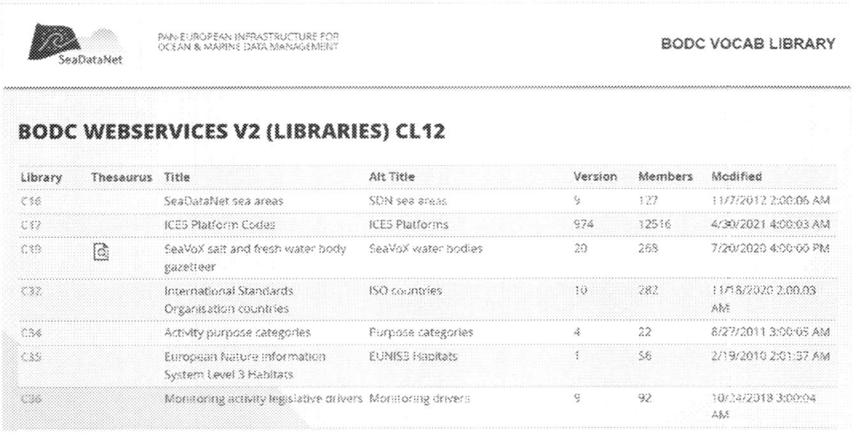

BODC WEBSERVICES V2 (LIBRARIES) CL12

Library	Thesaurus	Title	Alt Title	Version	Members	Modified
C16		SeaDataNet sea areas	SDN sea areas	9	127	11/7/2012 2:00:06 AM
C17		ICES Platform Codes	ICES Platforms	974	12516	4/30/2021 4:00:03 AM
C19	[icon]	SeaVoX salt and fresh water body gazetteer	SeaVoX water bodies	29	268	7/20/2020 4:00:00 PM
C32		International Standards Organisation countries	ISO countries	10	282	11/18/2020 2:00:03 AM
C34		Activity purpose categories	Purpose categories	4	22	8/27/2011 3:00:05 AM
C35		European Nature information System Level 3 Habitats	EUNIS3 Habitats	1	56	2/19/2010 2:01:37 AM
C36		Monitoring activity legislative drivers	Monitoring drivers	9	92	10/24/2018 3:00:04 AM

Figure 3.4 User interface at SeaDataNet portal to oversee and query all SeaDataNet common vocabularies.

To ease use, the P01 semantic model is exposed in its components by web services and on top of these a dedicated P01 facet search user interface has been deployed. This so-called "one-armed bandit" (https://vocab.seadatanet.org/p01-facet-search) can be reached from the overall SeaData-Net vocabularies interface by clicking on the "magnifying glass button" (Fig. 3.5).

This search tool makes it much easier to find relevant P01 terms by using the facets in combination with multiple free search terms, while P01 results can be exported in a CSV list for use in local mapping. In addition, a P01 vocabulary builder tool (https://www.bodc.ac.uk/resources/vocabularies/vocabulary_builder/) has been developed and deployed to facilitate data providers in composing and submitting requests for new P01 terms. The new terms can be built using semantic components. The following image illustrates how the P01 vocabulary builder can be reached.

The NERC Vocabulary Server (NVS) vocabularies underpin the mark-up of metadata and data within the SeaDataNet framework, thereby ensuring effective discovery and interoperability of its information assets. Many of these SeaDataNet-developed vocabularies are also very popular externally and support a broader community on a global scale. Greater up-take of NVS vocabularies is highly desirable, as it enables content enrichment while presenting opportunities to collaborate more closely with

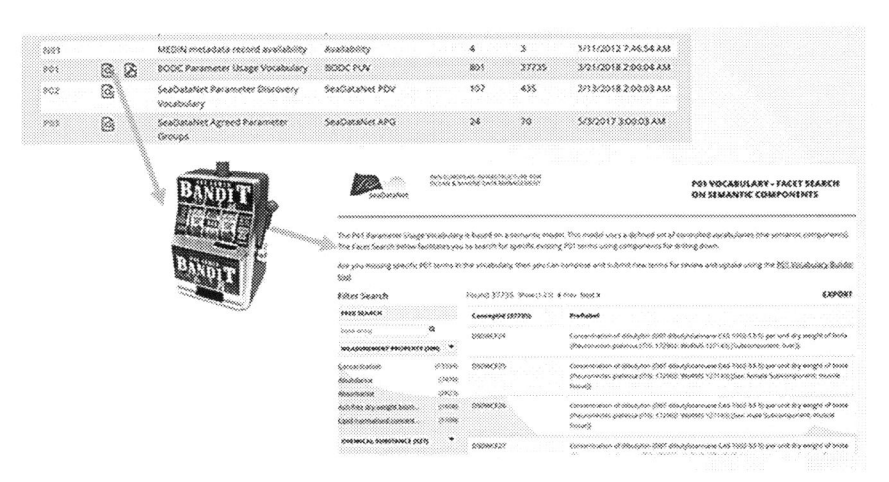

Figure 3.5 How to open the P01 semantic model facet search from the SeaDataNet vocabularies list user interface.

domain experts. Expansion of domain expertise is an ultimate goal for the governance of the vocabularies, as they become increasingly diverse and specialized, and maintaining user trust in content therefore becomes more challenging.

SeaDataNet data formats

Worldwide, data infrastructures mainly exchange data in two formats: NetCDF and/or ASCII. Within SeaDataNet, due to historical legacy also deriving from the MEDAR/MedAtlas project (Levitus, 2012), datasets are delivered to users, where possible, in common SeaDataNet data transport formats (https://www.seadatanet.org/Standards/Data-Transport-Formats), which interact with other SeaDataNet standards such as SeaDataNet-controlled vocabularies, SeaDataNet European directories, and SeaDataNet Quality Flag Scale, as well as with SeaDataNet analysis and presentation tools such as ODV and DIVA. The following SeaDataNet data transport formats have been defined:

- SeaDataNet ODV4 ASCII for profiles, time series, and trajectories;
- SeaDataNet NetCDF with CF compliance for profiles, time series, and trajectories;
- SeaDataNet MedAtlas as optional extra format;
- NetCDF with CF compliance for gridded datasets.

The first three formats have been extended with a SeaDataNet semantic header. The ODV4 format can be used directly in the popular Ocean Data View (ODV) analysis and presentation software package, which is maintained and regularly extended with new functionalities. The SeaDataNet NetCDF (CF) format for profiles, time series, and trajectories has been defined by bringing together a community comprising NetCDF and CF experts (such as from NCAR and UNIDATA), and many users of oceanographic point data. This NetCDF format can be used as an alternative for the SeaDataNet ODV 4 ASCII format, for profiles, time series, and trajectories.

SeaDataNet maintains and provides data providers with the NEMO software tool to convert from any type of ASCII format to the SeaDataNet ODV and Medatlas ASCII formats, as well as the SeaDataNet NetCDF (CF) format (for time series, profiles, and trajectory observations), which are then made accessible through the CDI service. Another SeaDataNet OCTOPUS software tool is used by data providers as a multiformat Checker, Converter, and Splitter tool.

Next to these SeaDataNet common data formats, several special SeaDataNet data formats are also used and documented for specific data types, such as:

- SeaDataNet ODV ASCII format for biodiversity data, as developed with EurOBIS;
- SeaDataNet ODV ASCII format for microlitter data, as developed with EMODnet Chemistry and TG-ML;
- ASCII data format for beach litter data, as developed with OSPAR, EMODnet Chemistry, and TG-ML;
- ASCII data format for seafloor litter data, as developed with ICES, EMODnet Chemistry, and TG-ML;
- SeaDataNet ODV ASCII format for flowCytoMetry data;
- NetCDF4 (CF) format for HF Radar data, as developed with Euro-GOOS and EMODnet Physics.

In addition, other common standards can be used for data formats. All applicable SeaDataNet data formats are included in the L24 controlled vocabulary.

Sensor web enablement (SWE)

Increasingly, marine data is collected by smart sensors and platforms. Several developments are ongoing in this field, with the development of new sensors for an expanding range of parameters, and new platforms that can carry a payload of multiple sensors and operate efficiently for long durations. These developments need to be accelerated, also for data management and data flow.

When dealing with (in-situ) observation data, there is a very large amount of different sensor data encodings and data models, as well as interfaces. This heterogeneity makes the integration of sensor data a very cumbersome task. For example, without a common standardized approach, it would be necessary to customize each application that consumes sensor data to the individual data formats and interfaces of all sensing devices delivering data.

To address this issue, the Open Geospatial Consortium (OGC), an international de-facto standardization organization in the field of spatial information infrastructures, has developed the Sensor Web Enablement (SWE – Fig. 3.6) framework of standards. The OGC SWE architecture comprises several specifications facilitating the sharing of observation data

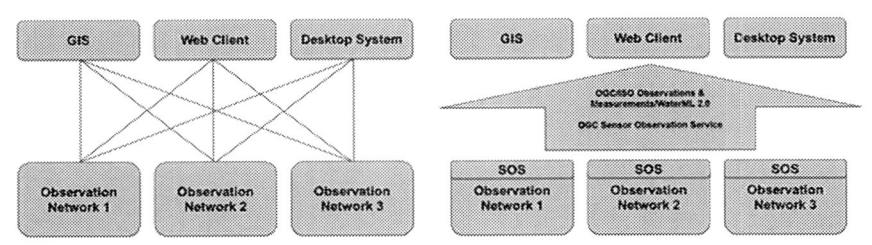

Figure 3.6 Left: no standards used, requiring individual integrating of all data sources; Right: reducing the integration efforts by using SWE standards, common interfaces and encodings.

and metadata via the web. Important building blocks are standards for observation data models (Observations and Measurements, O&M), for the corresponding metadata about measurement processes (Sensor Model Language, SensorML), and interfaces for providing sensor-related functionality (e.g., data access) via the World Wide Web (Sensor Observation Service, SOS).

The adoption of SOS SWE standards holds great promise as it facilitates data streamlining from platforms to receivers in real-time. It documents many relevant aspects of the sensors, platforms, and observations using marine SWE profiles and vocabularies, thus enriching the available metadata from observations at their origin, which will contribute to improving the FAIRness of datasets and documenting the provenance of observed data.

The key elements in this process are:

- SWE SOS server. The Sensor Observation Service provides a standardized interface for managing and retrieving metadata and observations from heterogeneous sensor systems.
- SWE Ingestion Service: This component aims to support sensor operators, researchers, and data owners when ingesting data and SWE metadata from operational observing platforms and sensors into a local storage system and publishing (selected) data streams from this database using SOS services to receiving servers. This facilitates operators when publishing streams of near-real-time and real-time observation data via SOS servers by first describing the structure of the observation network and data stream and then enabling automated data ingestion, storage, and publication process;
- SWE Viewing Services, based on the Helgoland Sensor Web Viewer, is an application for exploring and visualizing data streams from operational sensors and platforms.

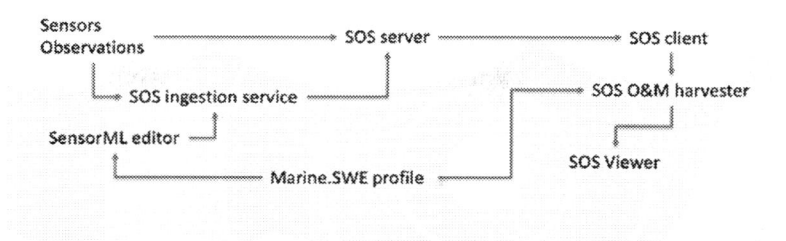

Figure 3.7 SWE demonstrator concept.

Fig. 3.7 shows the SOS SWE demonstrator concept where, starting from the observation, through the SOS ingestion service that implements standards for the interoperable integration of sensors and their observation data into web-based (spatial) data infrastructures (Bröring et al., 2011), it is possible to implement pull-based access to observation data as well as sensor metadata (SOS client) for data access and visualization (SOS Viewer).

One example of an open-source SWE Toolkit has been developed by the SeaDataNet consortium within the SeaDataCloud project. At the Sea-DataNet portal homepage, a link is given to an SWE Demonstrator (https://www.seadatanet.org/Software/Sensor-Web-Viewer) of the deployed SWE Toolkit. All of the software is published as open-source software. To assist uptake by operators of observing platforms, the Demonstrator page also includes information where all relevant documentation and GitHub resources can be found.

The marine data management landscape

Oceanographic and marine data include a very wide range of measurements and variables, originating from a broad, multidisciplinary spectrum of projects and programs. Oceanographic and marine data are collected by several thousands of research institutes, governmental organizations, and private companies. Various heterogeneous observing sensors are installed on research vessels, submarines, aircraft, moorings, drifting buoys, gliders, floats, fixed platforms, and satellites. These sensors measure physical, chemical, biological, geological, and geophysical parameters, with further data resulting from the analysis of water and sediment samples in a wide variety of parameters. Therefore, it is important to provide discovery and access to multidisciplinary and aggregated data sets and to provide analytical frameworks which support getting a better understanding of the ocean system and serving the large array of existing and potential applications.

Large amounts of multidisciplinary data need to be made available in an interoperable or harmonized manner in order to carry out the cross-domain analysis and processing of data that is necessary. Without a doubt, it is of great importance to ensure that maximum benefit can be derived from data once it has been acquired. The principles of "capture once — use many times" and achieving "FAIRness" are major targets for managing and serving the wealth of marine and ocean datasets to the existing and potential user community.

In the landscape of marine and ocean data management, great progress has been made over three decades with the development of standards, services, and establishing dedicated infrastructures. In 1961, the Intergovernmental Oceanographic Commission (IOC) of UNESCO established the International Oceanographic Data and Information Exchange (IODE) program, with the aim of enhancing marine research and meeting the needs of users for data and data products. Over the years, the IODE network of National Oceanographic Data Centers (NODCs) has collected and managed heterogeneous marine data from different sources. An IOC Ocean Data and Information System (ODIS) is today providing a "catalog of sources" of existing ocean related web-based sources/systems of data and information as well as products and services.

An important product of the international efforts in data archival is the World Ocean Database, a result of an IODE project carried out by NOAA's National Centers for Environmental Information (NCEI). The archival of data deriving from different sensors and divers databases required compatibility and comparability, as well as common data management practices (see also Chapter 1 and Chapter 4—the collaborative approach).

During recent decades, data and information have been collected, aggregated, and made available by infrastructures providing services for discovery and access and for ensuring long-term stewardship. Integrated marine observing systems have been developed worldwide (e.g., IMOS in Australia, the US Integrated Ocean Observing System) by consortia of institutions that established collaborative research infrastructures. Activities have been and are undertaken as part of international initiatives, led by the Intergovernmental Oceanographic Commission (IOC - http://www.ioc-unesco.org/), the World Meteorological Organization (WMO - https://public.wmo.int/en), the Food and Agricultural Organization (FAO - http://www.fao.org/home/en/), the Group on Earth Observations (GEO- http://www.earthobservations.org/index.php), the International Council for the Exploration of the Sea (ICES - https://www.ices.dk/), and others.

The European initiatives

Since the early 1990s this has been complemented by a wide range of EU initiatives, funded and/or supported by EU DG RTD (Research and Innovation), EU DG MARE (Maritime Affairs and Fisheries), EU DG DEFIS Defense Industry and Space (formerly known as EU DG GROW), EU DG ENV (Environment), and EU DG CONNECT (Communications Networks, Content and Technology), and aimed at developing a European capacity for collecting and managing marine in-situ and remote sensing data, while federating and interacting with national activities for developing data centers and data management systems. This has resulted in establishing leading European marine data management infrastructures, such as SeaDataNet (https://www.seadatanet.org: physics, chemistry, geophysics, geology, and biology), EurOBIS (https://www.eurobis.org/: marine biodiversity), Euro-Argo (https://www.euro-argo.eu/: ocean physics and marine biogeochemistry), EMODnet (https://www.emodnet.eu: bathymetry, chemistry, geology, physics, biology, seabed habitats, and human activities), ELIXIR-ENA (https://www.ebi.ac.uk/ena: biogenomics), ICOS-Ocean (https://otc.icos-cp.eu/: carbon), and Copernicus Marine Environmental Monitoring Service (CMEMS) (https://marine.copernicus.eu/: ocean analysis and forecasting).

These infrastructures are developed and operated by research, governmental, and industry organizations from European states, and in close interaction with the forementioned international initiatives. Each of these data infrastructures has established links to data originators and their data collection, facilitating supervision and engagement in the process that goes from collection to validation to storage and distribution, while several are also increasingly involved in generating data products and models, which are run by the infrastructure teams or made available as services for external users from research, government, and industry. These blue data infrastructures are also mostly complementary to each other, dealing with other data originators and/or different stages in the processing chains from data acquisition to data products to knowledge.

Moreover, they are working together in EU cluster projects such as ENVRI-FAIR (https://envri.eu/home-envri-fair/) and the Blue-Cloud (https://www.blue-cloud.org/). ENVRI-FAIR is aimed at analyzing and improving FAIRness of data services in four environmental subdomains, including marine, and also considering the requirements from the upcoming European Open Science Cloud (EOSC - https://www.eosc-portal.eu/).

The Blue-Cloud project is the data management component of the "The Future of Seas and Oceans Flagship Initiative" of the EU and brings together all mentioned blue data infrastructures and major e-infrastructures such as EUDAT (https://www.eudat.eu/), D4Science (https://www.d4science.org/), and WEkEO (marine DIAS) (https://www.wekeo.eu/). Blue-Cloud aims at establishing and offering researchers a smart federation of blue data resources, computing facilities, and analytical tools, and a Virtual Research Environment (VRE) for undertaking world-class science.

Cloud infrastructure

The Copernicus Data and Information Access Services (DIAS) (WEkEO - https://www.wekeo.eu/) is deploying a cloud-based platform to centralize access to all Sentinel satellite datasets (S1, S2, S3 Marine, S3 Land, S5P) and main Copernicus Service Data products from Copernicus Marine Service (CMEMS), Copernicus Atmospheric Service (CAMS), Copernicus Climate Service (C3S), and Copernicus Land Service (CLMS). The overarching objective of DIAS is to enhance access to Copernicus data and information for further use in an efficient computing environment implementing the paradigm of "bringing the user to the data" as one condition for unlocking the potential value of Copernicus for innovation, science, new business, implementation of public policies, and economic growth (Fig. 3.8).

WEkEO is the service for marine environmental data, virtual environments for data processing, and skilled user support (Fig. 3.9). WEkEO has

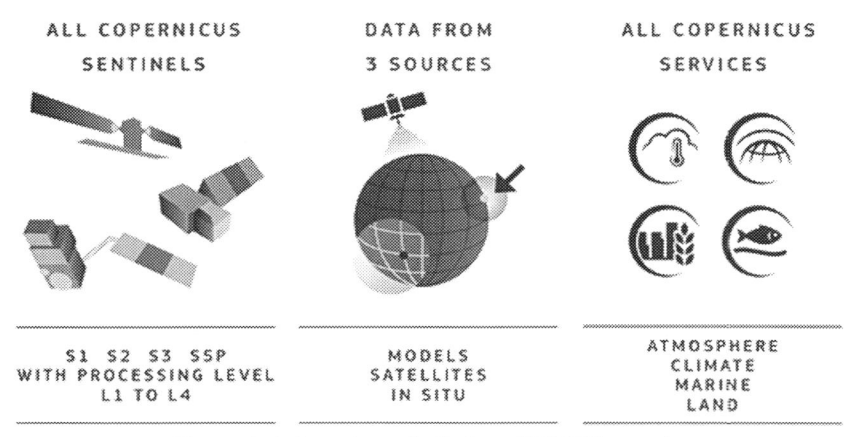

Figure 3.8 Overview of offering of WEkEO portal.

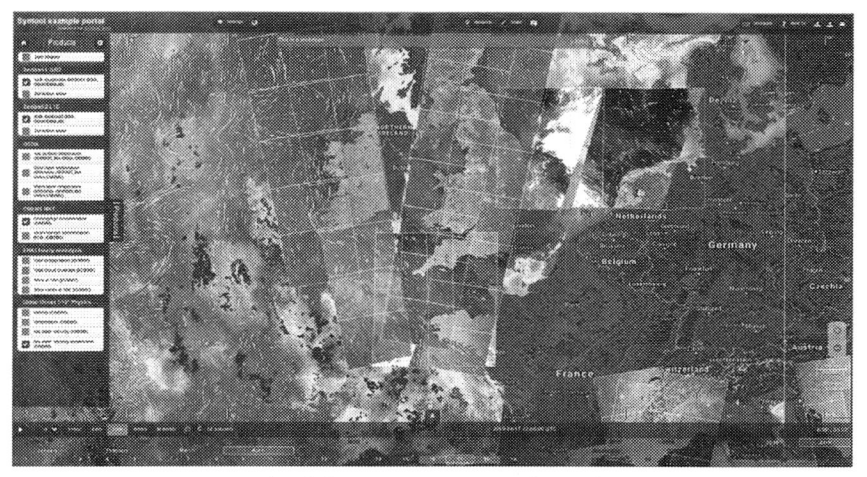

Figure 3.9 Example of data aggregation in the WEkEO user interface.

a public and free part for discovery and access to data and data products, while it also has a commercial part with various analysis applications and cloud space.

EMODnet—European Marine Observation and Data network

The European Marine Observation and Data network (EMODnet) (https://www.emodnet.eu) started operating in 2008 and is a long-term, marine-data initiative funded by the European Maritime and Fisheries Fund (managed by EU DG MARE), which, together with the Copernicus space program and the Data Collection Framework for fisheries, implements the EU's Marine Knowledge 2020 strategy. EMODnet nowadays connects a network of over 150 organizations, supported by the EU's Integrated Maritime Policy, who work together to observe the sea, process the data according to international standards and make that information freely available as interoperable data layers and in particular as pan-European data products. This "collect once and use many times" philosophy benefits all marine data users, including policymakers, scientists, private industry, and the public. EMODnet aims to increase productivity in all activities involving marine data, to promote innovation and to reduce uncertainty about the behavior of the sea. This will lessen the risks associated with private and public investments in the blue economy and facilitate more effective protection of the marine environment.

EMODnet provides easy and free access to marine data, metadata, and data products and services spanning seven broad disciplinary themes: bathymetry, geology, physics, chemistry, biology, seabed habitats, and human activities. Each theme is dealt with by a partnership of organizations that possess the expertise necessary to standardize the presentation of data and create data products. Each theme and its products and services can be reached via the central EMODnet portal. Moreover, for each of the themes, use is made of existing data management infrastructures, which deal with bringing data originators and data together, and which are providing relevant base data for developing EMODnet products and derived services. The synergy with EMODnet also provides a boost to the existing data management infrastructures as more data providers are stimulated to participate and share their data for EMODnet products. EMODnet turns marine data into maps, digital terrain models, time series and statistics, dynamic plots, map viewers, and other applications ready to support researchers, industries, and policymakers when tackling large societal challenges. In addition to the central portal and thematic portals and services, there is an EMODnet Ingestion portal, aiming at the uptake of data submissions from third parties that are not yet connected to the major European data infrastructures. A presentation of Bathymetry, Chemistry, Physics, and Ingestion portals is provided here. EMODnet Biology is discussed in another chapter of this book, EMODnet Physics is presented in a specific paragraph.

EMODnet Bathymetry

The EMODnet Bathymetry portal (https://www.emodnet-bathymetry.eu) is operated and further developed by a European partnership, comprising national hydrographic services, marine research institutes, industry, and SeaDataNet members. The partners combine expertise and experience in collecting, processing, and managing bathymetric data together with expertise in distributed data infrastructure development and operation and providing OGC services (WMS, WFS, and WCS) for viewing and distribution.

The main aims of EMODnet Bathymetry are:
- To bring together available bathymetric surveys and derived high-resolution composite DTMs (Digital Terrain Models)
- To produce and maintain the best Digital Terrain Model for European seas based on the gathered bathymetry data and with a grid resolution of 1/16 * 1/16 arc minute (circa 115 * 115 m)

- To publish and disseminate the EMODnet DTM widely with metadata, acknowledging used data and their data providers, OGC viewing services, and download services

EMODnet Bathymetry makes full use of the SeaDataNet infrastructure for managing the gathering of bathymetry datasets. References to the used data and their data holders can be found in the source references layer. Gathered survey datasets are described and included in the SeaDataNet Common Data Index (CDI) Data Discovery and Access service, metadata about composite DTMs are included in the SeaDataNet Sextant Catalog service for data products.

Chapter 4 gives additional information about the products of EMODnet Bathymetry.

EMODnet Chemistry

The EMODnet Chemistry (https://www.emodnet-chemistry.eu) consortium comprises data centers from the SeaDataNet network together with environmental monitoring agencies, regional sea conventions, ICES, chemical experts, and others. The partners combine expertise and experience in collecting, processing, and managing chemistry data together with expertise in distributed data infrastructure development and operation and providing OGC services (WMS, WFS, and WCS) for viewing and distribution.

The main aims of EMODnet Chemistry are to collate available chemistry observation datasets in order to produce and maintain validated aggregated and harmonized data collections and interpolated map products for eutrophication, contaminants, and marine litter, fit for support of the implementation of the Marine Strategy Framework Directive (MSFD). These data products are published and disseminated for viewing and downloading at the EMODnet Chemistry portal, while an INSPIRE compliant catalog is linked to providing metadata descriptions and DOI landing pages, acknowledging used data and their data providers, OGC viewing services, and download services.

For data gathering of most chemistry datasets, EMODnet Chemistry makes full use of the SeaDataNet standards, tools, and services of the CDI data discovery and access service. The gathering is done in direct communication with data originators to ensure the best sets of measured data and related metadata, and to prevent duplicates. Currently, the Chemistry CDI service provides online unified discovery and access to chemistry datasets from around one million entries, brought together by 65 data centers from >400 originators, for eutrophication (nutrients, oxygen, chlorophyll), contaminants, and marine litter (beach, seafloor, and microlitter), while data for beach litter and seafloor litter are gathered, managed, and published using two central European databases, which are developed and populated by EMODnet Chemistry in cooperation with the MSFD Technical Group on Marine Litter (TG-ML), EU JRC, Regional Sea Conventions, ICES, and several relevant EU projects, regional and local initiatives.

The gathered data are aggregated and validated by sea region. Therefore, a major challenge is to manage the heterogeneity, complexity, quality, and large volume of the gathered datasets and to process these into harmonized data collections. This is solved by regional coordinators, using SeaDataNet standards for vocabularies, QA-QC, and the SeaDataNet Ocean Data View (ODV) software, and supported by experienced chemistry experts from their institutes and the MSFD community. This results in harmonized validated data collections for each sea region, concerning eutrophication (MSFD indicator 5), contaminants (MSFD indicators 8 and 9), and European data products for marine litter (MSFD indicator 10) with focus on beach litter, seafloor litter, and microplastics. The eutrophication data collections are also used to generate a series of spatially interpolated maps of eutrophication parameters in time and depth per sea region.

EMODnet Chemistry undertakes close cooperation and tuning with the European Environment Agency (EEA) and the four Regional Sea Conventions (OSPAR, HELCOM, Bucharest Convention, and Barcelona Convention), JRC and ICES for making the data products fit for use in the MSFD process, while also an MoU has been established with Copernicus Marine Environmental Monitoring Service (CMEMS) for the exchange of validated data products for eutrophication to CMEMS, so as to further develop their ecosystem modeling and products.

EMODnet Ingestion

The EMODnet Data Ingestion portal (https://www.emodnet-ingestion.eu) was launched in February 2017. It aims at reaching out to organizations from research, public, and private sectors who are holding marine datasets and who are not yet connected and contributing to the existing marine data management infrastructures. Through a combination of central and national marketing activities, potential data providers are identified, encouraged, motivated, and supported to release their datasets through the EMODnet

Ingestion portal. The portal provides services that facilitate data holders when submitting their marine datasets for validation, safeguarding, and publishing by qualified data centers and subsequent distribution through European marine data infrastructures.

The Submission Service facilitates the submission of data files. A distinction is made in two phases in the life cycle of data submission (Fig. 3.10):

- Phase I: from data submission to publishing "as is"
- Phase II: further elaboration and integration (of subsets) in national, European, and EMODnet thematic portals.

The submission workflow is illustrated above and has the following steps:

- Step 1: Data submitter (possibly with help of EMODnet "ambassador") completes a number of key fields of the submission form and uploads a zip file with the datasets and related documentation;
- Step 2: Data Center is assigned who reviews and completes the submission form for publishing "as is" in Summary Service. Assignment goes by data theme and country;
- Step 3: Data Center elaborates, where possible, the datasets, resulting in availability in standard formats in the data center portal and European portals such as SeaDataNet, EurOBIS, a.o., and in thematic EMODnet portals.

The EMODnet network for validating and processing data submissions is recruited from the EMODnet Ingestion and EMODnet thematic portal consortia and at present comprises circa 50 qualified data centers for marine chemistry, physics, geology, bathymetry, biology, seabed habitats, and human activities data.

Since 2008, EMODnet has made huge advances in facilitating access to data from many sources. However, data still remain hidden or unusable because data holders do not share their data, due to restrictions in terms of

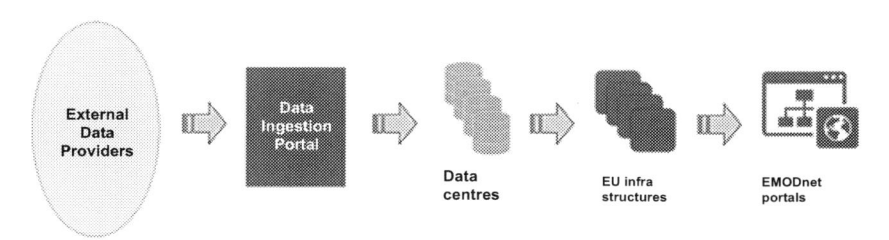

Figure 3.10 Data submission workflow.

resources, available time, or technical know-how. EMODnet's Data Ingestion facility tackles these problems, by reaching out to data holders and offering a support service to assist them in releasing their data for subsequent processing and quality control and ultimately publishing as open data.

Marketing and promotion have gained momentum and in October 2020 more than 700 dataset submissions were completed and published "as is" on the Ingestion portal and of which circa 300 have been elaborated and populated into SeaDataNet and EMODnet, representing thousands of new datasets. The majority of these submissions are from scientific institutes but some are from industry, related to offshore developments such as surveys for new pipelines, harbors, and wind farms. The principles of the EMODnet Ingestion service are illustrated very well in a special animation: "*Wake up your data! Set them free for Blue Society!*" (https://www.youtube.com/watch?v=p3vwngxyXuo).

Fit-for-use/fit-for-purpose infrastructure

The "fit-for-use/fit-for-purpose" concept has been well clarified in the initiative "EMODnet – Mediterranean CheckPoint," and its aim is to assess how and how much data meets applicable regulatory requirements and aims to enhance user satisfaction (Pinardi et al., 2017).

ISO19115-1:2014(E) and ISO19157:2013(E) provide basic definitions of the entities involved in data management; more specifically they define that datasets or dataset series is created by data producers and consumed by data users. The application of this generic ISO concept to the marine observation and the discussion about the difference between data producers and data users was already considered in the Data Adequacy Report (DAR) of EMODnet Mediterranean Checkpoint (Pinardi et al., 2017). This DAR elaborates the concepts of "fitness for purpose" (data producer needs) and "fitness for use" (data user requirements) in terms of "Universe of Discourse" (Moussat, 2014; Manzella et al., 2015; Pinardi et al., 2017) briefly: "Fitness for purpose" means that datasets or dataset series should be suitable for the intended purposes of the data producer. Its concept includes the quality assurance procedures, which is the set of planned and systematic actions necessary to provide appropriate confidence that a data product will satisfy the requirements for quality. Quality assurance includes management of the quality of instruments, materials, products, services related to management and inspection processes (checklists).

Information on quality and "fitness for use" requires the use of meaningful metadata and the involvement of technical arrangements that ensure interoperability. A data producer must apply actions for validating how well a dataset reflects its universe of discourse as defined in the data product specification.

Further attention to land-sea interface

The land-sea interface, i.e., coastal zone, is an area of high interest: around one-third of the European population lives close to a coastal zone, which hosts important commercial activities and also supports diverse ecosystems. Coastal zones are particularly vulnerable to climate change due to the combined effects of sea-level rise due to global warming and potential changes in the frequency and intensity of storms due to extreme weather events.

With its products, EMODnet is already providing tools and information for supporting the studies, research, monitoring, and assessment of the land-sea interface and in this project, there will be even more focus on it.

EMODnet is developing and updating a considerable number of products that can support land-sea interaction studies and land-sea interface management. These products range from sea level trends to river influence and contribution to land-sea exchanges to coastal sea surface currents to products on waves and winds, etc.

Support to Water Framework Directive

In the Water Framework Directive 2000/60/EC, member states commit to achieving the good qualitative and quantitative status of all water bodies. The criteria for assessing the ecological and chemical status of surface waters include the physical-chemical quality based on temperature, oxygenation, and nutrient conditions for drinkable waters as well as for fish waters, shellfish waters, bathing waters, and groundwater. It also covers marine waters up to one nautical mile from the shore. EMODnet is playing an increasingly important role: EMODnet Physics is well recognized as the primary integrator for river data and, in collaboration with EMODnet Chemistry, it is working on more and better river proxy products.

Support to the Marine Strategy Framework Directive

The EU's Marine Strategy Framework Directive has been in force since 2008. It requires member states to set up national marine strategies to achieve, or maintain where it exists, "good environmental status" by

2020. The MSFD promotes the ecosystem-based approach, which is a strategy for the integrated management of land, water, and living resources promoting conservation and sustainable use in an equitable way. The goal of ecosystem-based management is to maintain an ecosystem in a healthy, productive, and resilient condition so that it can provide the goods and services humans want and need. It considers the cumulative impacts on different sectors and aims to ensure that the cumulative pressures of human activities do not exceed levels that compromise the capacity of ecosystems to remain healthy, clean, and productive.

The MSFD is one of the most ambitious international marine protection legal frameworks, aligning the efforts of MS in coordination with non-EU countries, to apply ecosystem-based management and to achieve good environmental status. The scope of the directive stretches from the coast to the edge of member states' jurisdiction and protects the full range of marine biodiversity. The MSFD requires integrated planning (marine strategies) to be developed based on 11 descriptors and several criteria and parameters to be assessed by each member state. The MSFD descriptors set objectives such that biological diversity (D1), food web structure (D4), and sea-floor integrity (D6) are maintained, while the impacts from nonindigenous species (D2), fishing (D3), excess nutrients (D5), changes in hydrographical conditions (D7), contaminants in the environment (D8) and in seafood (D9), marine litter (D10), and underwater noise (D11) do not adversely alter the marine ecosystems. As the transboundary nature of certain pressures and ecosystems makes them very difficult to manage at the member state level alone, the directive states that regional sea conventions can aid cooperation.

In this framework, EMODnet is making available useful and complementary data for many of the indicators. For example, D1 combines "climatic conditions" (i.e., temperature, salinity, ice cover, light attenuation, …) with the information of the quality and occurrence of habitats and the distribution and abundance of species (https://ec.europa.eu/environment/marine/good-environmental-status/descriptor-1/index_en.htm); D2 states that "Food web components are also subject to environmental and climate variation"; D5 identifies in rivers and runoff of rainwater one primary source for the organic nutrients responsible for the eutrophication of marine waters; D6 needs the characteristics (physical, chemical, and biological) of the sea bottom to assess sea-floor integrity; D7 is about the hydrographical conditions that are characterized by the physical parameters of seawater: temperature, salinity, depth, currents, waves, turbulence, turbidity (related to the load of suspended particulate matter). Data on contaminants, eutrophication, and marine litter are provided by EMODnet Chemistry. For D11 (underwater noise), the work is implemented through the Technical Group on Noise (TG-NOISE), in close collaboration with the Regional Seas Conventions (RSC). EMODnet Physics distributes a harmonized pan-European Impulsive Noise Registry. Moreover, based on TG-NOISE guidance for underwater noise in European Seas, member states are starting to implement a monitoring program for continuous noise.

An operational fit-for-use infrastructure: EMODnet Physics

European repositories and infrastructures, i.e., CMEMS-INSTAC and SDN-NODC, are integrated with other available sources such as ICES, PANGAEA, PSMSL, SONEL, IOC, IOOS, IMOS, etc., are integrated into EMODnet Physics to make ocean physics parameters available.

EMODnet Physics (http://www.emodnet-physics.eu) is one of the seven domain-specific portals of the European Marine Observation and Data Network (EMODnet), and it is making available: temperature of the water column, salinity of the water column, horizontal velocity of the water column, changes in sea level and sea-level trends, wave height and period, wind speed and atmospheric pressure, water clarity (light attenuation), underwater sound (noise), inflow from rivers, and sea-ice coverage. Time series, profiles, and sampled datasets are made available, as recorded by fixed platforms (moorings, tide gauges, HF radars, etc.), moving platforms (ARGO, Lagrangian buoys, ferryboxes, etc.) and repeated observations (CTDs, etc.). Available products are collections of in-situ data, reanalysis, and trends from in-situ data, elaboration in space and/or time of in-situ data and model output for a given parameter.

Data collection

EMODnet Physics develops a common procedure to access ocean physics parameters and products from several sources: available products are collections of in-situ data, reanalysis, and trends from in-situ data, elaboration in space and/or time of in-situ data and model output for a given parameter. Data products time age ranges from real-time, near real-time, to validated long-term time series.

The synchronization process is based on smart adapters connecting data from sources to the system (Fig. 3.10), and apply procedures to harmonize information (common standards, common vocabularies, complete and

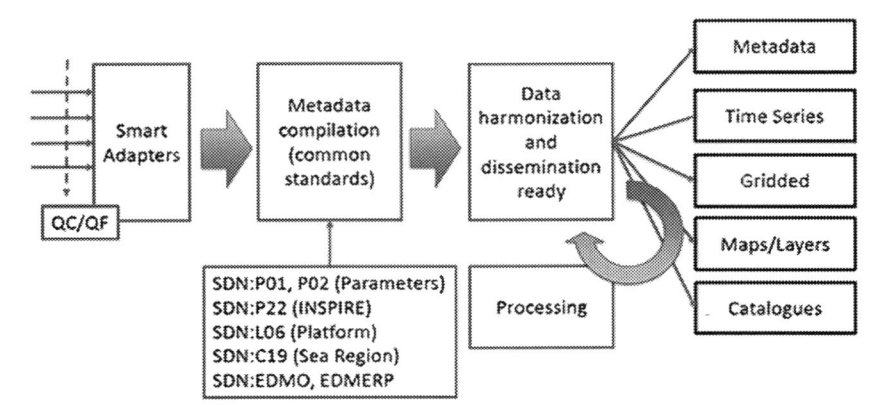

Figure 3.11 Data management flow in EMODnet Physics.

integrate metadata). At the end of the process, data is ready for dissemination through the EMODnet Physics dissemination channels/catalogs (i.e., map-viewer, ERDDAP, TDS, GeoServer, GeoNetwork catalog) (Fig. 3.11).

Real-time data acquisition and dissemination are based on the latest implementation of the Sensor Web Enablement (SWE) and Sensor Observation Service (SOS) standards. These interoperable interfaces permit the insertion and retrieval of georeferenced observation data in a standardized format. This new data stream management is performed in collaboration with EMODnet Data Ingestion and, besides developing and deploying the SOS, the two projects are working and contributing on a set of standards to implement ISO/OGC O&M features and SensorML for the marine domain (https://odip.github.io/MarineProfilesForSWE/).

The acquisition of near-real-time physical parameters is mainly an automated process: EMODnet Physics collects data from a network of providers and integrators (CMEMS InsTAC, global and regional assembly centers, regional ocean observing systems, etc.) and organizes products in the EMODnet Physics ERDDAP and hence in the map viewer. Typically, the transport format is NetCDF (CF Convention), as defined by the Euro-GOOS DATAMEQ working group and the SeaDataNet technical working team, and includes metadata and data quality flags.

Data quality is flagged according to an automatic—unsupervised procedure at the data source. EMODnet Physics is operationally processing this data flow to generate map layers (organized in the EMODnet GeoServer) and extract in-situ (monthly) trends, averages, and peak values of the parameters. Historically validated datasets are organized in collaboration with Sea-DataNet and its network of National Oceanographic Data Centers, which are supplying EMODnet Physics with products (climatology) on temperature and salinity of the water column. EMODnet Physics is also acting as an in-situ historical data collection broker between users and the NODCs. For historical, validated datasets (fixed stations—mooring, tide gauge), the metadata formats are the CDIs (common data indices) and the transport formats are ODV4 and NetCDF (CF convention).

For the parameters (and platforms) that are not managed by its pillars (e.g., river outflow, water noise, sea surface currents as recorded by HF radars, etc.), EMODnet Physics is supporting, promoting, and contributing to the development of the data management chain by applying common standards and procedures. Other aggregated and validated thematic collections/products (e.g., PSMSL, SONEL, SOCAT, etc.) are linked and ingested from sources (according to indications by the product principal investigator) by means of specific smart connectors.

Data sources

EMODnet Physics develops an ocean data information system based on a federated network of sources. These sources may be marine data integrators, marine data repositories, ocean observation programs, data assembly centers, marine institutes, oceanographic data centers, etc. The main (not exhaustive) list of sources is:

- CMEMS INSTAC (in situ measurement from EuroGOOS and ROOSs institutes)
- European Marine Institutes and National Oceanographic Data Centers (see annex)
- SeaDataNet (CDI and Climatology)
- GDAC (Coriolis)
- Global Sea Level Observing System (GLOSS)
- IOC Sea Level Station Monitoring (SLS)
- Permanent service for mean sea level (PSMSL)
- University of Hawaii Sea Level Center (UHSLC) – GLOSS Fast-Delivery Center
- Système d'Observation du Niveau des Eaux Littorales (SONEL)
- International Council for the Exploration of the Sea (ICES)
- Deep Ocean Multi-Disciplinary Ocean Reference Stations (OceanSITES)
- ARGO profiling float data
- Southern Oceans Observing System (SOOS)
- Global HF Radar Network
- Everyone's Gliding Observatories (EGO) and OceanGlider Network

- Marine Mammals Exploring the Oceans Pole to Pole (MEOP)
- Voluntary Observing Ship (VOS), Ship Of Opportunity Program (SOOP)
- Data Buoy Cooperation Panel (DBCP), Arctic Buoy Data (IAPB),
- Tropical Moored buoys: Pacific Ocean (TAO, TRITON), Atlantic Ocean (PIRATA), Indian Ocean (RAMA)
- PANGAEA – Data Publisher for Earth & Environmental Science
- European Multidisciplinary Seafloor and water column Observatory (EMSO)
- Global Ocean Surface Underway Data Pilot Project (GOSUD)
- US National Data Buoy Center (NDBC), Integrated Ocean Observing System (IOOS), National Oceanic and Atmospheric Administration (NOAA)
- Australian Integrated Marine Observing System (IMOS)
- Global Ocean Ship-Based Hydrographic Investigations Program (GO-SHIP)
- Global Ocean Data Analysis Project (GLODAP)
- Surface Ocean CO_2 Atlas (SOCAT)

Data processing

EMODnet Physics provides the user with information about the quality of data and the link to the applied quality check procedure. In general, if data is real-time and near-real-time data, the quality control procedure is automatic or semiautomatic. If data is a reanalysis or long-term series, the quality control procedure involves thematic expertise and supervision. Quality check procedures include consistency controls on time, location, range-spike, etc., and attach a Quality Flag to data (Table 3.6). The QC/QF method is defined according to the observation platform, the sensor type, and the parameter. Whenever possible EuroGOOS (https://archimer.ifremer.fr/doc/00251/36230/34790.pdf) and SeaDataNet (https://www.seadatanet.org/content/download/596/file/SeaDataNet_QC_procedures_V2_%28May_2010%29.pdf) recommendations are applied as close as possible to data origin.

Data publishing and data dissemination channels

EMODnet Physics develops and exploits various technologies and dissemination channels:

- ERDDAP—that implements FGDC Web Accessible Folder (WAF) with FGDC-STD-001-1998 and ISO 19115 WAF with ISO 19115—2/19139;

Table 3.6 Quality control/quality flags.

Code	meaning	comment
0	No QC was performed	—
1	Good data	All real-time QC tests passed.
2	Probably good data	These data should be used with caution.
3	Bad data that are potentially Correctable	These data are not to be used without scientific correction.
4	Bad data	Data have failed one or more of the tests.
5	Value changed	Data may be recovered after transmission error.
6	Not used	—
7	Nominal value	Data were not observed but reported. Example: an instrument target depth.
8	Interpolated value	Missing data may be interpolated from neighboring data in space or time.
9	Missing value	The value is missing, is not reported, is not applicable …

- THREDDS—that implements OpenDAP, NetCDF Subset Service, WCS 1.0 Service, WMS 1.3.0 Service, ncISO: Dataset Metadata Services, OAI Metadata harvesting;
- GeoServer—that implements several Open Geospatial Consortium protocols including Web Map Service (WMS), Web Feature Service (WFS), Web Coverage Service (WCS), and Web Map Tile Service (WMTS) and that was recently updated with the INSPIRE module;
- GeoNetwork—that implements WxS, OGC, ISO standards.
- web APIs, and web widgets

EMODnet Physics adopted ERDDAP as the core solution for data management and interoperability. ERDDAP supports both human interaction (e.g., OPeNDAP requests) and machine-to-machine interoperability. ERDDAP data server supports several common data file formats (HTML table, netCDF, csv, txt, mat, json, etc.) and output files are created on-the-fly in any of these formats. ERDDAP is a free and open-source code (JAVA program and source code are available in GitHub - https://github.com/BobSimons) that uses Apache compatible software licenses.

Data format reference

EMODnet Physics supports several data formats, ranging from csv, odv4, to netcdf. If the source is not implementing an ERDDAP data server to be linked to EMODnet Physics infrastructure (ERDDAP overcomes the main data format interoperability problems), in order to facilitate the ingestion/link process, it is recommended to use the Climate and Forecast convention NetCDF format version 4.

The CF metadata conventions (https://cf-trac.llnl.gov/trac) are designed to promote the processing and sharing of files created with the NetCDF API. The conventions define metadata that provides a definitive description of what the data in each variable represents and the spatial and temporal properties of the data. This enables users of data from different sources to decide which quantities are comparable and facilitates building applications with powerful extraction, regridding, and display capabilities. The standard is both mature and well-supported by formal governance for its further development. The standard is fully documented by a PDF manual accessible from a link from the CF metadata homepage (https://cf-trac.llnl.gov/trac). Note that CF is a developing standard; the current version is CF-1.6.

The European marine data infrastructures are based on the file format used to distribute OceanSITES data. Both CMEMS-InsTAC and SeaData-Net have developed some extensions to provide the user with more information. EMODnet Physics supports and recommends the inclusion of as many descriptive attributes as possible.

The EMODnet Physics Metadata format is based on an extension of the mandatory global attributes that are common between these three infrastructures.

New challenges
ENVRI-FAIR

As part of the EU ENVRI-FAIR project (https://envri.eu/home-envri-fair/), activities are undertaken in the marine domain to analyze and improve the FAIRness of several data Research Infrastructures in the marine domain. For their analyses, use is made of a methodology as developed and promoted by GO-FAIR (https://www.go-fair.org/). GO FAIR is a bottom-up, stakeholder-driven, and self-governed initiative that aims to implement the FAIR data principles. It offers an open and inclusive ecosystem for individuals, institutions, and organizations working together through Implementation Networks (INs). These implement clearly defined plans and deliverables to implement an element of the Internet of FAIR Data and Services (IFDS) within a defined time period. Moreover, they foster a community of harmonized FAIR practices, and finally, they communicate together on critical issues on which consensus has been reached and which are of general importance for the community. In the marine data community, for instance, SeaDataNet has become a GO-FAIR Implementation Network, and is active in improving its FAIRness of data and services, but also promoting solutions toward its members and other infrastructures, for example, as part of the ENVRI-FAIR project.

Several interesting scientific papers (Snowden et al., 2019; deYoung et al., 2019; Tanhua et al., 2019) have been composed, partly with the input of ENVRI-FAIR, about FAIRness in the marine domain, in the context of the OceanObs 2019 Conference.

European Open Science Cloud

The European Open Science Cloud (EOSC) is an initiative launched by the European Commission in 2016 as part of the European Cloud Initiative. EOSC aims to provide a virtual environment with open and seamless services for storage, management, analysis, and reuse of research data, across borders and scientific disciplines, leveraging and federating the existing e-infrastructures and thematic infrastructures. It aims to provide 1.7m EU researchers with an environment that has free, open services. EOSC will add value and leverage to infrastructure investment by member states and the EU. EOSC will provide a platform for European research, federating existing services, giving access to FAIR research data and services, a virtual space where science producers and consumers can come together, an open-ended range of content and services, and qualified European data and data products. Interesting sites for following the EOSC development are the EOSC portal (https://eosc-portal.eu), EOSC HUB (https://www.eosc-hub.eu/), and EOSC Secretariat (https://www.eoscsecretariat.eu/). The EOSC portal is set up as Europe's digital market for research data and services. It was launched at the end of 2019 and a range of pan-European e-infrastructure services can already be found within the EOSC providing data and computational solutions through a cloud-based environment. The development and sustainable operation of EOSC services are at the forefront of European funding priorities.

Pilot Blue-Cloud

In October 2019, the pilot "Blue-Cloud project" (https://www.blue-cloud.org) started, aimed at building and demonstrating a pilot Blue-Cloud as a thematic EOSC cloud to support research for improved understanding and management of the many aspects of ocean sustainability, ranging from sustainable fisheries to ecosystem health and pollution. The project is the data management component of the EU's "The Future of Seas and Oceans Flagship Initiative." Its concept is based on the work of the SeaDataNet network in its SeaDataCloud project. This Blue-Cloud pilot brings together leading European blue-data management infrastructures such as SeaDataNet (marine and ocean environment), EurOBIS (https://www.eurobis.org/) (marine biodiversity), Euro-Argo (https://www.euro-argo.eu/) (ocean physics and marine biogeochemistry), EMODnet (bathymetry, chemistry, geology, physics, biology, seabed habitats, and human activities), ELIXIR-ENA (https://www.ebi.ac.uk/ena) (biogenomics), Copernicus Marine Environmental Monitoring Service (http://marine.copernicus.eu/) (CMEMS) (ocean analysis and forecasting), and ICOS-Ocean (https://otc.icos-cp.eu/) (carbon) and horizontal e-infrastructures (EUDAT – https://www.eudat.eu, DIAS (WEkEO), and D4Science). It aims to deliver an initial "Blue-Cloud" framework, by a "smart" federation of "data resources," "computing resources," and "analytical service resources," to provide researchers with access to multidisciplinary data and a blue Virtual Research Environment (VRE) with dedicated virtual labs.

The project is a pilot and pathfinder, and from there a Blue-Cloud roadmap to 2030 is being prepared for expansion and sustainability of the infrastructure and services. The ambition is to mobilize input and support of all major stakeholders, be they data observing networks, marine data infrastructures, e-infrastructures, user communities, decision-makers, and funding agencies, for a larger and longer EU Blue-Cloud program. This program should aim at further developing, expanding, sustaining operations, and uptake by marine science of a European federated cyberinfrastructure for blue-data resources access and web-based marine science facilities. This will encompass investments for further developments and operation of each of the building blocks (existing and additional infrastructures), arranging inter alia more connected data providers and streamlined data submission, validation, storage, and distribution pipelines. Investments are also envisaged for further development, operation, deployment, and promotion of the federated blue cloud and cyberinfrastructure with a range of services by which many users should be reached and served, and which should lead to exceptional new knowledge and insights, relevant for policies, governance, management, economic activities, and science.

The objectives of the Blue-Cloud project are:

- To build and demonstrate a Pilot Blue Cloud by combining distributed marine data resources, computing platforms, and analytical services
- To develop services geared to support research to better understand and manage the many aspects of ocean sustainability
- To develop and validate a number of demonstrators of relevance for marine societal challenges
- To formulate a roadmap for the expansion and sustainability of the Blue Cloud infrastructure and services (Fig. 3.12).

A **Blue-Cloud data discovery and access service** is being developed as an overarching service to facilitate smart sharing of multidisciplinary datasets with human and machine users. A **Blue-Cloud Virtual Research Environment (VRE)** will orchestrate the computing and analytical services in specific integrated and managed applications that through the same VRE use the federated Blue-Cloud data resources in addition to external data resources (Fig. 3.13).

The Blue-Cloud innovation potential will be explored and unlocked by developing five dedicated Demonstrators such as Virtual Labs, co-designed with top-level marine researchers. Through the Blue-Cloud Platform, the

Figure 3.12 Federation of blue-data infrastructures and e-infrastructures.

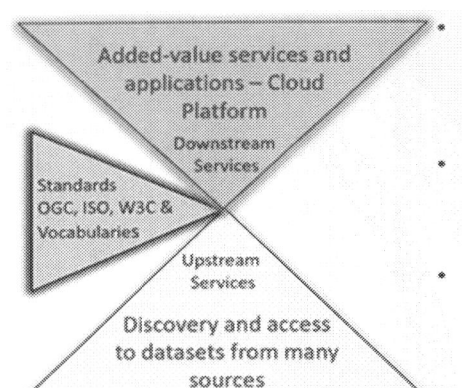

Providing a cloud platform with common services for data pre-processing, sub-setting, analyses, visualizations, publishing, DOIs...

Applying common standards and interoperability solutions for providing harmonised data and metadata

Providing harmonised discovery and access to data output from multiple sources, such as European research and monitoring data gathering, but also from other European and international data infrastructures

Figure 3.13 Leading concept for marine cyberinfrastructure for cloud-based science.

Virtual Labs will provide services such as access to datasets, products, and computation routines that can be exploited to analyze datasets and (re) generate research products.

Green Deal and Digital Twin of the Ocean

Fit-for-purpose and sustained ocean observations are an essential part of worldwide efforts to understand and protect the marine social–ecological systems while benefiting from their ecosystem services. Observations can be samples collected by ships, measurements from instruments on fixed platforms, autonomous and drifting systems, submersible platforms, ships at sea, or remote observing systems such as satellites and aircraft.

Marine data from these observations were difficult to find 10−20 years ago. Furthermore, they were accessible through long and sometimes costly negotiations and hard to put together to create a complete picture because of different standards, nomenclature, and baselines.

As part of the Green Deal strategy and the Digital Agenda for Europe, the EU has launched a new initiative for a Digital Twin of the Ocean. In two decades, the European Union invested in policies and infrastructures to make knowledge of the ocean central to environmental and climate policies as well as the blue economy. The EU member states, together with other countries (e.g., USA, Australia), have created an unrivalled marine data and forecasting infrastructure. These are working together under the principles of free and open access, interoperability, and "measure once, use many times."

The **Digital Twin of the Ocean** is the next step, filling the need to integrate a wide range of data sources to transform data into knowledge and to connect, engage, and empower citizens, governments, and industries by providing them with the capacity to make informed decisions. It will empower a shared responsibility to monitor, preserve, and enhance marine habitats, and support a sustainable blue economy (fishing, aquaculture, transport, offshore energy, etc.). It should allow assessment of the state of ecosystems, habitats, and the impact of human activities; forecasts of their short and long-term changes; development of biodiversity conservation strategies; management of sustainable economic activities; assessment of infrastructure vulnerability; development of mitigation, adaptation, and replacement plans to deal with climate risks and optimization of emergency responses to severe events such as storm surges. It will contribute to the development of digital interactive high-resolution models of the oceans, as part of the commitment to develop a very high precision digital model of the Earth (Destination Earth initiative). Building on the integration of existing EU leading-edge capacities in ocean observation (such as Eurofleets+, EuroArgo, Jerico, EMBRC, etc.), data infrastructures, and forecasting services (Copernicus, EMODnet, Blue Cloud, ERICs, -) through innovative IT technology, it will bring together infrastructures and communities in support to the EU Green Deal and societal transitions.

Conclusion and recommendations

The integration and exchange of heterogeneous data sources are key to managing big data applications in marine science. Over three decades, SeaDataNet has successfully established an operational pan-European infrastructure with more than 110 connected data centers that facilitate marine data discovery and accessibility, and a range of marine data management standards and vocabularies to facilitate marine data interoperability and reusability. Meanwhile EMODnet, since 2008, has made significant progress in making and publishing a series of thematic generic data products and services for many marine user communities, making use of available marine datasets as provided through SeaDataNet, EurOBIS, CMEMS, and other data infrastructures. These infrastructures represent very important building blocks for a unified European marine data infrastructure. However, not all data originators find their way to these infrastructures yet, with further work and additional funding being needed for these infrastructures to expand their capabilities for handling all kinds of datasets, to strengthen the interoperability between them and to develop cyber services. In that respect, progress is being made with developments by leading data infrastructures and e-infrastructures to provide cyber platforms with integrated data access and computational capabilities for the marine science community and wider research communities, in particular within the framework of the European Open Science Cloud (EOSC) with EU projects such as ENVRI-FAIR, Blue-Cloud, and LifeWatch, and in the framework of Copernicus with establishing the DIAS platforms with WEkEO for the climate and marine domains. These initiatives also feature developments for Virtual Research Environments (VREs), which promise wider uptake by marine research communities and implicate an important step toward marine science exploring EOSC opportunities.

An interoperable framework that builds on existing marine data management infrastructures and fully federate global sources of heterogeneous data is a priority for the Ocean Decade to achieve a "transparent and accessible ocean." During the All Atlantic Ocean Research Forum in February 2020, the need for a "Digital Ocean" or "DIGI TWIN for the Oceans," where all historical and current data about the ocean could be uploaded, accessed, updated in real-time, and used in decision-making, was confirmed by many stakeholders who recognized the need for regional funding to support such an initiative. This challenge is translated by the EU into the Green Deal strategy by launching a call for a Digital Twin of the Ocean (DTO) which could be considered as a further evolution of the Blue-Cloud development, expanding toward end-user engagement, and digital innovations for data collection, ingestion, processing, and visualization in support of many societal challenges. The DTO is also part of the larger Destination Earth initiative of the EU.

Recommendations

- Further development of the interoperability level of leading marine data management infrastructures for handling and exchanging multidisciplinary data between them and toward user communities and their applications, and expanding and upgrading their services with computational infrastructures for cloud computing, data storage, and access to big data analytical tools, in cooperation with leading European e-infrastructures.
- Incentivize and extensively promote the use of existing marine data management infrastructures by the marine science community as

providers and users of data to increase the current availability and accessibility of data in common standards, and subsequent use in Big Data applications.

- Promote significant participation of the marine domain in the development and operation of the Open Science Cloud and its services and similar initiatives at the international level. Entries are being made by leading data management infrastructures participating in Open Science Data projects (e.g., ENVRI-FAIR and Blue-Cloud) and joining consortia in international initiatives for further developments of Open Science (e.g., EOSC) core services and the market place. Furthermore, this should be done by informing and encouraging marine scientists to identify and explore marine science use-cases that can benefit from cloud-computing infrastructures. This will stimulate the creation and deployment of customized VREs for these use-cases that can show marine scientists and developers how to make optimum use of the new cyber opportunities. Training should be provided on the use of VREs. VREs should be interoperable to promote interdisciplinary collaboration and accelerate innovation. This will also foster more strategic partnerships between marine science and the ICT and data science communities.
- Marine science stakeholders are encouraged to contribute to the development of the Blue-Cloud roadmap to 2030 for expansion and sustainability of the infrastructure and services. The roadmap aims at mobilizing input and support of all major stakeholders (data observing networks, marine data infrastructures, e-infrastructures, user communities, decision-makers, and funding agencies), for dedicated initiatives. These initiatives should facilitate further development, expansion, sustained operation, and uptake by marine science of a federated cyberinfrastructure for access to blue-data resources and facilitating web-based science.
- Encourage cross-disciplinary fertilization of technologies between more advanced multimedia sectors and digital sectors with marine science to be able to scale-up cloud-computing initiatives for wider transdisciplinary applications.
- Further develop the cooperation, interoperability, and exchanges of data and services, including computational platforms, between leading European data infrastructures and international counterparts to facilitate

common access to ocean data on wider sea-basin and global scales and to work toward a "Digital Twin for the Ocean" that aligns with the objectives of the UN Decade of Ocean Science for Sustainable Development.

References

ARGO, 2021. Argo float data and metadata from Global Data Assembly Centre (Argo GDAC). SEANOE, doi:10.17882/42182. https://www.seanoe.org/data/00311/42182/.

Bradley, F., Fairall, C., 2007. A Guide to Making Climate Quality Meteorological and Flux Measurements at Sea. NOAA Technical Memorandum OAR PSD-311. https://repository.library.noaa.gov/view/noaa/17408.

Bröring, A., Maué, P., Janowicz, K., Nüst, D., Malewski, C., 2011. Semantically-enabled sensor plug & play for the sensor web. Sensors 11, 7568–7605. https://doi.org/10.3390/s110807568. https://www.mdpi.com/1424-8220/11/8/7568.

Bullister, J.L., Tanhua, T., 2010. Sampling and Measurement of Chlorofluorocarbons and Sulfur Hexafluoride in Seawater. The GO-SHIP Repeat Hydrography Manual: A Collection of Expert Reports and Guidelines. IOCCP Report No. 14. ICPO Publication SeriesN0, 134. https://www.go-ship.org/Manual/Bullister_Tanhua_CFCSF6.pdf.

deYoung, B., Visbeck, M., de Araujo Filho, M.C., Baringer, M.O., Black, C.A., Buch, E., Canonico, G., Coelho, P., Duha, J.T., Edwards, M., Fische, R.A.S., Fritz, J.-S., Ketelhake, S., Muelbert, J.H., Monteiro, P., Nolan, G., O'Rourke, E., Ott, M., Le Traon, P.Y., Pouliquen, S., Sousa-Pinto, I., Tanhua, T., Velho, F., Willis, Z., 2019. An integrated all-Atlantic Ocean observing system in 2030. Front. Mar. Sci. 6, 428. https://doi.org/10.3389/fmars.2019.00428.

Dickson, A.G., Sabine, C.L., Christian (Eds.), J.R., 2007. Guide to Best Practices for Ocean CO_2 Measurements. In: PICES Special Publication, 3. 191 pp. https://cdiac.ess-dive.lbl.gov/ftp/oceans/Handbook_2007/Guide_all_in_one.pdf.

Eikeseta, A.M., Mazzarella, A.B., Davíðsdóttire, B., Klingerb, D.H., Levinb, S.A., Rovenskayac, E., Stensetha, N.C., 2018. What is blue growth? The semantics of "Sustainable Development" of marine environments. Mar. Pol. 87, 177–179.

European Commission, 2019. The European Green Deal, Communication from the Commission to the European Parliament, the European Council, the Council, the European Economic and Social Committee and the Committee of the Regions, COM(2019) 640 Final. https://eur-lex.europa.eu/legal-content/EN/TXT/PDF/?uri=CELEX:52019DC0640.

Eurostat, 2018. Sustainable Development in the European Union. Monitoring Report on Progress towards the SDGs in an EU Context. Publications Office of the European Union, Eurostat statistical books, Luxembourg. https://doi.org/10.2785/221211.

Firing, E., Hummon, J.M., 2010. Shipboard ADCP Measurements. The GO-SHIP Repeat Hydrography Manual: A Collection of Expert Reports and Guidelines. IOCCP Report No. 14. ICPO Publication SeriesN0, 134. https://www.go-ship.org/Manual/Firing_SADCP.pdf.

Hydes, D.J., Aoyama, M., Aminot, A., Bakker, K., Becker, S., Coverly, S., Daniel, A., Dickson, A.G., Grosso, O., Kerouel, R., van Ooijen, J., Sato, K., Tanhua, T., Woodward, E.M.S., Zhang, J.Z., 2010. IOCCP Report No. 14. Determination of dissolved nutrients (n, p, si) in seawater with high precision and inter-comparability using gas-segmented continuous flow analyzers. The GO-SHIP Repeat Hydrography Manual: a collection of expert reports and guidelines, vol. 134. ICPO Publication SeriesN0. https://www.go-ship.org/Manual/Hydes_et_al_Nutrients.pdf.

IOC, 2008. Intergovernmental Oceanographic Commission 2008 IODE/JCOMM Forum on Oceanographic Data Management and Exchange Standards, IOC Project Office for IODE, Oostende, Belgium 21-25 January 2008. Oostende, Belgium: IOC/IODE Project Office, p. 45 (IOC Workshop Report No. 206) (English).

IOC, SCOR and IAPSO, 2010. The International Thermodynamic Equation of Seawater—2010: Calculation and Use of Thermodynamic Properties. Intergovernmental Oceanographic Commission, Manuals and Guides No.56, UNESCO(English), 196pp.

Intergovernmental Oceanographic Commission of UNESCO, 2013. IOC Strategic Plan for Oceanographic Data and Information Management (2013-2016). IOC Manuals and Guides, vol. 66, 45 pp. (English.).

Jenkins, W.J., Lott, D.E., Cahill, K., Curtice, J., Landry, P., 2010. IOCCP Report No. 14. Sampling and Measuring Helium Isotopes and Tritium in Seawater. The GO-SHIP Repeat Hydrography Manual: A Collection of Expert Reports and Guidelines, vol. 134. ICPO Publication SeriesN0. https://www.go-ship.org/Manual/Jenkins_TritHe3.pdf.

Kawano, T., 2010. IOCCP Report No. 14. Salinity Samples. The GO-SHIP Repeat Hydrography Manual: A Collection of Expert Reports and Guidelines, vol. 134. ICPO Publication SeriesN0. https://www.go-ship.org/Manual/Kawano_Salinity.pdf.

Langdon, C., 2010. IOCCP Report No. 14. Determination of Dissolved Oxygen in Seawater by Winkler Titration Using the Amperometric Technique. The GO-SHIP Repeat Hydrography Manual: A Collection of Expert Reports and Guidelines, vol. 134. ICPO Publication SeriesN0. https://www.go-ship.org/Manual/Langdon_Amperometric_oxygen.pdf.

Levitus, S., 2012. The UNESCO-IOC-IODE "Global Oceanographic Data Archeology and Rescue" (GODAR) project and "World Ocean Database" projects. Data Sci. J. 11, 46—71. https://doi.org/10.2481/dsj.012-014.

Malone, T.C., DiGiacomo, P.M., Goncalves, E., Knap, A.H., Talaue-McManus, L., de Mora, S., 2014. A global ocean observing system framework for sustainable development. Mar. Pol. 43, 262—272. https://doi.org/10.1016/j.marpol.2013.06.008.

Manzella, G., Pinardi, N., Guarnieri, A., De Dominicis, M., Moussat, E., Quimbert, E., Meillon, J., Blanc, F., Vincent, C., Kalllos, G., Kyriakid, C., Gomez-Pujol, L., Scarcella, G., Cruzado, A., Falcini, F., 2015. EMODnet MedSea CheckPoint First Data Adequacy Report. European Marine Observation and Data Network - EMODnet. https://doi.org/10.25423/cmcc/medsea_checkpoint_dar1.

McNichol, A.P., Quay, P.D., Gagnon, A.R., Burton, J.R., 2010. IOCCP Report No. 14. Collection and Measurement of Carbon Isotopes in Seawater Dic. The GO-SHIP Repeat Hydrography Manual: A Collection of Expert Reports and Guidelines, vol. 134. ICPO Publication SeriesN0. https://www.go-ship.org/Manual/McNichol_C1314.pdf.

McTaggart, K.E., Johnson, G.C., Johnson, M.C., Delahoyde, F.M., Swift, J.H., 2010. IOCCP Report No. 14. Notes on Ctd/o2 Data Acquisition and Processing Using Sea-Bird Hardware and Software (As Available). The GO-SHIP Repeat Hydrography Manual: A Collection of Expert Reports and Guidelines, vol. 134. ICPO Publication SeriesN0. https://www.go-ship.org/Manual/McTaggart_et_al_CTD.pdf.

Miola, A., Schlitz, F., 2019. Measuring sustainable development goals performance: how to monitor policy action in the 2030 Agenda implementation? Ecol. Econ. 164, 106373. https://doi.org/10.1016/j.ecolecon.2019.106373.

Moltmann, T., Turton, J., Zhang, H.-M., Nolan, G., Gouldman, C., Griesbauer, L., Willis, Z., Piniella, Á.M., Barrell, S., Andersson, E., Gallage, C., Charpentier, E., Belbeoch, M., Poli, P., Rea, A., Burger, E.F., Legler, D.M., Lumpkin, R., Meinig, C., O'Brien, K., Saha, K., Sutton, A., Zhang, D., Zhang, Y., 2019. A Global Ocean Observing System (GOOS), delivered through enhanced collaboration across regions, communities, and new technologies. Front. Mar. Sci. 6, 291. https://doi.org/10.3389/fmars.2019.00291.

Moussat, E., 2014. EMODnet Sea CheckPoint Methodology for Classifying the Existing Upstream Data According to Literature Survey. http://www.emodnet-mediterranean.eu/wp-content/uploads/2014/09/ANNEX1.pdf.

Nativi, S., Craglia, M., Pearlman, J., 2013. Earth science infrastructures interoperability: the brokering approach. IEEE J. Sel. Top. Appl. Earth Obs. Remote Sens. 6 (3), 1118–1129. https://doi.org/10.1109/JSTARS.2013.2243113.

Nativi, S., Mazzetti, P., Santoro, M., Papeschi, F., Craglia, M., Ochiai, O., 2015. Big Data challenges in building the global earth observation system of systems. Environ. Modell. Softw. 68, 1–26.

OECD, 2017. Measuring Distance to the SDG Targets: An Assessment of where OECD Countries Stand. Retrieved from: http://www.oecd.org/sdd/OECD-Measuring-Distance-to-SDG-Targets.pdf.

Pinardi, N., Simoncelli, S., Clementi, E., Manzella, G.M.R., Moussat, E., Quinbert, E., Blanc, F., Valladeau, G., Galanis, G., Kallos, G., Patlakas, P., Reizopoulou, S., Kyriakidou, C., Katara, I., Kouvarda, D., Skoulikidis, N., Gomez-Pujol, L., Vallespir, J., March, D., Tintoré, J., Fabi, G., Scarcella, G., Tassetti, A.N., Raichic, F., Cruzado, A., Bahamon, N., Falcini, F., Lilipot, J.-F., Duarte, R., Lecci, R., Bonaduce, A., Lyubartsev, V., Cesarini, C., Nuccetelli, M., Zodiatis, G., Stylianou, S., 2017. EMODnet MedSea CheckPoint – Second Data Adequacy Report. https://emodnet.eu/sites/emodnet.eu/files/public/D11.4-draft18.pdf.

Raymond, K., 1995. Reference model of open distributed processing (RM-ODP): introduction. In: Open Distributed Processing, Experiences with Distributed Environments. Proceedings of the Third IFIP TC 6/WG. Springer Nature. https://doi.org/10.1007/978-0-387-34882-7.

Snowden, D., Tsontos, V.M., Handegard, N.O., Zarate, M., O'Brien, K., Casey, K.S., Smith, N., Sagen, H., Bailey, K., Lewis, M.N., Arms, S.C., 2019. Data interoperability between elements of the global ocean observing system. Front. Mar. Sci. 6, 442. https://doi.org/10.3389/fmars.2019.00442.

Swift, J.H., 2010. IOCCP Report No. 14. Reference-quality Water Sample Data: Notes on Acquisition, Record Keeping, and Evaluation. The GO-SHIP Repeat Hydrography Manual: A Collection of Expert Reports and Guidelines, vol. 134. ICPO Publication SeriesN0. https://www.go-ship.org/Manual/Swift_DataEval.pdf.

Tanhua, T., Pouliquen, S., Hausman, J., O'Brien, K., Bricher, P., de Bruin, T., Buck, J.J.H., Burger, E.F., Carval, T., Casey, K.S., Diggs, S., Giorgetti, A., Glaves, H., Harscoat, V., Kinkade, D., Muelbert, J.H., Novellino, A., Pfeil, B., Pulsifer, P.L., Van de Putte, A., Robinson, E., Schaap, D., Smirnov, A., Smith, N., Snowden, D., Spears, T., Stall, S., Tacoma, M., Thijsse, P., Tronstad, S., Vandenberghe, T., Wengren, M., Wyborn, L., Zhao, Z., 2019. Ocean FAIR data services. Front. Mar. Sci. 6, 440. https://doi.org/10.3389/fmars.2019.00440.

Thurnherr, A.M., Visbeck, M., Firing, E., King, B.A., Hummon, J.M., Krahmann, G., Huber, B., 2010. A manual for acquiring lowered doppler current profiler data. The GO-SHIP Repeat Hydrography Manual: A Collection of Expert Reports and Guidelines, 134. IOCCP Report No. 14, ICPO Publication SeriesN0. https://www.go-ship.org/Manual/Thurnherr_LADCP.pdf.

Uchida, H., Johnson, G.C., McTaggart, K.E., 2010. IOCCP Report No. 14. CTD Oxygen Sensor Calibration Procedures. The GO-SHIP Repeat Hydrography Manual: A Collection of Expert Reports and Guidelines, vol. 134. ICPO Publication SeriesN0. https://www.go-ship.org/Manual/Uchida_CTDO2proc.pdf.

UNESCO, 2006 - IODE Steering Group for MEDI, 2006. Intergovernmental Oceanographic Commission Reports of Meetings of Experts and Equivalent Bodies. Drexel University Philadelphia, USA. https://www.jodc.go.jp/info/ioc_doc/Experts/SGMedi3.pdf.

Wilkinson, M.D., Dumontier, M., Aalbersberg, I., Appleton, G., Axton, M., Baak, A., Blomberg, N., Boiten, J.-W., Bonino da Silva Santos, L., Bourne, P.E., Bouwman, J., Brookes, A.J., Clark, T., Crosas, M., Dillo, I., Dumon, O., Edmunds, S., Evelo, C.T., Finkers, R., Gonzalez-Beltran, A., Gray, A.J.G., Groth, P., Goble, C., Grethe, J.S., Heringa, J., 't Hoen, P.A.C., Hooft, R., Kuhn, T., Kok, R., Kok, J., Lusher, S.J., Martone, M.E., Mons, A., Packer, A.L., Persson, B., Rocca-Serra, P., Roos, M., van Schaik, R., Sansone, S.-A., Schultes, E., Sengstag, T., Slater, T., Strawn, G., Swertz, M.A., Thompson, M., van der Lei, J., van Mulligen, E., Velterop, J., Waagmeester, A., Wittenburg, P., Wolstencroft, K., Zhao, J., Mons, B., 2016. The FAIR Guiding Principles for scientific data management and stewardship. Sci. Data 3, 160018. https://doi.org/10.1038/sdata.2016.18(2016).

Wright, G., Rochette, J., Greiber, T., 2016. Sustainable development of the oceans: closing the gaps in the international legal framework. In: Mauerhofer, V. (Ed.), Legal Aspects of Sustainable Development. Springer, Cham. https://doi.org/10.1007/978-3-319-26021-1_27.

PART THREE

Society-driven data and co-production

A collaborative framework among data producers, managers, and users

S. Simoncelli[1], Giuseppe M.R. Manzella[2,3], A. Storto[4], A. Pisano[4], M. Lipizer[5], A. Barth[6], V. Myroshnychenko[7], T. Boyer[8], C. Troupin[6], C. Coatanoan[9], A. Pititto[10], R. Schlitzer[11], Dick M.A. Schaap[12], S. Diggs[13]

[1]Istituto Nazionale di Geofisica e Vulcanologia, Sezione di Bologna, Italy
[2]The Historical Oceanography Society, La Spezia, Italy
[3]OceanHis SrL, Torino, Italy
[4]Consiglio Nazionale delle Ricerche - Istituto di Scienze Marine (CNR-ISMAR), Rome, Italy
[5]Istituto Nazionale di Oceanografia e di Geofisica Sperimentale — OGS, Trieste, Italy
[6]University of Liege, Liege, Belgium
[7]Middle East Technical University, Institute of Marine Sciences, Erdemli-Mersin, Turkey
[8]National Centers for Environmental Information, National Oceanic and Atmospheric Administration, Asheville, NC, United States
[9]Ifremer Centre de Bretagne, Plouzané, Brest, France
[10]COGEA, Rome, Italy
[11]Alfred Wegener Institute, Bremerhaven, Germany
[12]Mariene Informatie Service MARIS B.V., Nootdorp, the Netherlands
[13]Scripps Institution of Oceanography, University of California San Diego, La Jolla, CA, United States

Introduction

The advent of the United Nations Ocean Science Decade 2021—30 for Sustainable Development (Ryabinin et al., 2019, https://www.oceandecade.org/) will bring about a blue revolution with the aim of boosting ocean knowledge and literacy and serve the whole society in driving informed decision-making processes and the blue growth. The continuous observation of the ocean is a crucial activity in this process that allows monitoring the ocean state, its functioning, and evolution in a changing environment and climate, but nowadays the sustainability of the observing network is at risk. Beside the efforts to design and maintain at global and regional scale an adequate integrated ocean observing system, the need to manage and exploit at best the deriving huge amount of data is equally important, especially due to the increasing data flow originated from new autonomous platforms. This necessitates the implementation of a global ocean data management system made by the interconnection between different existing

Ocean Science Data
ISBN: 978-0-12-823427-3
https://doi.org/10.1016/B978-0-12-823427-3.00001-3
© 2022 Elsevier Inc.
All rights reserved.
197

infrastructures, evolved in the past decades in different marine disciplines and thematics: physics, chemistry, biology, ecology, geology.

New data management paradigms emerged in the past decades, such as the development and adoption of common standards (INSPIRE Directive 2007/2/EC of the European Parliament), the open data policy (Open Data Directive (EU) 2019/1024), and the FAIR principles (Findable, Accessible, Interoperable, and Reusable - Wilkinson, 2016), which are changing the way science is advancing but also how it is conceived. Data reuse is the basic concept behind data sharing and open data, which was supported by the former DG-MARE Commissioner, Maria Damanaki, in the communication "Marine Knowledge 2020" (European Commission, 2010).

Since the pioneer analysis made by Redfield (1934), the need for standards was addressed as one of the most important actions for data use and data sharing, that became evident when, during the World Ocean Circulation Experiment (WOCE, e.g., Ganachaut and Wunsch, 2003) and the Joint Global Ocean Flux Study (JGOFS, e.g., Fasham et al., 2001), researchers were discussing about the need of integrating and comparing nutrient measurements made by different laboratories. An offset was found in nutrient data by comparing measurements made in deep waters (depth over 3500 m) during different cruises and measurements made at nearby stations. The identified discrepancy was indicating inconsistencies in the preparation of calibration standards and use of different methodologies. Agreed-on practices, methodologies, intercalibrations, Certified Reference Materials (CRMs), clean room techniques, qualified personnel, etc., were not sufficient to assure good quality of data. Sampling strategies and data collection methodologies, sample pretreatment, transport, and storage were identified as integral part of the quality assurance and good measuring practices (see also Chapters 2 and 3). Since then, various Quality Control (QC) and Assurance (QA) procedures have been established, for example, by National Oceanographic Data Centers, international projects (e.g., SeaDataNet — Fichaut et al., 2013; WOCE - World Ocean Circulation Experiment — Wunsch, 2005; GTSPP - Global Temperature and Salinity Profile Program — Wilson, 1998; GOSUD - Global Ocean Surface Underway Data — GOSUD, 2016), and expert groups (e.g., Bushnell et al., 2019). An overwhelming number of methodological documentations, from manual Standard Operating Procedures (SOP) to community agreed best practices, have been created by individuals, institutions, and expert groups. In order to overcome the fragmentation of existing documentation and material and to guide community efforts in the creation of SOPs and best practices, the "Ocean Best Practices System" (OBPS, Pearlman et al., 2019) has been endorsed and supported by the UNESCO's Intergovernmental Oceanographic Commission (IOC, http://msp.ioc-unesco.org/) as a central hub for universal access.

Therefore, the societal benefit of science can be achieved through a community effort that collects, manages, curates, and shares the data acquired with specific purposes and enables multiple stakeholders from both public and private sectors to use and reuse them many times and for different purposes. Transparency and quality are the basic requirements underlying this process to promote efficiency, competition, and innovation and reduce the uncertainty in knowledge. Many marine observation and data collection initiatives are promoting data sharing and open data with the main goal to support research, to make data available to the wide public, to assist informed policies, and to support the blue growth.

New societal needs and the emerging blue economy sectors (marine transportation, aquaculture, fishery, tourism, renewable energy) necessitate scientific sound information to minimize the impact of human activities on the marine environment and guarantee sustainable growth. In fact, the sustainable development of the ocean and seas (Agenda 2030, SDG 14 UN General Assembly, 2015) can be achieved only by preserving a good environmental status (MSFD — Marine Strategy Framework Directive 2008/56/EC, e.g., Borja et al., 2010) and assuring that natural resources are not depleted. Increasing pressures like acidification, eutrophication, biodiversity loss, pollution, overexploitation, and illegal activities are also affecting the oceans, and an integrated and ecosystem-based management of the marine environment is the only practicable solution to become a resilient society. An example is the ongoing development and progressive implementation worldwide of the Marine Spatial Planning (MSP) approach through the joint MSP Roadmap adopted by UNESCO-IOC and the Directorate-General for Maritime Affairs and Fisheries (DG MARE) of the European Commission (Directive 2014/89/EU).

Ocean knowledge and environmental assessment are based on an integrated multidisciplinary and multiscale ocean observing and monitoring system (Fig. 4.1) consisting of three pillars: observations, numerical models, and data assimilation techniques. In-situ and satellite observations, within their accuracy, provide the reference ("truth") information on the physical and biogeochemical ocean state and are then essential to develop, calibrate, and validate predictive models, as well as constrain them toward reality through data assimilation techniques. Deriving huge amounts of data and

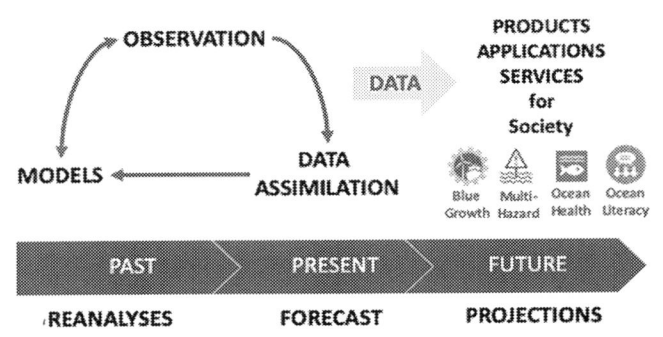

Figure 4.1 Integrated observation and monitoring system schema that shows the interconnection between ocean observation and modeling to run models to reproduce the past, predict the near future, and simulate the future state of the ocean, creating a massive amount of data that are used to develop products, applications, and services for society.

products can drive the development of new knowledge but also downstream application and services. Such a system enables us to understand past and present ocean conditions and forecast future ocean conditions thanks also to the advent of retrospective analyses and future scenario applications, which enable to design mitigation and adaptation strategies within an evolving marine environment. Accurate predictive capabilities permit the implementation of services for real-time decision-making, multihazard warning systems, and anticipatory marine spatial planning.

The optimization of the existing observing and prediction system relies on the engagement of multiple stakeholders along with the many phases of the marine data value chain to identify monitoring gaps and continuously tailor data products requirements to the emerging societal needs. This integrated assessment process passes through the generation of progressive value-added products, which synthesize information through the creation of various monitoring indicators of ocean and climate conditions, ecosystem health, or human impacts. Operational oceanography community effort allowed the progressive implementation of fit-for-purpose and functional monitoring indicators, adequate to the multiple reporting activities like the IPCC Assessment Reports, the EU Copernicus Ocean State Reports, the European Environment Agency, and the Bulletin of the American Meteorological Society (BAMS) State of the Climate.

Among future priority areas of research (Ryabinin et al., 2019) and technical development there are the comprehensive mapping of the ocean, describing the physical, biological, chemical, geological variables, ecosystems, natural resources, cultural objects, marine and maritime human activities, etc., and the improvement of modeling capabilities oriented toward the creation of a digital twin of the ocean that represents the socioecological dimensions of the ocean through digital means. The mapping of global ocean depth (bathymetry) is, for example, still fragmented and nonhomogeneous in terms of resolution; the effort of assembling different sources of data or data types at both regional and local scales is still ongoing to match the theoretical requirements. Other examples are the mapping of seabed habitats, marine litter, chemical pollutants, and human activities, which are all very important to assess the sources and cumulative impact of multiple stressors of the marine environment. Data and information technology will be the key enabler of virtual representation of the ocean because it will allow us to collate, preserve, manage, and exchange data for their rapid massive usage and transformation into information and knowledge. Big data analytics and artificial intelligence are also fertilizing ocean science creating expectations on new capabilities and findings.

A *"transparent and accessible"* ocean (Ryabinin et al., 2019) is thus one of the societal outcomes expected from the Decade to provide rapid access to ocean data and information technologies, enabling the capacity of environmental assessment and informed actions to protect and restore the ocean health, to use ocean space and resources sustainably. The vision consists of a global ocean data system like a digital ecosystem made by multiple, interoperable and scalable components, which necessitates continuous international codesign and coproduction efforts. It should build upon the existing data infrastructures that allow interoperable data sharing and stewardship in tight collaboration between data providers, producers, and users. It should continuously expand and upgrade, taking into account the most recent community best practices and user requirements. Likewise, it should progressively incorporate diverse thematic ocean data, with main focus on Essential Variables (https://earthdata.nasa.gov/learn/backgrounders/ essential-variables) but extending to socioeconomic disciplines and humanities, including cultural heritage to fully understand the ocean as a key component of the Earth system and human life. A "fitness for purpose" digital data ecosystem should deal with in situ, remotely sensed, and model data, integrating historical and contemporary observations. The codevelopment of software and tools should set up collaborative virtual environments, where users can access data without limitations and create innovative solutions to face future societal challenges.

The Agenda 2030 (UN General Assembly A/RES/70/1, 2015) calls for codesign, coproduction, codelivery, solution-oriented research as a necessary step forward to fully implement an efficient data cycle able to transform all existing data into value for society. An important element is the move from distinct integrated/coordinated observing systems and data services to ocean data and information systems that encompass the whole data value chain. The marine data value chain will be the backbone of this chapter, which Fig. 4.2 summarizes together with the supporting activities needed to realize it and the guiding principles lying behind it. The realization of the data value chain implies a huge collaborative effort to run and sustain the technology infrastructure and develop the services, together with high coordination and communication skills. Key component of the data infrastructure is the system monitoring dedicated to routinely control the performance of the implemented services and the infrastructure content.

The objective is to overview how to progressively usher ocean data into information products, leading to a greater understanding of ocean status, functioning and evolution and how the resulting knowledge can further be synthetized through indicators with the multiple purpose of reporting the scientific outcome and support informed decision-making for human well-being.

The chapter develops following the progressive complexity and information content of ocean data products, and it is articulated as follows: data cycle and data collections; gridded data products, in particular synoptic and climatological maps, satellite products, ocean reanalysis, new challenges data products and the complexity of deriving them. Finally, the chapter discusses the quality of data products before drawing conclusions and recommendations for the Decade to come.

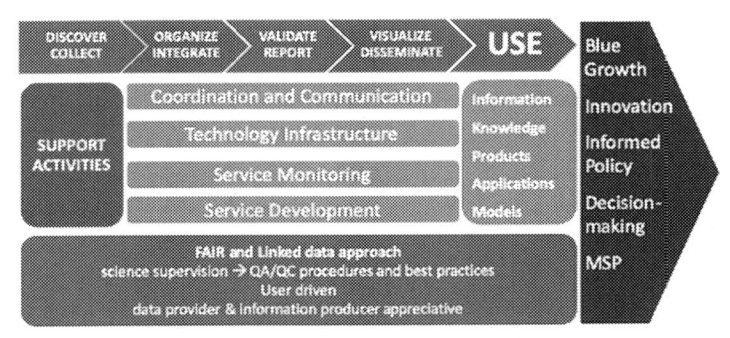

Figure 4.2 The marine data value chain: it includes the data life cycle, the necessary supporting activities to fulfill it, and the underlying principles.

Data cycle and data collection

The physical and chemical parameters of the ocean's subsurface to this day have not been fully measured nor routinely monitored. Space-based measurements provide a high resolution view of the surface ocean but can give only some insight to the physical nature of the subsurface ocean, in an integrated fashion without revealing the actual structure and functioning. The difficulty of sending ships to all corners of the global ocean has always made manual subsurface measurements high cost and low yield—point by point across vast areas. Even in the dawning age of autonomous subsurface measurements (profiling floats, gliders, ...), positioning and then maintaining measurements across the ocean is a large and incomplete undertaking. As for covering the full depth of the ocean, the mean ocean depth is around 3800 m. The core Argo program (Roemmich et al., 2019; Wong et al., 2020) measures the top 2000 m, leaving more than half of the ocean without regular monitoring. Deep Argo is meant to remedy the lack of deep monitoring, but is an ongoing effort, not global (or even basin-wide) yet. Outside of global monitoring systems such as Argo, there is a robust monitoring of subsurface ocean conditions for the management of fisheries and other economic activities, as well as environmental monitoring to assess the health of areas of national and international interests. Further there are regular and irregular research cruises to areas of the ocean to study specific phenomena or simply to map the physical, chemical, and biological conditions.

The difficulty and expense of subsurface ocean monitoring make it imperative to gather all ocean data from all sources within and without the standard global ocean observing system, and provide a more fully measured and monitored ocean. Ensuring the quality and usability of the aggregated data is also of primary importance, as the expense and effort of gathering the information is wasted if the data are not sufficiently described or of insufficient quality for environmental monitoring. Further, given that the ocean changes slowly over time, and it is the changes that are expected to reveal themselves during monitoring, integrating ocean data from all possible sources, both historic and current, is indispensable for long-term monitoring of the global, regional and even the local ocean.

Historical data: the GODAR project

In the early nineties, some initiatives started with the aim to identify and rescue historical oceanographic data, highlighting the need to manage data with a coordinated and harmonized approach. The "Global Oceanographic Data Archaeology and Rescue" (GODAR) project started in 1993, under the auspices of the UNESCO Intergovernmental Oceanographic Commission (IOC), to locate and rescue historical oceanographic data for the pre-1992 period, safeguarding an important amount of data types. The World Ocean Database Project (WOD), an international project of the IOC International Oceanographic Data Exchange (IODE), managed by the National Oceanographic and Atmospheric (NOAA), was in 2001 an important follow-up. The purpose of the WOD was (and is) to aggregate all available ocean profiles (subsurface ocean variables vs. depth; multiple depths at the same location/date/time) across instruments, platforms, and measurement methods. At its beginning in the early 1990s, the WOD was envisioned simply as a means to make available the input data and quality control for the World Ocean Atlas (WOA) series. The WOA is a set of gridded climatological mean fields of ocean variables (temperature, salinity, oxygen, nutrients, etc.) calculated from in-situ ocean profiles. Basic quality control procedures for the WOA (Boyer and Levitus, 1994) were specifically designed and implemented to flag anomalous data. To ensure reproducibility of the WOA, the WOD was (and is) released in concert with the WOA and includes every quality flag which represents a data point, which was either used or not for the WOA.

In Europe, two pilot projects (Mediterranean Oceanic Data Base—MODB—and MedAtlas) started in 1994 to rescue temperature, salinity, and bio-chemical data in the Mediterranean and Black Seas. The partners of the two projects joined a concerted EU funded action called briefly MEDAR/MedAtlas (Fichaut et al., 2003). This project started in 1998 and was endorsed by IOC and GODAR. A complete history of GODAR/WOD and MEDAR/MedAtlas is presented in the paper by Levitus (2012). GODAR/WOD and MEDAR/MedAtlas gave priorities also to the preparation of catalogs, development of metadata, and implementation of quality control procedures by using common protocols for data formatting/management/exchange that were demanded by the operational oceanography community to answer to the emerging ocean forecasting needs. The main goals were including, since then, the preparation of value-added products such as objective analyses and climatological gridded statistics to be used as initial and boundary conditions in modeling, the identification of variabilities from seasonal to interannual/decadal time scales, and the ocean data assimilation efforts.

Evolution of quality paradigms

The successive research and technology developments in ocean science have been running in parallel to the increasing data management need to collect and provide ready-to-use data for both near real-time applications (e.g., ocean forecasting) and delayed mode analysis. The data use, sharing, and exchange require the integration and harmonization of data coming from heterogeneous sources into consistent long data records, which are the main target of marine and climate data infrastructures from which to derive ocean information products. Fichtinger et al. (2011) describe the data harmonization and interoperability aspects, introduced by the EU INSPIRE Directive - 2007, which made possible datasets to be combined and data infrastructure services to run without manual intervention. The contextual introduction of Data Specifications allowed data providers to transform their data according to precise encoding rules.

The traditional view of data, focused on collecting, processing, analyzing, and publishing of results, has been definitely substituted by a life cycle view that highlights the importance of finding, storing, and sharing data (Griffin et al., 2018). This radical change requires the execution of architectures, policies, practices, and procedures encompassing the full data life cycle in order to satisfy the needs of data producers, intermediaries, and end users. As a consequence, the prior planning of any data gathering activity has become mandatory to document data creation, content, and context, but also to fulfill data quality requirements.

Data quality requires predefined Quality Assurance (QA) strategies based on the selection of internationally validated methodologies for sampling and analysis, the mandatory use of reference materials, and the participation in "blind" international intercomparison exercises. Data providers must follow specific QA procedures and protocols applied before and during the dataset creation (e.g., Bushnell et al., 2019). The Commission of the European Community (CEC) and IOC (1993) manual first listed the data gatherer responsibilities in terms of documentation, calibration/intercalibration exercise, sampling strategy, admissible ranges of data, algorithms used, corrections, and flags. Most recently, QA protocols, Standard Operating Procedures (SOPs), and best practices have been developed from global expert groups, considering international agreed policies and standards that are accessible via the OBPS (https://www.oceanbestpractices.org/).

The application of QA guidelines into the specific fieldwork is guaranteed by a list of actions to be performed and checked routinely. The preliminary information to achieve the target of a Data Quality Management System consists of an overview of the sensors and the methodologies adopted, in particular (Mora, 2014):

1. the knowledge of the sensor's accuracy;
2. the calibration and intercomparison of sensors;
3. the adequacy of the sampling strategy to the scope;
4. Quality Assurance (QA) of field work;
5. Quality Control (QC) of collected data;

Records of monitored parameters should include a minimum set of information mapped through metadata: the description of the sensors and platforms configuration, the measurement position, the measured quantity units, and the processing date and time. Other information useful to allow data intercomparison and reuse includes the measuring system adopted and its accuracy.

QC is defined as the process of defect detection and rejection, which is essentially focused on the "process output," i.e., it regards procedures to be applied after data gathering. The concept of data QC is vital for data reuse, and without it data from different sources cannot be combined to gain value. Scientific, analytical, and statistical evaluations must determine if data present adequate quality to support the intended data usage, resulting in labeling each numerical value with a Quality Flag (QF) and avoid modifying the original data record following a harmonized scheme of QFs (IOC, 2013). Examples of QF scale are available from the SeaDataNet Common Vocabulary (IOC, 2013; IOC/UNESCO, 2019) and from the WOD (IOC, 2013). QFs ensure that the quality of the data is apparent to the user, who holds sufficient information to decide the suitability for a specific task and can apply the proper data filtering. QC practices include data integrity checks (e.g., format), data value checks (threshold checks, minimum/ maximum rate of change), neighbor checks, climatology checks, and highly depend on the data thematic, sensor type, and the available amount of time for the analysis (latency).

When prime importance is that the data are provided without delay, a few simple checks are performed in Real Time (RT), and it is not always possible to carry out instrument calibrations, corrections to times for clock errors, or more than the most rudimentary position checks. Data going through the GTS (Global Telecommunication System) are an example of RT data transmission. Their QC is done quickly with automated procedures without final annexed indication on the quality of the data being distributed. Where quality control is carried out, observations that fail the tests are usually removed from the data stream. Near Real Time (NRT) data are characterized by latencies normally between 1 day and several weeks, allowing the implementation of automatic QC. The NRT QC consists basically of: platform identification, data and location check, position on land test, regional range test, pressure increasing test, spike test, bottom spike test, gradient test, digit rollover test, stuck value test, density inversion, gray list, gross salinity or temperature sensor drift, frozen profile test, deepest pressure test. The Delayed Mode (DM) QC requires more complete crosschecking and analysis procedure, including visual inspection and comparison with climatologies.

It is important to distinguish here the difference among QA/QC and quality assessment process (Bushnell et al., 2019), which consists of evaluating if data meet the necessary quality requirements or are adequate for an intended use. Quality assessment is treated in detail in next sections. Data accuracy is another core attribute of quality dimension, it represents one of the quality elements of data reliability (Cai and Zhu, 2015), and it can be evaluated by analytical comparison with reference data. Data uncertainty determination and its propagation along with the data value chain represents an important present challenge, since it influences the user selection of a dataset or a data product among others or even decision–making. The first version (v0.1) of International Quality Controlled Ocean Database (IQuOD) released in 2018 is the first example of ocean subsurface temperature database with uncertainty information associated with each measurement (Cowley et al., 2021).

Several actors apply QA/QC procedures along with the data cycle and value chain (Fig. 4.3): the data originator before the data entry in a database, the data manager during the data ingestion, the intermediary to verify/ improve data integrity/consistency/completeness in a multisource dataset and, the information producer to generate derived products, such as, for example, gridded fields, ocean reanalysis, or more complex indicators. Data consistency within a single dataset or within a collection of datasets is an important quality element to preserve within a data infrastructure and in data products. Data experts apply secondary QA/QC procedures to analyze the whole database data and metadata content to guarantee high quality standards and products.

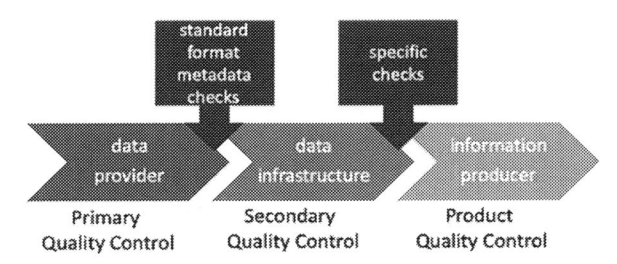

Figure 4.3 Schematic of the different quality control phases along with the marine data and production cycle.

Data infrastructures

Nowadays sound research outcomes must rely on the most recent, extensive, and best quality datasets, collected either by specific monitoring campaigns or assembled from data infrastructures. Many marine data infrastructures give insight through web portals or catalogs of the available marine data products, ensuring their free, consistent, and reliable access. Consolidated services, capable of handling parts or the whole marine data value chain, are allowing ocean science to advance toward effective decision support, informed policy, and blue growth. The upstream, midstream, and downstream of ocean services, schematized in Fig. 4.4, are mapped correspondingly atop the marine data and information flow. The upstream is linked to the ocean data records creation and preservation pillar. The midstream corresponds to the pillar on the application-oriented ocean information products, which might integrate non-ocean data and models. The

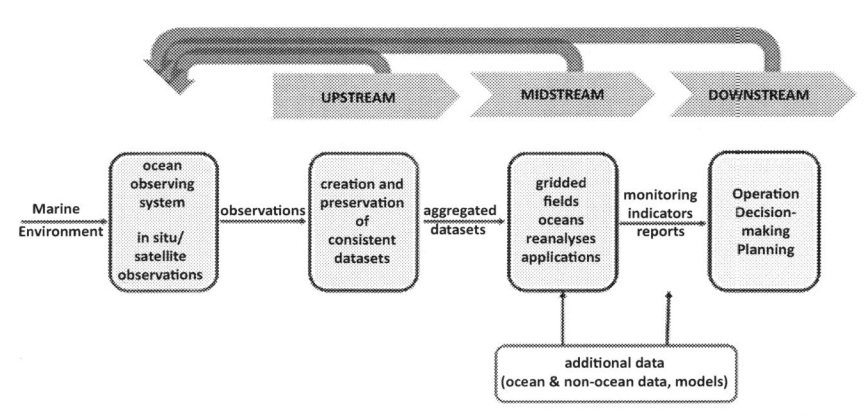

Figure 4.4 Ocean data flow and the service compartments build on top of it. *Adapted from (Zeng et al., 2019).*

downstream includes the transition between "Products and Applications" pillar and "Decision-Making" pillar, as connected by the "Indicators and Reports" flow, as well as the "Decision-Making" pillar itself. Furthermore, a feedback mechanism through multiple users' engagement allows to improve and further develop not only the services to be more effective but also to optimize the underlying ocean observing system.

Data infrastructures, which constitute the upstream service, build on the discovery of data sources or providers and their connection to a network. Data sources can be: research institutes which hold data from scientific expeditions; environmental protection agencies that perform routinely monitoring surveys; research programs and infrastructures that manage marine monitoring systems (e.g., Argo, OceanSITES, GO-SHIP) or observatories and networks (e.g., EMSO, LTER, GEO-BON), private companies that work in the blue economy sector.

The access to data sources to collect the data is a critical step that relies on two main approaches (Nativi et al., 2012):

1. the data provider becomes a stable node of a network (virtual database); a central system user submits queries by means of a shared intermediary interface;
2. the data infrastructure takes care of the data ingestion building up ad hoc solutions for each data provider (see Chapter 3).

In the second approach, the data infrastructure organizes the data adopting common data and metadata models (vocabulary, format, and standards) that facilitate the integration of different data sources and data types. A full metadata description (data provider, instrument type, parameters measured, measurement units) of the data is mandatory in this phase, since it assures to archive, preserve, and analyze the data for further reuse.

Additional crucial metadata information has to be defined once the data are ready to be ingested in order to uniquely identify the data provider, the data curator, or holding center; the program and project frameworks in which the data have been collected; the cruise and station identifiers; the data access restrictions. This metadata information follows after the definition of specific vocabularies, maintained and managed as complementary services from the data infrastructure.

The advent and adoption of metadata standards allow to uniquely identify a data record, its processing phases, and to acknowledge all the actors of the data life cycle (data providers, managers, and information producers), assuring full process traceability and transparency. The enabling power of a full metadata description is constantly increasing at different levels, giving the possibility to have a global view of the observing networks in operation and available datasets through dashboards (e.g., https://www.ocean-ops.org/, https://map.emodnet-physics.eu/, http://www.marineinsitu.eu/dashboard/) and data access web portals. It enables the monitoring of the data-sharing landscape per country/ocean region, the overview of the data types, and their space-time coverage. The constant monitoring of space-time distribution of a specific data type or essential variable permits the identification of observing system gaps and the subsequent formulation of strategies for the generation of new data according to the evolving societal needs or instead prioritizing the use of existing data depending on the user requirements. Another issue to consider is the omission of existing data and metadata not yet shared and available, which demands specific attention and actions in order to be gradually limited.

The rapid availability of overview characteristics of the observing system at global and basin scale enables their codevelopment through user engagement at different levels, depending on the kind of user considered, which can be policymakers, nongovernmental organizations, forecasting centers, scientists, students, and so on. Moreover, some data infrastructures consent to apply a flexible data access policy according to data provider wishes and promote data sharing in whatever way. In this case the monitoring of data access metadata information, free access data versus restricted data, allows also to check the compliance with the defined data policies.

The publishing and dissemination of data and products are then crucial for users' uptake and feedback. The general characteristics and the reliability of the datasets, together with the methodologies applied and the format adopted, are usually delivered to the users via product information documents, manuals, or user guides. Harmonized data and derived products can be visualized through static images, geoportals, dashboards, which could permit their tailored extraction too. Aggregated data collections and derived products can be also released through web catalogs landing pages (e.g., https://www.seadatanet.org/Products#/).

Harmonized data records or aggregated collections can be considered the basic data product from which to derive information through further analysis and processing. They can be used to generate higher level data products, both in terms of complexity and deriving information, like maps (gridding). They are also needed to drive and validate predictive or reanalysis models and to derive monitoring indicators.

Some key examples of European open-source data integrators and service providers financed by the EU (Martín Míguez et al., 2019) are

SeaDataNet, the European Marine Observation Data network (EMODnet), and the Copernicus Marine Environment Monitoring Service (CMEMS, https://marine.copernicus.eu/). The WOD project, managed at NOAA National Centers for Environmental Information (NCEI) and the IQuOD (http://www.iquod.org/) IODE project, are other international examples. These initiatives cope with the common objective to collect and integrate all available marine data from different instruments, platforms, and measurement methods and present to the public completely free and without restriction data collections in a uniformly formatted and quality-controlled manner. They originated with different purposes that led to different data flows, data population characteristics, and timing of data processing and release.

The CMEMS was conceived to "provide regular and systematic reference information on the physical and biogeochemical ocean and sea-ice state for the global ocean and the European regional seas" (Le Traon et al., 2019), through the combined use of observations (in-situ and satellite) and numerical model analyses. Observations are managed by the Thematic Assembly Centers (TACs) to provide high-level observation-based products (as e.g., temperature, salinity, current, sea level, waves, chlorophyll, oxygen) and used by the Monitoring and Forecasting Centers (MFCs) to validate/assess their model-based products and constrain the underlying ocean forecasting systems. Observation and model-based products are typically provided in NRT and DM, the latter as reprocessed (for observations) and reanalysis (for model) consistent and stable, namely long-term, datasets for climate studies.

WOD and SDN are instead dealing mainly with the management and preservation of DM observations, while EMODnet is a network of organizations that work together to aggregate and process marine data from diverse sources and generate data products. It provides a gateway to marine data and products through a number of thematic portals, and each thematic group builds on various data initiatives (data infrastructures, networks, projects, data assembly centers) and adds value by (1) facilitating access to the data and (2) generating new products from them (Martín Míguez et al., 2019).

SeaDataNet represents one of the data infrastructures behind EMODnet (Bathymetry, Chemistry and partially Physics lot). It is a first example of a distributed data management system where data are stored and handled in a network of National Oceanographic Data Centers (NODC) and research institutes. Data originators are interconnected to a central Common Data Index (CDI) management infrastructure and access portal (https://cdi.

seadatanet.org/) and provide validated, standardized, and formatted data thanks to common guidelines and specific software developed by expert data managers. The distributed system has advantages such as scalability and interoperability but gives flexibility to data providers in maintaining and updating their data. The drawback of such a solution is the higher probability of duplicates and data anomalies than a single database as well as the necessity to perform aggregation of data on the fly while processing user requests.

Data flow

Data discovery and assembly is the first and crucial step to populate a database, but common standards and formats need to be adopted by the data providers that connect to and constantly feed the database or through an intensive body of work in understanding incoming datasets. This process ensures that all information is present and properly formatted for appropriate use of the data, and that the data are actually the reporting of the measurements intended.

The incoming data can derive from ocean data systems, such as the Global Temperature and Salinity Profile Program (GTSPP), the Global Ocean Surface Underway Data Pilot Project (GOSUD), moored buoys (i.e., TAO, TRITON, PIRATA, RAMA), Deep Ocean Multi-Disciplinary Ocean Reference Stations (OceanSITES), Argo profiling float data (GDAC), Sea Level data (GLOSS), from National Oceanographic Data Centers (NODCs), research institutes, environmental protection agencies monitoring arrays, projects, and research institutes. A constant monitoring of the Global Ocean Observing System (GOOS) status and growth and its timely exchange of data and metadata is available thanks to the dashboard (https://www.ocean-ops.org/) implemented by the World Meteorological Organization (WMO)-IOC Joint Commission for Oceanography and Marine Meteorology (JCOMM) in-situ Observations Programs Support (OceanOPS).

It is a vital and time-consuming phase to QC data integrity during the ingestion process, which involves a great deal of computer programming and knowledge of formats, and requires intense communication with data originators. This QC phase permits to detect missing mandatory information, errors made during the data transfer or reformatting and eventual duplicates, meaning that a copy of the submitted dataset is already available in the database.

The analysis of metadata of the database content allows to monitor its population in terms of parameters' availability, spatial and temporal distribution, originators, platforms, projects, programs. The duplicate check instead avoids unnecessary growth of the database, and guarantees the preservation of the best data version. Duplicates can skew the statistics and bias the successive analysis thus affecting the quality of derived products. In order to avoid this, the duplicates have to be identified and resolved: the best copy should be kept while others should be removed or labeled somehow. Different types of duplicates can reside in a database, and an exhaustive metadata description is the first requirement to identify the best version. Identical metadata records (longitude, latitude, time, instrument, data provider) could be associated with different measurement data due to different processing or different associated quality flags. On the opposite the same measurement values could be associated with divergent metadata description. Duplicates might occur because:

- same data are submitted through different paths, e.g., directly by originator and also by national and by regional data centers or by several partners of a project;
- real-time data are resubmitted in delayed mode with better precision and quality;
- historical data are combined from multiple archives.

The exact duplicates—those having the same metadata (location, time) and data—are easy to identify and reject during the ingestion process. Another type of duplicates are the identical profiles (i.e., containing the same variables with identical values at each depth) that are attributed to different position/time. The causes for this can be either human mistakes, rounding errors, or reporting local time instead of GMT in one of the duplicates, or this can be a "stuck" profile from a drifter that is repeatedly transmitted from different locations. Finding such profiles in a database is also simple but identification of the correct copy requires analysis of supplementary information and, most probably, communication with data providers for clarification. The most difficult problem is the detection and resolving of "near" duplicates that happen when the same data received through different paths were treated differently, for example, the first version contains original raw data, the second contains data interpolated to standard depth levels, the third has adjusted coordinates and time and assigned quality flags, and so on. Upon detecting potential "near" duplicates, the decision on which one to keep and which to remove should be based on the objective criteria, e.g., retain a profile having more depth levels and better quality scores. If the criteria do not allow to identify the best copy, the decision can be to retain the one with the latest updated date. The large oceanographic databases have well-established procedures for identification of duplicates in the incoming datasets internally and against the whole content of the database. However, not all duplicates can be resolved in automatic mode; some cases need expert involvement, which is time-consuming and costly. The problem of duplicates arises once again when there is a need to merge data from different databases for purposes of joint analysis or product, for example, for calculation of climatic fields based on the most complete dataset. For example, according to the assessment done within the SeaData-Cloud project the overlapping between SeaDataNet, WOD, and CMEMS vary from 22% to 42% in the Black Sea region (Myroshnychenko and Simoncelli, 2020).

Successively, the data need to be further validated applying for each parameter the commonly recognized Quality Control (QC) procedures. Commonly the data ingested within the database have been already quality assessed by the data providers (see Fig. 4.3 schematic) that assign QFs according to the adopted data representation but a secondary validation is essential at the data infrastructure level to guarantee the harmonization, consistency, compatibility, and comparability of data integrated from many originators and instrument types over time and validated with different methodologies. This secondary validation loop usually represents a phase of a wider QA strategy. Eventually, in this phase the QFs of the detected anomalous data can be changed.

SeaDataNet developed a Quality Assurance Strategy (QAS), schematized in Fig. 4.5, in order to achieve the desired quality of data and derived products. It consists of five main phases:

1. data harvesting of all data files contained in the network of data centers (NODCs);
2. file and parameter aggregation to generate metadata-enriched data collections;
3. secondary QC analysis at regional level;
4. feedback to data providers on data anomalies;
5. analysis/correction of data anomalies and update of the respective CDI records.

This iterative approach results in versioning of data collections that after each loop extend their time coverage and advance in quality. Each QAS loop involves many actors and processes, which require a big communication and collaboration effort to progressively optimize and automate the

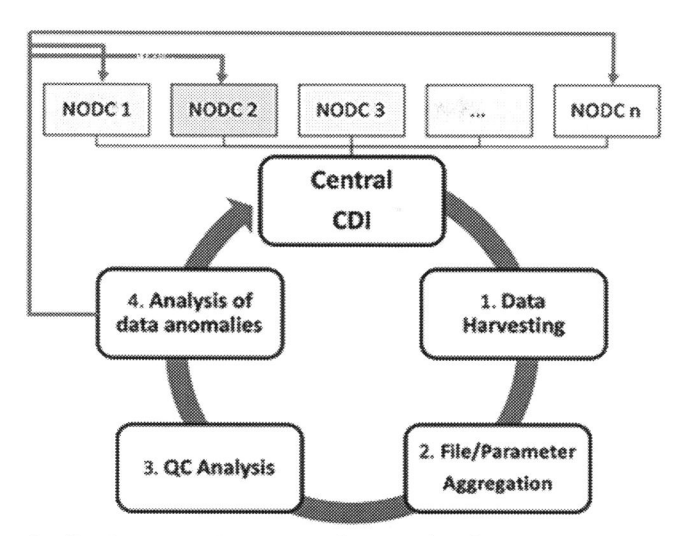

Figure 4.5 Quality Assurance Strategy implemented within SeaDataNet infrastructure.

workflow. The secondary QC is conducted by experts at regional level in order to harmonize data quality mainly by visual inspection thanks to the *Ocean Data View* software (ODV) (Schlitzer 2002; https://odv.awi.de/). The first resulting aggregated datasets for sea temperature and salinity covering all the European sea basins (Arctic Sea, Baltic Sea, Black Sea, Mediterranean Sea, North Sea, and North Atlantic) were released for the first time in 2014 and have undergone several updates within QAS cycles, the last two (Simoncelli et al., 2019, 2021) in the framework of SeaDataCloud project (https://www.seadatanet.org/About-us/SeaDataCloud). These data collections form the basis for the generation of added value and synthesis products, as the SeaDataCloud temperature and salinity climatological mean fields (Simoncelli et al., 2020a,b), and are available through the Sextant web catalog (https://www.seadatanet.org/Products#/) together with their associated Digital Object Identifier (DOI) and Product Information Document (PIDoc) reporting the products' generation methodology, main characteristics, validation, usability, and technical specifications to facilitate users' uptake.

Fig. 4.5 can also be applied to the WOD model, with some differences. The main difference is that the WOD is not a distributed system. So, in a first stage (1) data harvesting is actualized through domestic (U.S.) and international data exchange mechanisms. Data exchanged from NODCs through the IOC IODE as well as other sources, domestic and international, are archived in their original form at NOAAs National Center for Environmental Information (NCEI). This ensures that the data as close to original (first recorded/calculated) form are preserved for reference and verification. This also ensured the long-term archival of data not specifically the province of any NODC, such as from the CLIVAR and Carbon Hydrographic Data Office (CCHDO) and the Global Temperature and Salinity Profile Program (GTSPP). In a second stage (2), data aggregation becomes a large task of format conversion and metadata extraction and interpretation. The nondistributed nature of stages (1) and (2) comes at a cost in complexity and resources devoted to data conversion as depicted in Fig. 4.6. Stages (3) and (4) are performed within the WOD system for the specific goal of calculating the World Ocean Atlas climatological mean fields of ocean variables. There is also a community-wide effort to provide stage (3) and stage (4) automated and expert quality control and anomaly detection following (and setting) international best practices. This effort of the International Quality Controlled Oceanographic Database (IQuOD) is systematically and statistically setting the criteria for the best set (and sequence) of automated quality control for ocean profile data (Good et al., 2020), setting uncertainty values on each measurement (Cowley et al., 2021), intelligent metadata (essential metadata not included in the original data files, but discerned from the nature of the data and included metadata (Palmer et al., 2018), and a system and network for expert anomaly detection with a

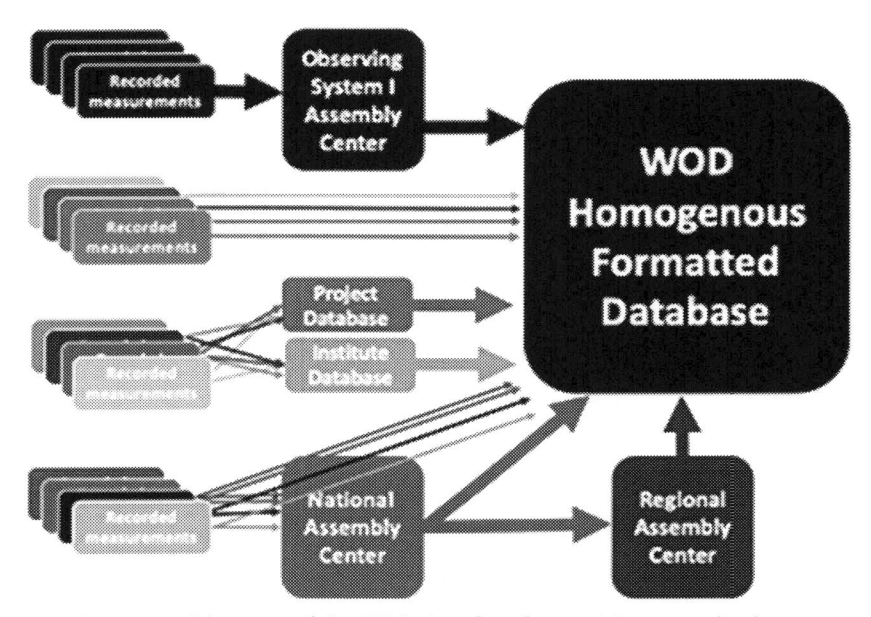

Figure 4.6 Schematic of the WOD data flow from originator to database.

machine learning element (Castelao, 2020). The Central CDI for the WOD is somewhat independent from the system and there are multiple public access points for the data, including the WOD select system and the NOAA Big Data Program cloud service (https://registry.opendata.aws/noaa-wod/).

Online data services and coproduction

Recent years have seen rapid and unprecedented advances in network and cloud technologies, both, in terms of network hardware and storage systems as well as in terms of new software tools, protocols, and transmission standards. This has led to huge increases in network bandwidth, cloud storage capacity, and, most importantly, new online software tools. One such new technology that is changing the way we are using the internet are Websocket connections providing permanent connections between a server and a client that allow sending messages in both directions with extremely low latency. This allows, for the first time, online services with truly interactive user experience.

These technological advances drive a fundamental change in the way we are managing and using data. While applying to all areas of research and society, this is especially true for the environmental and marine sciences. Largely driven by the need to document and understand the ongoing environmental and climate changes, oceanographers and IT specialists are building new systems that give online access to very large and high-quality datasets to wide user communities, ranging from scientists to policymakers and the general public.

The overall aim is to create and release large aggregated datasets based on many multinational data streams and containing data for a wide range of parameters, including classical hydrographic data, nutrients, carbon, and a variety of tracers and contaminants. As detection of temporal changes is a key research goal, extending the temporal coverage is crucial. This might require tedious and time-consuming browsing of archives, and discovery and rescue of valuable historical data. Examples of such data aggregation activities are the pan-European hydrographic, eutrophication, and contaminants data collections created within the SeaDataCloud and EMODnet Chemistry (https://www.emodnet-chemistry.eu/) projects, the WOD and various CMEMS datasets. In addition, many large thematic aggregated data collections are created and maintained by groups of scientists and computer experts in self-organized, bottom-up initiatives. Examples are (1) the GLODAP hydrographic and carbon data for the global ocean (Olsen et al., 2020), (2) the MEOP Marine Mammals Temperature and Salinity

Data 2004–17 (Treasure et al., 2017), (3) the SOCAT global fCO2 dataset 1957–2020 (Bakker et al., 2016), and (4) the Global Transmissometer Database V3.2020 (Gardner et al., 2020). All these datasets are invaluable resources for environmental and climate change research.

The typical, traditional use of these community datasets has been to download the respective aggregated data collection to the user's computer and analyze the data locally with installed software. As many of the datasets are very large, this approach has serious drawbacks and involves long and tedious downloads as well as significant storage resources on the local machine. Even worse, whenever a dataset changes (many data products are updated annually) the local copies become outdated and need to be replaced with downloads of the updated dataset.

With recent technological advances in place, there is now a better approach: store and maintain a single copy of the dataset in the cloud and provide online services for data access, subsetting, extraction, analysis, and visualization. This relieves the user of the tedious download and software installation tasks and reduces the hardware requirements on the user side enormously, thus allowing using these services on new types of hardware, such as tablets and smartphones. Updating a dataset is now as simple as replacing the single dataset copy on the server with the new version.

Following this new strategy, scientists at the Alfred Wegener Institute in Bremerhaven, Germany, have developed webODV, an online version of the popular ODV software (Schlitzer 2002; https://odv.awi.de/). As its main components, webODV provides two services: (a) data exploration and (b) data extraction.

The webODV data explorer service uses the new ODV–online interface, which mimics the user interface of the ODV software in the web browser. Look-and-feel of the browser interface matches the ODV interface, and virtually all functionality as well as graphics types of ODV are supported. This makes it very easy for the large group of previous ODV users to get started with ODV-online and the webODV data explorer. An example image of the webODV data explorer browser window with explanations of the interface elements is shown in Fig. 4.7.

The *webODV* data extraction service allows users to subset aggregated datasets in many different ways and to extract these data subsets in a variety of formats. The data extractor guides users through a sequence of steps: (1) select the station subset, (2) select the variables to be included in the output, (3) visualize the selected data, and (4) perform the download of the selected data. Stations can be selected by cruise name, geographical

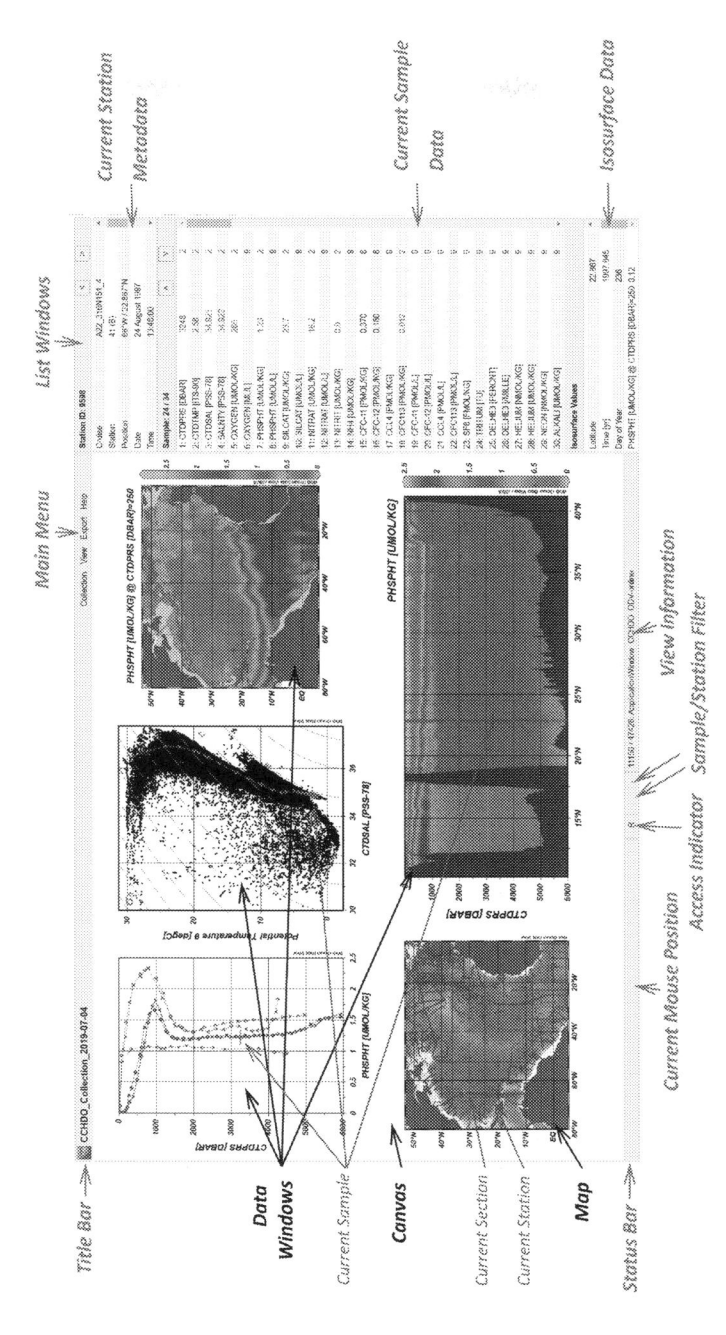

Figure 4.7 Example image of the webODV data explorer browser window with explanation of the interface elements.

domain, date/time, and availability of data for specific variables. Supported output formats are spreadsheet (ASCII text), ODV collection, and netCDF.

The *webODV* services are operational and employed in the *SeaDataCloud* Virtual Research Environment (VRE;https://vre.seadatanet.org/), in the EMODnet Chemistry project (https://emodnet-chemistry.webodv.awi. de/), the GEOTRACES data services (https://geotraces.webodv.awi.de/), and on the *webODV* Explore website (https://explore.webodv.awi.de/) that lets users explore many popular environmental datasets.

The SeaDataCloud cloud environment (Fig. 4.8) hosts a replication of SeaDataNet database content which improves data access service, and it offers a collaborative environment to perform data-driven research. Its pilot version has been codesigned in order to offer an efficient working experience through expert users' engagement that provided recommendations about the VRE and its auxiliary software functioning, such as *webODV* for supporting online QA/QC and DIVAnd (Barth et al., 2014) for computing climatological gridded fields. The potential is to facilitate database monitoring, QA/QC processes, workflow automation, publication, and speed up thanks to the codeveloped softwares. The VRE will also permit the application of machine learning algorithms for supporting expert's QC of increasing amounts of data, whose visual inspection is becoming unfeasible. Mieruch et al. (2021) is an example of such application.

A coproduction process with the additional possibility to integrate external datasets will deliver higher quality products. The VRE final user will benefit from a new working experience with ready-to-use high quality data, softwares, and workflows to train and further advance research.

In parallel to the SeaDataCloud, the WODcloud (WODc) project aims to incorporate the international community into this vital step, allowing data

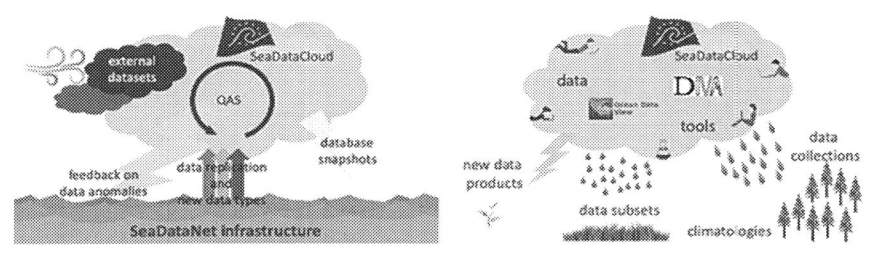

Figure 4.8 Schematic of SeaDataCloud Virtual Research Environment: (left) the SeaDataNet infrastructure content is replicated in a cloud environment for data access service optimization; (right) customized data and tools for data analysis, visualization, and products' generation provide the users a new working experience.

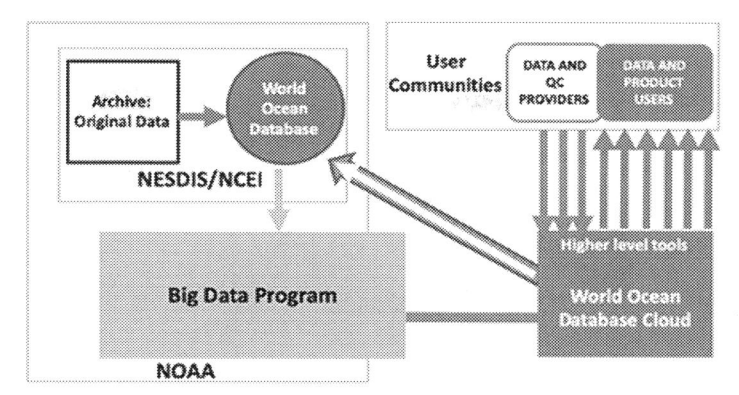

Figure 4.9 Schematic of the WOD cloud.

providers to upload their data to the cloud in a form which can be both easily incorporated into the WOD and which can also be immediately used in concert with the WOD within tools for oceanographic and related research. The WOD will remain an international resource managed at NCEI, and final quality checks and metadata will be provided at NCEI, but data will be available for use jointly with the WOD immediately upon upload to the WODc and the community will not need to wait for quarterly (3-month) updates of the WOD or for the deeper quality control involved in the release of WOD versions in concert with release of WOA versions.

WODc (Fig. 4.9) is a system in development which will allow the user to upload their dataset into the cloud and utilize them immediately along with the actively managed WOD and any other datasets similarly uploaded to the cloud space. The cloud space will contain open-source tools for analysis and also be accessible to the users. Data uploaded to the cloud space will be aggregated in the actively managed WOD following aggregation procedures (see above). So, the WOD data ecosystem, the IQuOD, and the WODc will combine to realize the initial goal of the WOD—complete easy, free, and open access to the most comprehensive uniformly formatted, quality-controlled, historic and recent oceanographic profile data, enhancing the utility and coverage of the global ocean observing system.

Gridded products

The level of complexity and the information content of data products increase moving along the marine data value chain (Fig. 4.4). Once a validated dataset or data collection has been assembled and harmonized, it

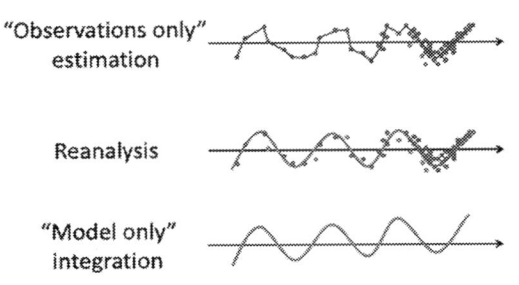

Figure 4.10 Schematic view of past ocean reconstruction methodologies.

can be used for multiple purposes and applications. Many applications require mapping sparse data in space and time onto a regular 2D or 3D grid, eventually embedding gap-free algorithms. This procedure can be applied to in-situ and/or remotely sensed observations through different techniques, obtaining synoptic or climatological gridded fields, which can be used for ocean monitoring, process-oriented studies, model initialization, or validation. Monthly objective analyses for the global ocean computed from in-situ temperature and salinity profiles are example products from CMEMS and Met Office Hadley Center (Good et al., 2013). These kinds of products can be discontinuous and noisy, highly depending on the space-time distribution of observations; however, the associated uncertainties tell the users where the products are more or less reliable.

Validated in-situ or remotely sensed data records are then used for model assessment or can be assimilated in ocean general circulation models in near real time for forecasting activities or in delay mode within retrospective experiments designed for climate applications, such as ocean reanalyses. Reanalysis products are characterized by dynamical balances, higher temporal resolution, up to daily scales, and temporal continuity, since the model hydrodynamic equations project forward in time the initial information and data assimilation constrains the model solution close to the available observations (Fig. 4.10).

The output of both observations-only estimates and reanalyses is used to derive ocean monitoring indicators and investigate the evolution of the ocean state.

Observation-only gridded products

In order to derive gridded products, observations must be generally aggregated over a chosen time scale. Depending on the application and data density, different options are possible. For some measurement campaigns (or permanent measurement networks) a very high spatial coverage can be achieved over a local area. In such areas a quasi-synoptic view (all observations are thus assumed to represent the same time instance) can be derived. For larger areas the spatial resolution is often not sufficient and averages have to be computed over a larger time span. Common choices are monthly or seasonal fields averaging the data from all years but handling the different months or seasons separately. Long-term averages (e.g., decades) are usually referred to as climatologies. Beside assessing the mean ocean state or a climatological average year, there is also often the need to assess long-term changes. This can, for example, be addressed by combining data over several years for a given month or season (using, e.g., a sliding window of decade). This allows us to isolate the effects of the seasonal cycle and long-term changes.

Climatologies allow us to assess the mean state of the ocean and to detect anomalous events relative to the mean state. It also allows to decompose the variability into different time scales (such as seasonal and decadal), which can more easily be attributed to the dynamics of the ocean and the atmospheric forcing action on the ocean.

Ocean variables sampled at sparse locations need to be mapped on a regular grid for many applications, as if the user needs to compute derivatives (e.g., to compute geostrophic currents) or integrals of the field (such as the ocean heat content). The gridded ocean fields are also used in the context of ocean modeling to initialize ocean models and to validate them. One can detect biases in ocean models due, e.g., to their limited resolution or inappropriate parameterization. It is also common to reduce drift (biases increasing over time) in models via nudging (or other form of data assimilation).

Challenges to create gridded data products

Several challenges are to be faced when creating gridded ocean data products. In particular, ocean observations tend to be relatively sparse and with inhomogeneous distribution (more data near the coast than in the open ocean). Also, the presence of (undetected) outliers and sampling biases (due to inhomogeneous spatial/temporal sampling) do have in general a detrimental effect on the gridded data product. Beside measurement errors, the observations typically do not represent the same spatial and temporal scales of a climatological mean field. In particular, observations are generally instantaneous measurements while climatologies are long-term means. This discrepancy is generally often called representation or representativity errors (Daley, 1993).

To some degree sampling biases can result in a representativity error but outliers should be typically discarded in the derivation of the gridded data product. High-resolution dataset (e.g., a time series at a high temporal frequency) might have correlated representativity error. Either this correlation is taken into account explicitly or the weights of such observations have to be reduced by artificially increasing their expected error variance. Alternatively, these high-resolution datasets can be subsampled or binned (i.e., reducing the resolution by averaging).

When merging different data collections in order to increase spatial and temporal resolution and coverage, another issue to be considered is the handling of duplicates that may substantially alter the quality of a gridded product. In general, as stated before, the measured values, the coordinates, the time, and other metadata are sufficient to detect if two observations are duplicates. However, this might not be so straightforward if observations have been interpolated or rounded to different decimal values.

General concepts and methods

Different methods exist to generate gridded fields, but they share some common concepts. Generally, the techniques use a first guess or a background estimate, which can be an average of all observations or a reference climatology. The gridding method then uses anomalies (observed value minus the background estimate) and the background estimate is added to the interpolated field to obtain the final field. The observed values are typically spread over a given length-scale which defines the correlation length. Sometimes, this parameter is only defined implicitly. This length-scale can be either estimated from theoretical consideration by choosing, for example, the Rossby radius of deformation as the dominant length-scale or by fitting an assumed covariance function (e.g., Thiebaux, 1986; Emery and Thomson, 2001) as illustrated in Fig. 4.11.

Various methods exist in oceanography to generate gridded data products (in particular climatologies) from in-situ observations. Among the most widespread are the Cressman analysis, optimal interpolation, and the variational inverse method with different degrees of complexity. The Cressman method (Cressman, 1959) performs a weighted sum of the observation (or anomalies of observations) for every grid point. The weights depend on

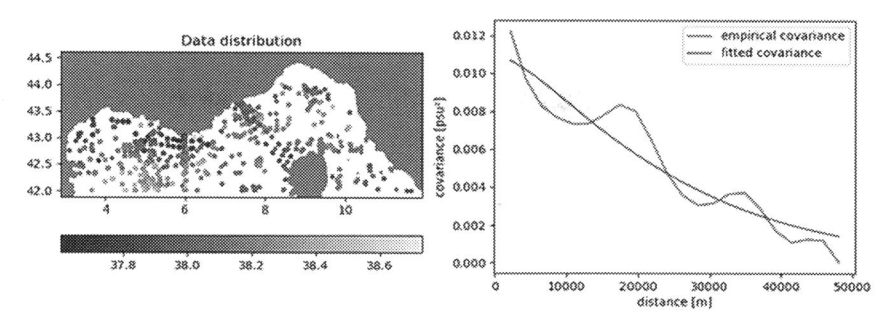

Figure 4.11 (Left panel) Data distribution of surface salinity observations extracted from the World Ocean Database 2018 (WOD18, Boyer et al., 2018) for January in the Gulf of Lion (North-west Mediterranean Sea). (Right panel) The pairwise covariance of data points grouped by distances (empirical covariance) and the fitted covariance model to estimate the correlation length.

the distance between the observation and the considered grid cell. In the Barnes scheme (Barnes, 1964), these weights decrease following a Gaussian radial basis function based on a given length scale. The Cressman method with the Barnes scheme is, for example, used in the World Ocean Atlas 2018 (Locarnini et al., 2018; Zweng et al., 2018).

Optimal interpolation (sometimes also called statistical interpolation) aims to combine in an optimal way a background estimate and observations (Gandin, 1965; Bretherton et al., 1976; Carter and Robinson, 1987). Assuming the expected error covariance of the observations and the background estimate are known, one can derive the optimal linear combination of both minimizing the expected error variance.

Variational inverse methods aim to minimize a cost function requiring that the interpolated field is close to the observations, close to a background estimate, and "smooth" (which can be formalized by derivatives of the field). This technique is implemented in tools like DIVA (Troupin et al., 2012; Beckers et al., 2014) and DIVAnd (Barth et al., 2014). The variational inverse method is for appropriately chosen error covariances equivalent to optimal interpolation, despite the difference of the initial formulation. In optimal interpolation the expected error variance of the analysis can be computed efficiently than for the variational inverse method. The variational inverse method can more easily integrate additional constraints, for example, advection constrain (isolines of tracers approximately aligned with ocean currents) and decouple water masses based on the land–sea mask. Dynamical constraints have been examined in the context of creating gridded data products of ocean surface currents from high-frequency radar current (Barth et al., 2021). While the formalism can be extended by adding linear constraints, nonlinear constraints such as the inequality constraints (e.g., water column stability requiring that the density should increase monotonically with depth) or positivity (concentration should be positive) cannot be easily implemented in this framework. For concentrations of chemical parameters one can however transform the data using a nonlinear function such as the logarithm (or an empirical determined function) so that the transformed variable is approximately Gaussian distributed. This approach is referred to as *Gaussian Anamorphosis* (Wackernagel, 2003), which is often used for, e.g., chemical parameters.

Uncertainty of gridded products

Besides measurement errors, there is also a representativity error affecting the observations. For example, when a monthly climatology is created, the difference between an instantaneous value (of a particular year) and the monthly mean (for all years) is called *representativity error*. These errors in the observation (potential measurement errors and representativity error) have an impact on the uncertainty of the gridded products.

Another source of uncertainty, in most cases an even larger contribution, is due to gaps in the spatial or temporal data distribution. The expected error variance can be derived from optimal interpolation and by extension also from variational inverse methods. However, it should be noted that several relatively strong assumptions are necessary (mainly about the background error covariance) to derive fields representing the uncertainty (Beckers et al., 2014). The expected error fields give a qualitative indication of the accuracy of the interpolated field and can be used for instance to mask the areas where the analysis is above a certain error threshold (usually 30% or 50%) and thus is less reliable. A more robust way to assess the accuracy is to validate the data products with independent data. In the cross-validation method a (random) subset of observations is set aside to perform the validation (Wahba and Wendelberger, 1980; Brankart and Brasseur, 1996; Beckers et al., 2014). It is necessary to ensure that this data subset is independent of the data used for the analysis. The choice of how this subset is generated can have a significant impact on the validation statistics. For instance, if a fraction of a high-resolution vertical profile is set aside for validation, then the RMS error of the analysis influences the retained data points from the high-resolution profile and the validation data would be artificially low.

However, it is also important to keep in mind that the cross-validation error includes the contribution for the error of the analysis and the error of the observations (with the representativity error). Only for synoptic observations, the latter can be considered as small.

The validation of climatological gridded fields can be conducted also through consistency checks or comparison with reference products (e.g., Troupin et al., 2010; Iona et al., 2018), providing users with information about the suitability of a specific product for an intended use. A consistency check can be conducted qualitatively by visual inspection or quantitatively by computing metrics, e.g., the Root Mean Square Difference (RMSD) and the bias. Fig. 4.12 provides an example of consistency analysis between the SeaDataCloud Mediterranean Sea climatology (Simoncelli et al., 2020), computed over the 1955–2018 period at 1/8th of degree of resolution and the reference global product WOA18 at 1/4th degree (WOA18 averaged decades, Locarnini et al., 2018) in which bias and RMSD give quantitative indication in which months and on which levels the two products exhibit the largest differences, and the maps give visual insights of where the main differences arise. In this example the main differences can be ascribed to the smaller scales and coastal gradients that the regional SDC climatology can resolve with respect to the global WOA18 climatology.

Example workflow for gridded product generation in SeaDataNet and EMODnet

A climatology can be defined as a set of gridded fields for different variables, typically sea water temperature and salinity, over a period covering several years. The creation of gridded fields from in-situ measurements constitutes a good example of added-value product useful for a wide range of users. The DIVAnd tool has been developed (Barth et al., 2014) and successively uptaken in the framework of SeaDataCloud and EMODnet Chemistry projects to generate climatologies. Thanks to DIVAnd implementation in conjunction with Jupyter notebooks (https://jupyter.org/) and its codevelopment via GitHub collaborative coding, it is possible to continuously advance the workflow efficiency, transparency, and the derived products' quality. In fact, the whole workflow, from data to product, can be reproduced thanks to the different persistent unique identifiers (PIDs).

Starting with the input data and metadata, PIDs are attributed to each measurement, allowing one to easily retrieve the information relative to the data provider, cruise and project names, institute, principal investigator, among other fields. The benefit is twofold: firstly, data providers know if

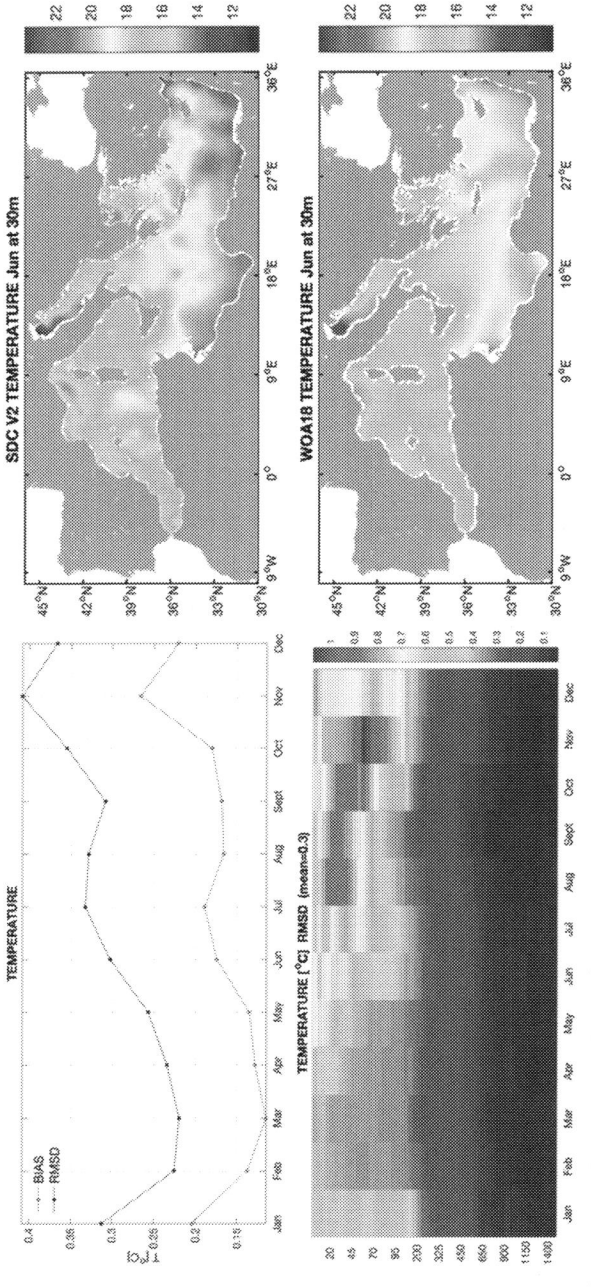

Figure 4.12 Consistency check of the SeaDataCloud Mediterranean Sea temperature climatology 1955—2018 (Simoncelli et al., 2020) the World Ocean Atlas 2018 (WOA18 averaged decades, Locarnini et al., 2018): (top left panel) basin averaged bias and Root Mean Square Difference (RMSD) monthly time series; (bottom left panel) Hovmoller plot of basin averaged RMSD computed along the water column on the overlapping vertical levels in the whole Mediterranean Sea; (right top panel) SDC temperature climatology at 30 m of depth in June; (bottom right panel) WOA18 temperature climatology at 30 m of depth in June.

their data are used in a given product, thus giving them more visibility and acknowledgment; secondly, the final product users can provide a feedback on the data quality, for instance if suspect measurements are detected, and this feedback is brought to the corresponding data provider.

Ideally, the software tool employed to generate the products should also be citable; this is why a Semantic Versioning Specification is used. In addition, a DOI is minted for each new release of the software tool. This is done through the Zenodo platform, which allows an automatic management of the DOI for software code stored in GitHub. Concretely, this translates to a new metadata field for the products, which indicates the version of the software tool that was run on the input data.

The resulting fields are stored as Network Common Data Form (netCDF, Rew et al., 1990) files. The choice of such a format was obvious for several reasons: from a programmer perspective: the availability of libraries in various programming languages (Fortran, Python, Julia, …) to read and write netCDF files; from a final user perspective: the variety of software tools to visualize netCDF files (ncview, ncBrowse, Panoply, FERRET, …); from a data manager perspective: the ease to distribute the files with frameworks such as OPeNDAP or ERDDAP. Web Map Services (WMS) can also be added to the data server so that users can visualize the content of the files on an interactive map (Leaflet, OpenLayers).

In the context of recent European initiatives, *Virtual Research Environment* (VRE) has been set up and deployed. Their goal is to allow users to access data, tools, and computing resources in a seamless manner. In the frame of SeaDataCloud project, users are provided with the regional data collections of temperature and salinity (https://www.seadatanet.org/Products, Simoncelli et al., 2021) that they can combine with their own sources of data. In addition, tools such as ODV (Schlitzer, 2002) or DIVAnd (Barth et al., 2014) are deployed in the VRE (Fig. 4.7), thus making it possible for users to perform advanced analysis on the data and create ad hoc products.

Satellite products

Earth observation (EO) satellites have the unique capability to allow a synoptic and regular observation of the Earth surface. Ocean remote sensing (also referred to as *satellite oceanography*) can provide a systematic global mapping of the whole ocean surface at relatively high spatial and temporal resolution covering the blue (physics), green (carbon and biogeochemistry), and white (sea ice) ocean. Infrared and/or microwave satellite observations can supply global (and regional) mapping of ocean variables as, e.g., sea surface temperature and chlorophyll-a concentration up to twice a day, a characteristic not feasible by using in-situ measurements only.

Currently, EO satellite sensors allow the measurement of five primary variables: sea surface temperature, surface roughness, sea surface elevation (or sea surface height), sea surface salinity, and ocean color. Additionally, a wide range of high-level physical and biological quantities can be derived. Ocean color, i.e., the visible radiation (400–700 nm) emerging from the sea surface (or water-leaving radiance), allows the retrieval of biological variables as, e.g., the concentration of chlorophyll-a and the colored dissolved organic matter. The measurement of the sea surface elevation, based on radar altimeters, allows determining surface geostrophic currents and significant wave high. Surface roughness, namely the measurement of the backscattered radar signal, allows wind speed estimation at the sea surface and waves' direction.

The provision of these (and new) satellite-based products has been a demanding task in the past decades, responding to the increasing emerging policy and user needs. On the one hand, near-real-time satellite observations are essential elements of integrated (typically with in-situ and model data) operational Earth observing systems, used, e.g., for maritime safety, marine resource, and marine and coastal environment monitoring. On the other hand, the availability of long-term satellite-based data records allows the analysis and monitoring of the present state of the ocean with respect to the past, interpreting changes and trends in the marine environment. Last, but not least, satellite observations are needed to validate model analysis and reanalysis, constrain and/or initialize meteorological and ocean forecasting systems.

Satellite products are classified according to their processing level that ranges from the so-called Level-0 to Level-4. Level-0 (L0) are raw (unprocessed) sensor data, as received from the satellite. L0 data are then converted by the agencies responsible for the sensor (upstream data providers) to higher levels, namely into more useful variables. Level-1 (L1) are image data in sensor units, as, e.g., top-of-atmosphere radiance, brightness temperature. The next processing step converts L1 data into derived geophysical variables (L2), as, e.g., water-leaving radiance, chlorophyll-a concentration, sea surface temperature. L2 provides geolocated data but usually given in swath coordinates. Level-3 (L3) provides image data generated by regridding L2 data onto a regular latitude-longitude grid. Regridding can be performed in different ways, as, e.g., through interpolation of the L2 observations found within the final L3 grid cell. Collating (merging) single-sensor or multisensor L3 data provides the so-called Level-3 Collated (L3C) and super-Collated (L3S) products, typically available as daily (day and night) files (Fig. 4.13). Since each sensor has its own accuracy and retrieval characteristics (such as spatial resolution, viewing geometry, etc.), a bias-adjustment procedure is

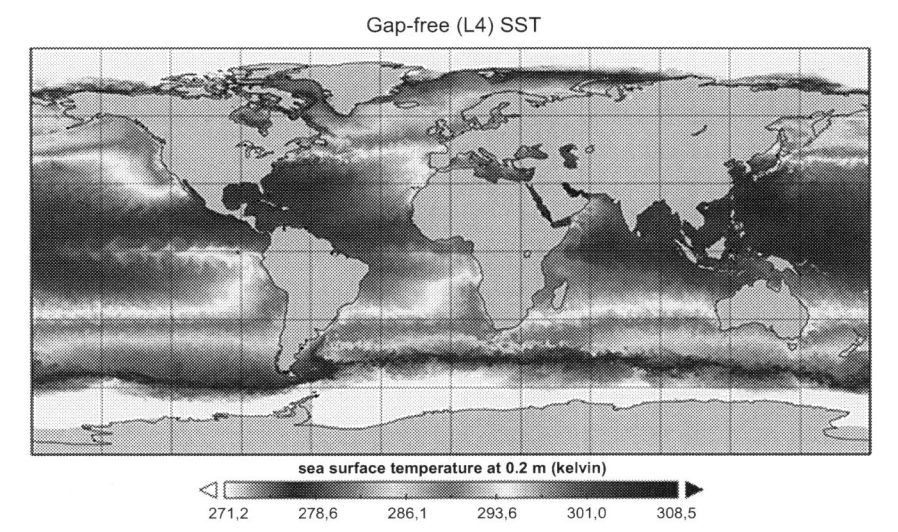

Figure 4.13 Examples of AVHRR-19 global daily composite (L3C) and gap-free (L4) SST data.

generally adopted when creating a L3S file. In practice, the most accurate sensor is chosen as reference while the others are bias-adjusted to the reference (e.g., Buongiorno Nardelli et al., 2013). Finally, since many satellite products are generally affected by several data voids (due to clouds, rain, land, sea-ice, etc.), the last processing step builds gap-free (L4) data, by using lower level (L2 and/or L3) data (Fig. 4.13). Gap-free data are typically obtained by optimal interpolation (Buongiorno Nardelli et al., 2013) or variational assimilation techniques (e.g., Good et al., 2020).

The CMEMS represents one of the main examples of how satellite observations have been exploited to derive and disseminate high-level user-driven products (Le Traon et al., 2019). Observations (including in-situ data) are a fundamental pillar of the value-added chain of CMEMS, which derives from upstream satellite observations (L1 and/or L2 products) downstream high-level (L3 and/or L4) products. CMEMS satellite-based products include sea surface temperature and salinity, sea surface height, ocean color, sea ice, winds, and waves, covering the global ocean and the European Seas. Compared to global products, regional products typically provide added values, as, e.g., enhanced spatial resolution and/or improved quality, due to local (regional) algorithms' parametrization. All these products are generally provided as near-real-time (e.g., provided on a daily basis) and/or multi-year (long-term) L4 data in order to be directly usable for downstream applications.

In any case, the timely availability and quality of satellite products are highly dependent on upstream data providers. Among them, ESA, EUMETSAT, NASA, and NOAA process on the ground and deliver worldwide satellite data in real time. For a comprehensive ocean remote sensing review, we refer to Robinson (2004).

Satellite-based sea surface temperature retrieval

Sea surface temperature (SST) is an Essential Ocean Variable (EOV), since it is at the base of many physical processes and operational applications, and plays a key role in regulating the Earth's climate system. Indeed, SST modulates and responds to the exchanges of heat, momentum, and freshwater at the interface between the ocean and the atmosphere, which in turn modifies the horizontal and vertical temperature gradients of the upper ocean. SST responds to climate variability (see, e.g., Deser et al., 2010) becoming an Essential Climate Variable (ECV) that allows to assess, monitor, and predict the impact of climate change. SST is also required in weather (meteorological and oceanographic) and climate forecast systems, used, e.g., as boundary conditions and/or assimilation schema (e.g., Chelton and Wenz, 2005). It is therefore essential to have accurate SST data at both global and regional scale, and available as near-real-time and long-term datasets to accomplish both operational and climate applications.

SST has been routinely measured from space since the late 1981 by a variety of satellites and sensors. The basic principle of satellite-derived SST observation is based on the measurement of the radiation spontaneously emitted by the sea surface that, in the ideal case of a blackbody, would follow Planck's law. Then, from a measure of the radiation emitted by the sea surface it is possible to derive the brightness temperature (Tb), which differs from SST by the emissivity of the sea surface and the effects of the atmosphere, which attenuates the radiation through the processes of scattering and absorption. Specific atmospheric correction procedures are required when deriving SST from Tb. This is the processing step that converts brightness temperature data (L1 product) to SST data (L2 product).

SST is typically measured by infrared and microwave radiometers. Infrared radiometers, such as the AVHRR (Advanced Very High Resolution Radiometer) and AATSR (Advanced Along Track Scanning Radiometer) sensors, can provide accurate (up to $0.1-0.3K$) SST measurements at high spatial resolution (up to about 1 km) in clear sky conditions (see, e.g., Embury et al., 2012). Indeed, cloud coverage does not allow any SST measurement in the infrared channels. This limitation is overcome by the all-weather microwave sensors, which can "see" through clouds (with the limitations imposed by land and rain, see for example, Gentemann et al., 2009).

There are different definitions of SST according to the instrument used to measure it. Among them, we recall the skin, subskin, and foundation SST. The skin SST is the water temperature measured by an infrared radiometer and is representative of the first microns from the sea surface. The subskin temperature is typically associated with the measurement of a microwave radiometer and representative of the first millimeter from the surface. Beneath the subskin, SST is referred to as depth SST (where depth takes the value of the depth). Finally, the foundation SST is the temperature free, or nearly free, of any diurnal temperature variability (diurnal cycle). For more detail about SST definitions, we refer to the Group for High Resolution SST (GHRSST) (http://www.ghrsst.org/), and for a comprehensive SST review to Minnett et al. (2019).

SST products examples

There is a variety of satellite-based SST datasets, characterized by different spatial and temporal resolution, temporal coverage, archived and distributed by several institutions and international agencies. Examples of well-known long-term SST datasets covering the satellite era (1981—present) include Global Ocean OSTIA Sea Surface Temperature and Sea Ice Reprocessed (Good et al., 2020); ESA_CCI and C3S Reprocessed Sea Surface Temperature Analyses (Merchant et al., 2019); NOAA Daily OISST v2.1, previously known/referred to as Reynolds SST analysis (Huang et al., 2021); and Hadley Center Sea Ice and Sea Surface Temperature (HadISST1) (Rayner et al., 2003).

These datasets combine satellite only (ESA CCI-C3S SST), and satellite along with in-situ temperature measurements (OSTIA, NOAA OISST, HadISST) to generate gap-free SST fields at global scale. All these datasets are designed to provide accurate SST estimates at high spatial and temporal resolutions to fulfill the requirements of a large variety of users. Table 4.1 summarizes the basic characteristics of each dataset.

OSTIA and ESA CCI-C3S SST reprocessed products provide daily gap-free (L4) maps of foundation and 20 cm SST, respectively, and sea ice concentration at 0.05 degrees × 0.05 degrees regular grid (approximately 5—6 km) from the end of 1981 to 2019 using in-situ and infrared and microwave data. While OSTIA is built from infrared and microwave satellite data and in-situ data, ESA CCI-C3S makes use of infrared satellite observations only, namely the (A)ATSR, SLSTR, and the AVHRR series of sensors. These are operational products developed by the UK Met Office and distributed through CMEMS and the Climate Data Store (CDS) of the Copernicus Climate Service (C3S).

The NOAA Daily OISST dataset, previously known as Reynolds SST analysis, consists of global daily gap-free maps of SST at 20 cm depth at 0.25 degrees × 0.25 degrees regular grid from 1981 to present. This dataset is routinely produced by NOAA/NESDIS/NCEI and publicly provided at https://www.ncdc.noaa.gov/oisst/. Finally, the Hadley Center Sea Ice and Sea Surface Temperature dataset (HadISST1) provides global gap-free maps of monthly SST and sea ice concentration data at 1 degrees × 1 degrees regular grid from 1870 to present (https://www.metoffice.gov.uk/hadobs/hadisst/, Rayner et al., 2003).

These datasets can be used for fundamental climate applications compatible with the length of each time series, such as long-term monitoring of

Table 4.1 Descriptive product comparison summary for satellite-based SST datasets. Input observations are derived from satellite infrared (IR), microwave (MW), and/or in-situ measurements.

Dataset	Institution	Type of product	Time range	Observation input	Type of SST	Horizontal grid spacing	Temporal resolution
OSTIA	Met office	L4	1981/10/01—2019/12/31	IR + MW + in situ	Foundation SST	Global 0.05 degrees × 0.05 degrees	Daily
ESA CCI-C3S SST (v.2.0)	Met office	L4	1981/09/01—2019/12/31	IR	SST at 0.2 m	Global 0.05 degrees × 0.05 degrees	Daily
NOAA daily OISST (v.2.1)	NOAA	L4	1981—Present	IR + in situ	SST at 0.2 m	Global 0.25 degrees × 0.25 degrees	Daily
HadISST1	Met office	L4	1870—Present	IR + in situ	—	Global 1 degrees × 1 degrees	Monthly

SST changes, as for example, trends (see e.g., Pisano et al., 2020) and comparison to or boundary condition for numerical models. Other target applications include the use of these datasets in the definition of climatic indices (e.g., ENSO indices), assessment, and monitoring of weather extreme events, including marine heatwaves (e.g., Bensoussan et al., 2019) and their impact on marine ecosystem and related services.

Users are encouraged to consider also the type of SST offered by each producer, distinguishing between, for example, skin, subskin or 20 cm depth SST, and foundation SST according to the specific application for which the data are meant to be used.

The new generation of satellite sensors, as for example, those on-board the European Space Agency (ESA) Sentinel satellites (e.g., Donlon et al., 2012), and future missions, as for example, the Copernicus Imaging Microwave Radiometer (CIMR) mission (Kilic et al., 2018), allow continuity and evolution of satellite-based applications providing data of improved quality and spatial resolution.

An important aspect related to satellite remote sensing is the validation of its measurements. Huge effort is dedicated to properly estimating uncertainties in satellite products (Loew et al., 2017). A common validation framework is to use in-situ data as reference measurements. Specifically, a matchup dataset of colocated (in space and time) satellite and in-situ measurements is built. Then, a set of statistical metrics, such as mean bias and rootmean square difference (RMSD), is computed from the differences between satellite and in-situ matchups, and used to quantify the accuracy of the satellite product.

Overall, satellite-based observations are an essential tool for both scientific and operational applications. Though satellite observations are representative of the sea surface only, the combined use of in situ and observations along with, e.g., artificial intelligence algorithms and/or modeling, and data assimilation allows the reconstruction of three-dimensional ocean fields (see, e.g., Sammartino et al., 2020; Bell et al., 2015).

Ocean reanalysis

Another family of gridded ocean product is the ocean reanalysis (i.e., ocean retrospective analysis, also known sometimes as ocean synthesis). Ocean reanalyses are climate-oriented tools, which combine oceanic observations with an ocean general circulation model through data assimilation (e.g., Stammer et al., 2016). Indirectly, reanalyses also account for atmospheric observations through the sea surface boundary conditions, usually provided by atmospheric reanalyses, which are the analog of ocean reanalyses in meteorology.

The fundamental tools of the ocean reanalyses (observation preprocessing, data assimilation schemes, numerical ocean prediction models) are shared with operational oceanography. However, as reanalyses aim at reconstructing the past climate of the oceans in a way as temporally consistent as possible, their uniqueness consists in the adoption of the same version and configuration of such tools during the simulation of the reanalyzed period. Reanalyses therefore differ from operational oceanography that aims at having the best performing analysis and forecast possible at present through incremental implementation of model and data improvements, and the ingestion of all newly available observations (see schematic in Fig. 4.10). Reanalyses also differ from model-only simulations (hindcasts), which are blind to the oceanic observations but preserve full temporal continuity of output fields, and thus conservation of the seawater properties, and to the statistical mapping of observations (objective analyses), which usually exhibit stronger adherence to the observational datasets but fail in depicting physically balanced multivariate fields.

The popularity of ocean reanalyses has increased over time, due to several methodological advances and the recent possibility to reach eddy-resolving resolution even in global products, thanks to the increase in computational power. Both these factors in turn increased the confidence the climate community has on them. Ocean reanalyses thus have joined the catalog of climate services in several national and international programs. Reanalyses are being updated in near real-time, and there is a plan for instance within the CMEMS to transition toward real-time production of reanalyses (with 1 month or even shorter period of latency), possibly updating the reanalysis once delayed time data become available. This is an important step forward toward the routine use of reanalyses for climate monitoring, compared to the production of reanalyses in the 2000s and 2010s, generally performed on a "once in a while" basis. However, processing of high-quality in-situ and satellite observations has intrinsic delays; for instance, accurate orbital data for the processing of sea level anomalies from altimeter missions are generally estimated with delays of several months, and delayed-time products become available about 6 months after the observations. This suggests that interim reanalysis products, which timely use real-time data and are later rerun once the high-quality data are available could likely combine the need for timeliness and maximum accuracy of different stakeholders.

Possibly, ocean reanalyses use delayed-mode and research-quality observational data and boundary forcing datasets (namely atmospheric reanalyses for the sea surface boundary conditions and, in case of regional products, larger-scale ocean reanalyses for the lateral boundary conditions) to ensure the highest quality level and accuracy stability. There exist a large variety of reanalysis products, either global or regional, performed at coarse or eddy-resolving horizontal resolution, spanning only the altimetry satellite era or covering the entire 20th century. Recent experimental applications can reach even longer time periods, owing to the ingestion of proxy data rather than observations from instrumented platforms (e.g., Widmann et al., 2010). Historically, the production of reanalyses has been fostered by their use for the initialization of coupled models in long-range (subseasonal to multiannual) prediction systems, and now they became a commonly adopted climate tool, used also to force downstream applications such as off-line biogeochemical simulations, Lagrangian dispersion models, etc.

Among the goals of ocean reanalyses, the correct reproduction of low-frequency climate signals—e.g., interannual trends of key climate variables, changes in large-scale circulation features, etc.—stands as a crucial objective, at the same time complex and challenging, which in general operational oceanography is not concerned with. The success of reanalyses in these regards is hampered principally by irregular and intermittent observational sampling, causing spurious and thus nonphysical variability, and systematic errors in the ocean model configuration and/or in the forcing datasets. In order to attenuate the impact of the irregular observational sampling, conservative solutions are often adopted in the configuration of the ocean reanalysis, considering that the majority of data assimilation schemes are nonconservative and may thus accommodate sharp changes in the reanalyzed variables. For instance, the majority of reanalyses ingest gridded (gap-filled) sea surface temperature (SST) products (L4, see earlier) rather than swath or regridded data (L2 or L3, respectively) in order to preserve temporal and spatial continuity of the assimilated SST dataset. SST observations are in turn important to maintain a temporally homogeneous correction of the radiative fluxes at the sea surface and the implied mixed layer variability. Indeed, several ocean reanalyses exploit SST data to correct the air-sea heat fluxes rather than assimilating directly such data to correct the interface temperature whose constraining effect could last shorter. In such a case, it is critical to avoid temporal discontinuities in the heat flux constraint that could amplify in the subsurface.

Another example is the use of model-derived mean dynamic topography (MDT) in the assimilation of altimetry data for ocean reanalyses spanning periods longer than the altimetry era (Storto et al., 2011). The MDT is the height that sea level anomalies are referenced to, namely the stationary component of the ocean topography due to permanent circulation patterns. Using a realistic MDT but inconsistent with that of the ocean model may cause abrupt changes, especially in the barotropic circulation, when altimetry data start to be available at the end of 1992. More, in general, transient processes may arise when new observation types are assimilated, leading in turn to aforementioned spurious variability, notably evident in ocean transport and eddy kinetic energy diagnostics whose temporal variability is generally small in reality (Yang et al., 2017). For instance, to overcome such kind of issues within 20th century reanalyses when bathythermograph profiles become available slowly after World War II, it was proposed (Giese et al., 2016) to assimilate only sea surface observations—whose availability dates back to the 19th century.

An important ingredient of ocean reanalyses for mitigating systematic errors in ocean general circulation models or forcing datasets is the use of bias correction schemes that, unlike most atmospheric data assimilation systems concerned with biases in the observations, are targeted to the correction of model biases. Obviously, estimating such biases is not straightforward because of the lack of reference data; bias correction can be achieved for instance by using either climatological or time-varying gridded observational products (Balmaseda et al., 2007), or the iterative use of analysis increments from previous reanalysis runs (Canter et al., 2016). Bias correction shall be tuned to correct systematic errors without flattening out the climate signal, which in turn requires dedicated sensitivity analyses (Storto and Masina, 2016). Model bias correction is particularly relevant when deep ocean biases and drifts originate from suboptimal background-error vertical covariances (e.g., Lellouche et al., 2018), used in data assimilation to spread bottomward the informative content of upper ocean observations, or in case of spurious stratification/enhanced convection coming from the intermittent application of the analysis increments. As the ocean below 2000 m of depth is practically unobserved and unconstrained, bias-correction is the only possible approach to keep realistically low variability in the deep and abyssal oceans. Bias correction is also important to prevent that systematic errors in the atmospheric forcing (e.g., Janowiak et al., 2010) cause long-lasting misrepresentation of the upper ocean water masses.

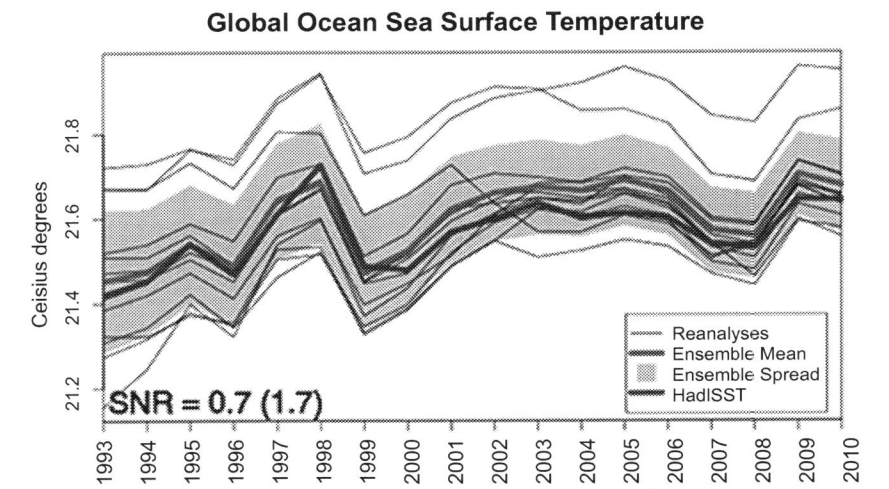

Figure 4.14 Time series of global ocean sea surface temperature from the ORA-IP inter-comparison project (Balmaseda et al., 2015). The plot shows the ensemble mean and standard deviation and the individual ocean reanalyses. The reference HadISST (Rayner et al., 2003) dataset is shown in blue (black in printed version). The signal-to-noise ratio (SNR) is defined here for simplicity as the ratio between the temporal standard deviation of the ensemble mean divided by the temporal mean of the ensemble standard deviation. The numbers in parentheses refer to the SNR ratio computed from anomalies. *From (Storto et al., 2019). https://doi.org/10.3389/fmars.2019.00418.*

Validating ocean reanalyses has often been performed through cross-comparison between several reanalysis products, where the underlying assumption is that the consensus of the products indicates reliability. The consistency check can include a reference dataset as in Fig. 4.14 extracted from Storto et al. (2019). These intercomparison exercises showed indeed that parameters directly observable and well-sampled by the global ocean observing network are in good agreement across the reanalyses, while those that are scarcely observed are not (Balmaseda et al., 2015).

Unfortunately, there exist very limited (independent) verifying datasets for reanalyses, which generally consist in local data from observational campaigns, observations from drifters, tide gauges or high frequency radars that may not be assimilated because of their sparseness for basin-scale or global applications. There is however an increasing request for fit-for-purpose validation of ocean reanalyses, generally fostered by local communities or specific downstream applications, which calls in turn for a closer cooperation between global and basin-scale observational communities and coastal observatories. The assimilation of synthetic observations, a technique inherited by observing system simulation experiments (OSSE, e.g., Halliwell et al., 2017) to assess the impact of forthcoming observations through extraction of synthetic observations from another realistic realization of a reanalysis, is another possibility to explore the performance of a reanalysis system with respect to generic or specific target diagnostics. The use of synthetic observations is increasingly used in the ocean reanalysis community to shed light on strengths and weaknesses of single or multiple products. In a different context, reanalyses are also used as synthetic ground truth in experiments where the effect of different mapping strategies on objective analyses is assessed (e.g., Allison et al., 2019).

Partly linked to the reanalysis validation challenge, the adoption of the ensemble approach (single- or multimodel) emerges as a sensible way to provide climatic time series with associated uncertainty ranges, estimated from the ensemble dispersion. Provided that differences in ensemble member formulations (e.g., model physics, assimilated or forcing dataset, etc.) span several sources of structural uncertainty of reanalyses, the ensemble average operation is in general able to reduce systematic errors that can compensate each other (Palmer et al., 2017). To this end, multimodel ensemble means have proved superior to any individual member in a number of global ocean studies (e.g., Storto et al., 2017). Recently, the adoption of super-ensemble analyses, where several reanalyses are combined with observation-only products, has been used in the context of ocean monitoring indexes, providing encouraging results for some targeted applications (von Schuckmann et al., 2018; Storto et al., 2019).

Societal challenges products

The global increase in sea-based activities, the overall expansion in maritime transport, the growing coastal urbanization, and the foreseen increase in offshore activities may threaten marine ecosystems through several kinds of physical, chemical, and biological disturbances and through contamination by hazardous substances. The need to effectively manage growing human activities in order to achieve and preserve Good Environmental Status (GES *sensu* MSFD, Marine Strategy Framework Directive, European Commission, 2008) and to guarantee a sustainable use of the maritime space, requires the correct understanding of ocean conditions as well as of the pressures and impacts of human activities. Multidisciplinary information is required to address the multiple societal needs of guaranteeing sustainability, safety, and security of human activities, as well as good environmental conservation. The development of midstream and downstream services and applications, from either public bodies and private enterprises in support of decision-making processes, frequently needs the contribution of non-ocean data and/or information products, as schematized in Fig. 4.4. The integration of ocean products with multidisciplinary information strengthens the necessity of data and metadata harmonization for their rapid uptake and exploitation. This section presents some examples of societal challenges products with the aim of discussing the difficulties and the way forward of collecting and integrating some key ocean and non-ocean data for societal benefit.

Bathymetry is the basic information to support planning of most maritime activities, to assist navigation safety, to monitor coastal erosion, sea level changes. It is also crucial to define high resolution models' topography and increase their reliability, especially in the coastal area. However, notwithstanding its consolidated importance, there are still many marine areas that are poorly mapped (Ryabinin et al., 2019), representing a serious risk for navigation accidents and consequent pollution dispersion, but also a limitation for the predictability of hydrodynamics and multihazards in coastal areas.

Marine pollution from land-based as well as from sea-based sources is also a long-known problem, tackled within multiple national, regional, and global frameworks (OSPAR, HELCOM, UNEP/MAP, MEDPOL, MARPOL). Nevertheless, growing number of substances released in the environment, the fate of persistent organic pollutants (POPs), and emerging contaminants still represent challenges for pollution assessment, as required by Regional Sea Conventions, by European environmental directives (e.g., European Commission, 2000 - WFD, European Commission, 2002 - ICZM, European Commission, 2008 - MSFD, European Commission, 2014 - MSP) and by the European Environmental Agency (EEA), at least in the European framework. Assessment of marine pollution at subregional and regional scale involves the use of data from multiple sources and the possibility to combine diverse datasets to obtain practical and understandable data products required by different kinds of stakeholders and decision-makers.

Monitoring of contaminant levels in the marine environment is essential for the evaluation of environmental status and to identify pressures to the marine ecosystem, and thus to enable management and reduction of pollution sources. Observations must provide information on spatial and temporal variability and trends in concentration. The objective is to verify progress toward GES (European Commission, 2008) and to evaluate if pollution is progressively being phased out, i.e., the presence of contaminants in the marine environment and their biological effects are kept within acceptable limits, so as to ensure that there are no significant impacts on or risk to the marine environment. Ensuring GES, however, has more than just environmental implications. It is a crucial requirement for a healthy ocean economy as well. However vast, the ocean and its resources are still finite. Therefore, preservation also responds to the objective of ensuring "the sustainable use of marine goods and services by present and future generations".

The possibility to map, at the same time and with the same approach and tools, pollution status (e.g., from data on anthropogenic litter or on chemical contaminants) and sources of pollution (i.e., distribution of human activities possibly responsible for pollutant release such as maritime traffic, touristic activities) may greatly support identification of the main hazards for marine ecosystem status and may support correct management procedures (Liubartseva et al., 2018; Prevenios et al., 2018). Furthermore, integrating data products of marine pollution with products on the spatial and/or temporal distribution of human activities together with spatial information on meteo-oceanographic conditions (e.g., winds, ocean currents) can greatly assist to identify sources, predict the fate of discharge and pathways of anthropogenic impacts, and to design better pollution prevention measures (Sepp Neves et al., 2016, 2020).

Hence, mapping human activity in the ocean has become especially important. On the one hand, it consents to gain a better understanding of pollution sources and their cumulative impact; on the other, it enables a better understanding of the complex set of interactions and synergies between competing uses of the ocean to which the MSP initiative intends to respond. The overlapping of different thematic layer maps would provide an optimal tool for anticipatory planning and decision-making. Unfortunately, while over the past few decades considerable progress has been made toward mapping of physical, chemical, biological, and geological parameters, mapping human activity is not as equally advanced. This might be largely due to the fact that there seems to be no or very few uncoordinated, international, national, or regional communities that act as a point of reference for human activity in the ocean. While physics, chemistry, geology, biology, hydrography, oceanography, etc., are well-established scientific disciplines with a long-standing tradition of cooperation between scientists and institutions, the same cannot be said for human activities. As a consequence, mapping efforts are more recent and largely uncoordinated. Recently, initiatives such as EMODnet, which includes a thematic group on human activities (https://www.emodnet-humanactivities.eu/), have considerably improved the situation in terms of agreed standards and common vocabularies. At the same time, the road ahead is still long and more effort is needed to obtain a comprehensive mapping of human activity in the ocean.

Bathymetric maps

Although knowledge of bathymetry has always been considered important, many marine regions are still not adequately mapped in the year 2020 as required by science and industry, based on the General Bathymetric Chart of the Oceans (GEBCO) data and products (https://www.gebco.net). The Baltic Sea was mapped at 90% of its area, but some portion of East Asia was mapped at 8% (https://www.gebco.net/about_us/committees_ and_groups/scrum/mapping_projects/). The lack of bathymetric data with an adequate high spatial resolution compromises the achievement of UN's Sustainable Development Goals n. 14 (https://sdgs.un.org/goals) and to achieve the ambitious goal of having a 100% mapped ocean with high spatial resolution, a collaborative project between Nippon Foundation of Japan and GEBCO was established in 2018 (https://seabed2030.org/). The project is also conducted under the auspices of the International Hydrographic Organization, UNESCO-IOC, and sees the participation of Europe through the EMODnet program.

An historical overview of sea-sounding has been provided in Chapter 1. Knowledge of bathymetric maps was required by the *Directions for Observations and Experiments to Be Made by Masters of Ships, Pilots, and Other Fit Persons in Their Sea-Voyages* (Moray and Hook, 1667): "Though the art of navigation, one of the most useful in the world, be of late vastly improved, yet remain there are many things to be known and done, the knowledge and performance whereof, would tend to the accomplishment of it: as the making of exact maps of all coasts, ports, harbours, bays, promontories, islands, with their several prospects and bearings; describing of tides, depths, currents, and other things considerable in the seas: turnings, passages, creeks, sands, shelves, rocks, and other dangers."

In the modern view, Moray and Hooke's sentence contains many of the existing problems for the production of a good bathymetric map. There is a need to have deep ocean maps with sufficient accuracy for modern science and industry, but to have much more precise maps in coastal areas, up into estuaries, deltas, and rivers, considering intermittently flooded coastal/river areas and tidal variations.

The ancient method to measure bathymetric data was the sea-sounding with lead lines or premeasured cables described in Chapter 1. The bathymetric survey datasets originate from national hydrographic services, research institutes, governmental departments, and companies. In the era of electronic maps, there is a need to digitize hydrographic information that in many cases has been acquired over many years or even centuries.

Today, different devices are used to measure depth of the sea, from remotely operated near-bottom vehicles to satellites. The sensors on board the vehicles include acoustic, optic, or radar systems, and they can provide data at different spatial resolutions. However, Dierssen and Theberge (2014) underlined that bathymetry data collected using "remote sensing" methods are characterized by not static measurements, since tides in different locations and times can change bathymetry data up to several meters. Another problem with bathymetry maps is arising from different "chart datum," a reference water level generally derived from tide phases. Lowest Astronomical Tide (LAT) and Mean Lower Low Tide (MLLT) are the most common chart datums, but also other reference levels are used, such as Mean Higher High Water (MHHW). In seas having small tidal excursions, Mean Sea Level is adopted. Additionally, maps of surface elevation of rivers can be provided as percentile of water surface elevation (e.g., https://waterdata.usgs.gov/nwis/rt/).

An effective data management system and production workflow have to be designed and deployed in accordance with international standards and common practice to overcome the inherent diversity in terms of data management practices and the heterogeneity in bathymetry input data. This is the case of the European EMODnet Bathymetry project (https://www.emodnet-bathymetry.eu) initiated in 2008, which addresses the whole life cycle of the bathymetric information: from the elementary observation (soundings) up to the Digital Terrain Model (DTM) product delivery, as illustrated in Fig. 4.15.

Different approaches can be adopted to the creation of high-resolution bathymetric maps, the most common starting from integration of all available data. Traditionally, bathymetric maps are produced from data collected by ships with specialized hardware and time-consuming and expensive methodology. New digital image processing techniques have been developed, by employing "super resolution" methods, i.e., enhancing the resolution of maps by recovering missing details with "deep-learning" software (e.g., Sonogashira et al., 2020).

EMODnet Bathymetry adopted an innovative approach and combines satellite data, processed bathymetry datasets, and auxiliary data, providing from 2018 a high-resolution bathymetry (about 115 m × 115 m) for the

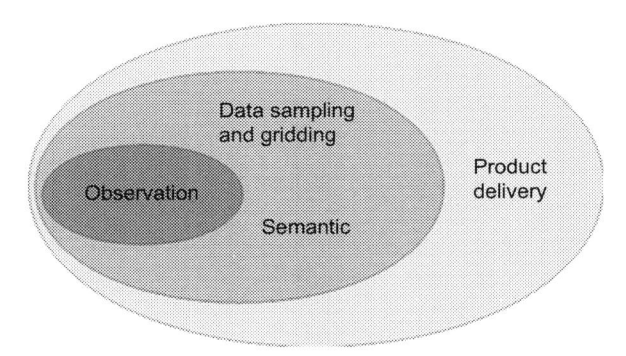

Figure 4.15 Life cycle of bathymetric information from observation to product delivery, as followed by the EMODnet Bathymetry consortium (https://www.emodnet-bathymetry.eu).

European seas. The DTM is based upon more than 9,600 bathymetric survey datasets, composite DTMs, and satellite-derived bathymetry data products, altogether from 48 data providers. Gaps are completed by using the GEBCO (General Bathymetric Chart of the Oceans) DTM which has a lower grid resolution of circa 450 m × 450 m. Each DTM grid cell contains a reference to the predominant source dataset used and its INSPIRE compliant metadata.

Fig. 4.16 shows the different steps of the generation of the EMODnet DTM and the inventory/distribution of individual sources of bathymetric information. Read the figure from the top, where source data are described (metadata production) and sampled using dedicated software (https://www.seadatanet.org/Software/MIKADO). Metadata are made available for population into common catalogs while sampled data are made available for compilation in a series of regional DTMs for (parts of) sea basins, prior to the validation and integration in the complete DTM and publishing on the web portal. Quality assessment and quality control are carried out at regional level by the groups of data providers. The regional DTMs are produced using a standard software (Globe, GLobal Oceanographic Bathymetry Explorer, Poncelet et al., 2020) that provides a complete data processing chain, from acquisition to generation of digital elevation models.

The resulting EMODnet DTM format has attributes per grid cell for minimum, maximum, average standard deviation of water depth in meters, number of values used for interpolation over the grid cell, and reference to the prevailing source of data with metadata. The final step from regional DTMs into overall EMODnet DTM is not only an integration but also a

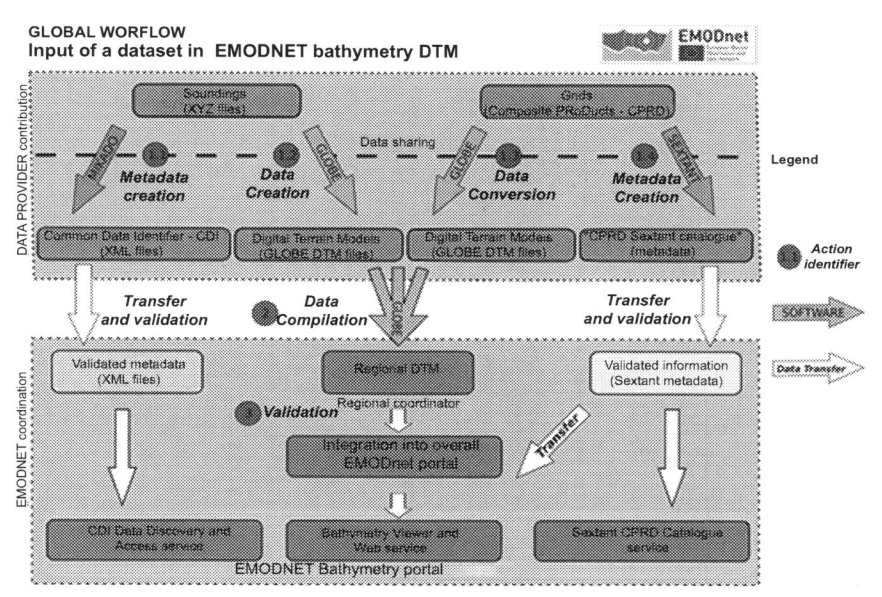

Figure 4.16 Overall workflow for the generation and publication of the EMODnet Bathymetry Digital Terrain Model (DTM).

final QA/QC to achieve a high-quality bathymetry product that can be made public at the portal for viewing and for downloading. It starts with a number of QA/QC actions on the received regional DTMs, followed by integrating the various regional DTMs in the overall products, which requires activities for spotting and avoiding edge-effects by local smoothing and other activities.

The S-44 document of the International Hydrographic Organization (IHO, 2008) describes the specifications of bathymetric surveys, with respect to the precision needed to ensure safety of navigation. However, many data providers of EMODnet Bathymetry are not hydrographic services and thus not subject to the use of the IHO S-44 standard as a support for the acquisition and processing of their bathymetric data. Therefore, EMODnet Bathymetry agreed on a classification of qualitative aspects of the source data, which provide the so-called quality index. Those aspects basically rely on the precision reached by the individual components of the system, characterized by the horizontal indicator (QI_Horizontal) and the vertical indicator (QI Vertical). Associated with these elements are the age of the survey indicator (QI_Age) and the purpose of the survey indicator (QI_purpose).

Pollution products

One of the most consolidated approaches to manage marine contaminant data in Europe relies on the pan-European infrastructure for ocean and marine data management (SeaDataNet, https://www.seadatanet.org/), which has developed standard vocabularies (e.g., Parameter Usage Vocabulary, P01 - see Chapter 3) implemented by the British Oceanographic Data Center (BODC), metadata profiles and formats, shared data quality control procedures, standard file formats, as well as dedicated software to guarantee data interoperability according to the FAIR principles (Wilkinson, 2016). In the framework of parallel initiatives, particularly focused on pollutants, e.g., EMODnet Chemistry (https://www.emodnet-chemistry.eu/) and Harmo-NIA (https://harmonia.adrioninterreg.eu/), an in-depth analysis of marine pollution data revealed the high complexity of this kind of variables, which implies the need of accurate and detailed metadata related to the matrix (e.g., water, sediment, biota) being monitored, to sampling methods and analytical procedures and quality assurance/quality control (QA/QC) protocols and, lastly, the need to harmonize measurement units, in order to allow data comparability and fitness for visualization purposes (Molina Jack et al., 2020; Berto et al., 2020). The standard vocabulary (P01), based on a semantic model which uses a defined set of controlled vocabularies (the semantic building blocks), allows keeping relevant metadata connected to the data. However, the high heterogeneity and complexity due to multiple combinations of chemical substances—matrix properties—sampling and analytical characteristics result in a huge number of P01 terms for the same chemical substance. As an example of data heterogeneity and complexity, 43 different terms (standard P01) are currently available to indicate the water concentration of one single substance, e.g., cadmium, one of the priority substances listed in most environmental directives (Directive 2013/39/EU — European Commission, 2013). Given this high heterogeneity and complexity, pollution data comparability and, ultimately, their suitability for regional and subregional data products are really challenging.

In order to tackle the complexity of data on marine pollutants, the implementation of pollution products requires the following approach (Fig. 4.17):

- Data and metadata standardization
- Parameter harmonization
- Data QC
- Mapping of spatial and/or temporal trends of comparable variables

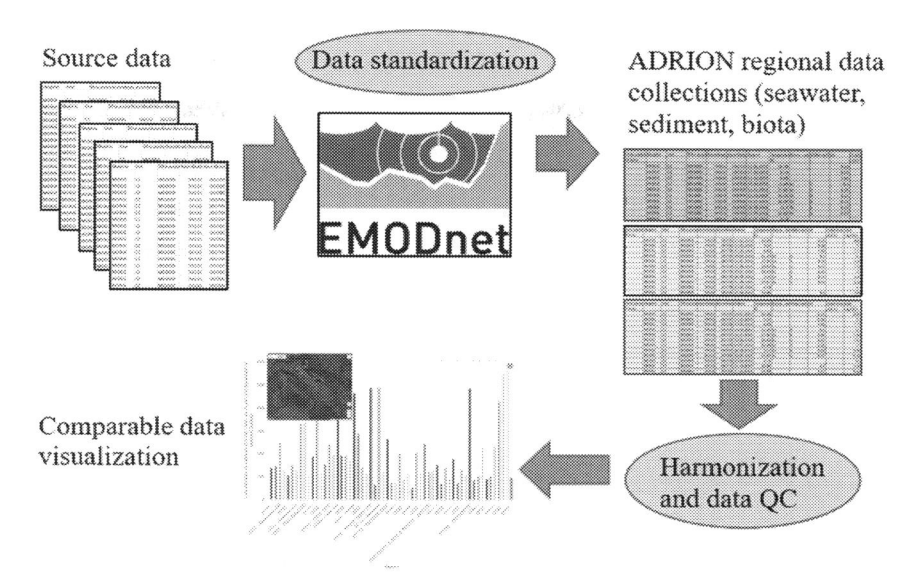

Figure 4.17 Approach for harmonization in data management and visualization. *Adapted from-(Lipizer et al., 2021).*

Data and metadata standardization implies use of consolidated protocols (Giorgetti et al., 2020) to produce interoperable metadata and data. Parameter harmonization includes measurement unit conversions (when possible) toward a limited set of common units, and harmonization of parameter coding, with P01 vocabulary.

While primary QA/QC is the responsibility of the data provider, the secondary QC of datasets provided by multiple sources is conducted at regional level (in analogy with SeaDataNet, see Fig. 4.5), relying on regional expertise of QC operators (Berto et al., 2020). The secondary QC comprises a preliminary format check of metadata completeness, correctness, and measurement units' consistency. Successively the regional aggregated datasets are subject to comparison with method Limit of Detection and Limit of Quantification, comparison with regional or subregional ranges of concentrations (when available), check for clearly impossible data ranges (e.g., different orders of magnitude) (Giani et al., 2020; Lipizer et al., 2020).

Due to the heterogeneity in chemical substances monitored in different countries, to the limited spatial availability of data and to the patchy and "hot-spot" distribution of marine contaminants, mapping marine pollution is extremely critical. Driven by the need to compare concentrations of pollutants in a region which hosts several, often overlapping, human activities, and in particular a large number of offshore platforms, the distribution of concentrations of hydrocarbons in the surface sediments, fluoranthene per unit of dry sediment is provided in Fig. 4.18, together with a map of monitoring stations and positions of offshore platforms as an example of a multidisciplinary data product, useful to detect pressures (offshore platforms distribution) and impacts (contaminant concentrations) on the marine environment.

With regard to data products on marine contaminants, there are not, at the moment, commonly agreed standards but, rather, highly heterogeneous and user-driven visualization outputs, targeting specific and diverse needs. In addition, heterogeneity and complexity of data of marine contaminants still limit data comparability and the possibility to merge data provided by multiple sources to produce regional and subregional scale products. Nevertheless, harmonized regional scale information, together with geospatial information on human-driven pressures (e.g., maritime activities), is highly needed for a sustainable management of the maritime environment, particularly in transnational areas such as in the European Seas; therefore, new types of products to further expand the marine environment monitoring capability and their usage are strongly needed.

Cross-discipline mapping, which takes into account socioeconomic and environmental aspects, provides an effective visual representation for decision-makers, managers, and policymakers on which to base their decisions and anticipatory planning (Noble et al., 2019; Lonsdale et al., 2020).

Human activity products

While standardized data management procedures are well consolidated in the fields of physical and chemical oceanography, this is not the case for human activities' data, which are crucial to identify the sources of pollution and to assess its impact on marine environment. Community efforts in established scientific disciplines have a long track record, but this is not the case for human activities, which in fact include a plethora of economic and noneconomic activities, not necessarily related to each other, which are grouped together for the simple reason that they all take place in the ocean. EMODnet Human Activities (https://www.emodnet-humanactivities.eu/) aims to facilitate access to existing marine data on activities carried out in EU waters, by building a single-entry point for geographic information on (at the time of writing) 17 different themes. An online portal makes available

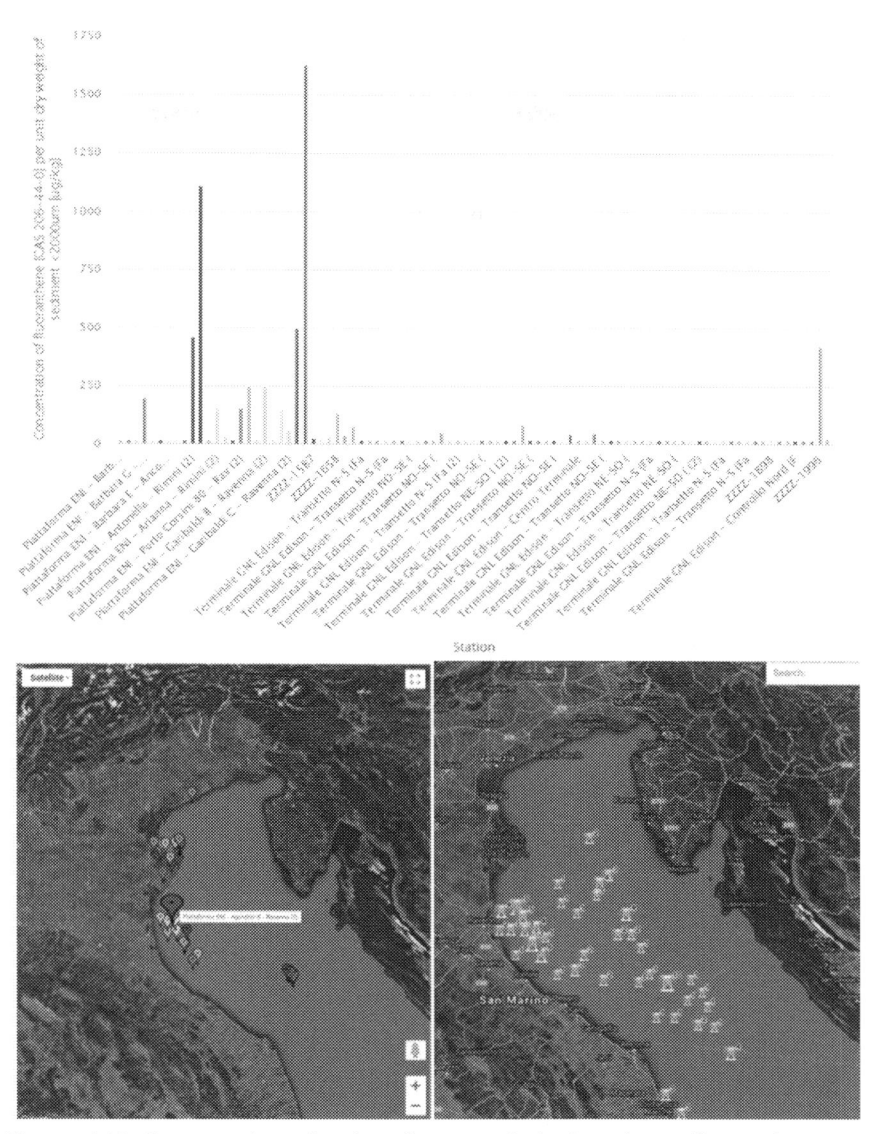

Figure 4.18 Concentration of polycyclic aromatic hydrocarbons (fluoranthene, μg kg^{-1}) per unit dry weight of sediment (top panel) in the stations indicated in the map (bottom left). The bar plot represents the median concentrations if multiple data are available for the same station. An interactive system at https://vrtlac.izor.hr/ords/harmonia/H_VIZUAL (last accessed: January 27, 2021) allows to visualize contaminant concentrations navigating on the stations. The distribution of offshore platforms from EMODnet Human Activities is shown on the bottom right map. *From https://www. emodnet-humanactivities.eu/view-data.php, last accessed: February 05, 2021.*

information on their geographical position, spatial extent, temporal variation, time when data was provided, and attributes to indicate the intensity of each activity.

EMODnet Human Activities does not carry out its own data collection: it simply integrates existing resources made available from current initiatives, and harmonizes data in terms of parameters, units of measurement, vocabulary, and language. In the framework of SeaDataNet, a P08 (see Chapter 3) vocabulary for human activities has been set up and is regularly updated. Moreover, EMODnet Human Activities performs QA/QC checks on data in terms of metadata curation, data format, data standards compliance, geographic location, coordinate system, error detection, accuracy, data aggregation. The data are aggregated and presented so as to preserve personal privacy and commercially sensitive information, a peculiarity of this kind of data. The data also include a time interval so that historic as well as current activities can be included.

Vessel density maps are an example of a data product that can respond to many societal challenges, because the oceans are a hub of human activity, and maritime transport is so widespread that it inevitably affects anybody working in or with the ocean. While maritime transport is believed to be the most environmentally friendly transport mode per ton/kilometer, it also is a source of pollution by petroleum hydrocarbons, antifouling biocides, litter, noise pollution, air pollution, introduction of alien species, etc. In addition, the "ubiquitous" presence of ships in the ocean is such that any other human activity needs to take them into account. Laying pipelines and cables, developing offshore wind farms, installing aquaculture cages are all processes that require careful planning, lest ship routes might interfere with them.

Vessel density maps show the distribution of ships (i.e., of maritime traffic), based on the instantaneous number of vessels per unit area, such as a square kilometer, a square degree, etc., and are often generated starting from ship positions retrieved from the Automatic Identification System (AIS). AIS is an automatic ship transponder system used on board certain vessels and is used by Vessel Traffic Services (VTS). The system was conceived to assist vessels' watch standing officers and allow maritime authorities to track and monitor vessel movements for purposes such as collision avoidance, maritime security, aid to navigation, search and rescue, etc. After its introduction it has found further uses, e.g., in fishing fleet monitoring and control and maritime traffic studies, including production of vessel density maps. Vessels fitted with AIS transponders can be tracked by AIS base stations located along coast lines or, when out of range of terrestrial networks, through a growing number of satellites that are fitted with special AIS receivers which are capable of deconflicting a large number of signatures.

EMODnet Human Activities maps were created by ingesting terrestrial and satellite AIS data, and calculating the time spent by each single ship in each cell of a grid (details on the method at https://www.emodnet-humanactivities.eu/documents/Vessel%20density%20maps_method_v1.5.pdf). Fig. 4.19 shows the vessel density in the Adriatic Sea, the same region of Fig. 4.18, as a complementary information on the probable pollution provenance. Depending on the area to be analyzed, the process itself requires considerable computing power, as well as a set of AIS data which in many cases has to be purchased from a commercial provider. For these reasons, researchers and/or private companies might find it considerably expensive and time-consuming to create a vessel density map; hence they often resort to prepackaged products available on the market. In this sense, the freely available EMODnet vessel density maps marked a milestone achievement, in that they enabled researchers and businesses to

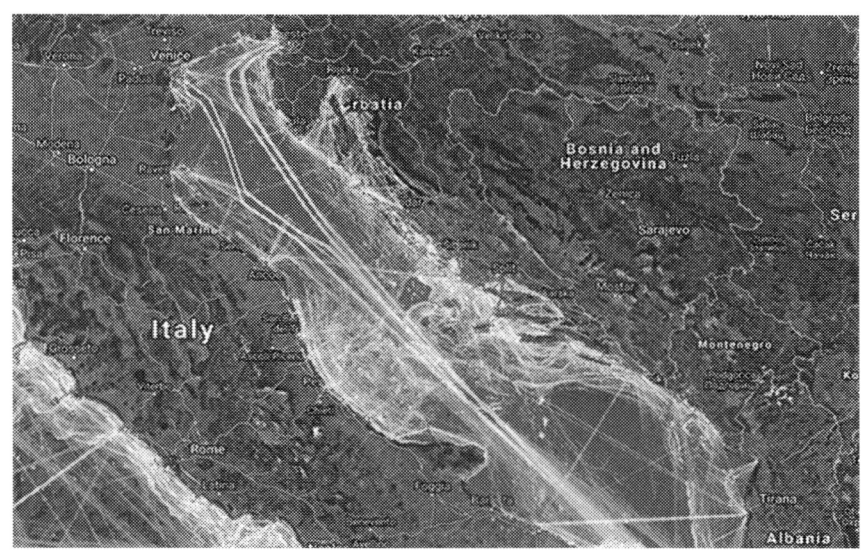

Figure 4.19 Vessel density in the Adriatic Sea, August 2019. The darker the color, the higher the density. Density is expressed in number of hours spent by ship in a 1 km² grid cell in a month. *From https://www.emodnet-humanactivities.eu/view-data.php, last accessed: February 18, 2021.*

rely on an open and well-documented product that serves many ocean research fields. For instance, EMODnet Human Activities' users have combined vessel density maps with a wide range of oceanographic and physical parameters to improve their modeling. By analyzing biological observations, currents, and ship traffic patterns, it is possible to gain a better understanding of how alien species are introduced in a marine region; ship traffic also influences the distribution of marine litter; noise data from stations can be combined with estimations of noise from shipping, so to interpolate noise data at a higher resolution; existing vessel traffic was also looked at as a potential vehicle for lime disposal to tackle ocean acidification in the Mediterranean (Butenschön et al., 2021).

Products quality and transparency

The quality of information and knowledge in ocean science is based on the adequacy of the integrated ocean observing and monitoring system to the societal needs. This concept encompasses the whole marine data value chain and all the actors playing a role along with it: data providers, data managers, information producers, and the users. Shepherd (2018) outlined that reuse of data collected by both private and public organizations can increase economic competition and innovation in emerging maritime sectors. The reuse of spatial data is easy when information about their quality and fitness for use is available, and when technical and legal barriers for integrating these into the user systems are removed.

The process of democratization of data, described by Buck et al. (2019), due to the public and free availability and accessibility of massive amounts of ocean data from many infrastructures, is changing the way data are converted into information and knowledge and how people use and gain value from them. An efficient exploitation of data content is mandatory to capitalize the data gathering investment; however, this necessitates an effective data management system. This also reverses the way how data and products are traditionally assessed, passing from a provider/producer-driven to a user-driven approach. The challenge is how to translate the societal needs into user requirements on the base of which to design/optimize a fit-for-purpose observing and monitoring system and derive fit-for-use data products. The user should also be made aware of data or product's properties and quality in order to properly use them for the intended purpose.

The users' community is vast (academia, public authorities, private sector, nongovernmental organizations) and the needs are very diverse depending

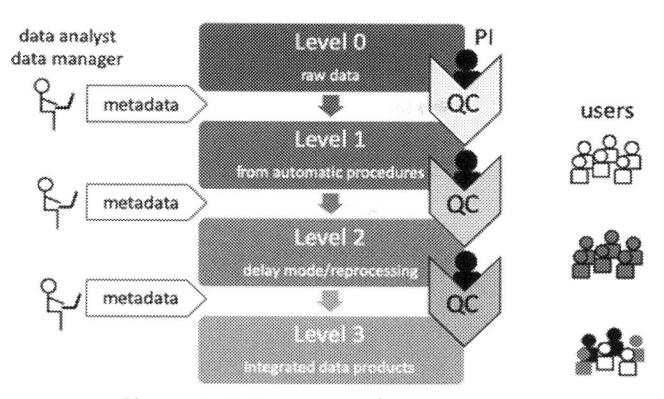

Figure 4.20 Taxonomy of data products.

on the level of complexity of the demanded products and their timeliness. Fig. 4.20 presents a tentative schematic of data and products taxonomy, which shows the many processes and the roles that enter into play along with the data life cycle and value chain according to the level of processing and latency. This schema has to be adapted to each data domain; it starts from the raw data, gathered from the data provider (level 0), to which a data manager associates a metadata description related to acquisition process (creation, content, and context) and the quality control procedure conducted by automatic procedures and/or the data expert.

Different types of users may access diverse level products and become intermediary, but all of them need the standardized descriptive metadata associated to both data and the processing in order to establish if products are suitable for a particular application. Data provenance and lineage are thus crucial information that guarantees full transparency along with the data life cycle and the traceability of data sources and producers. In fact, indexed data through unique persistent identifiers can be shared, integrated, and made publicly available through web services to be reused many times independently of their original purpose. The adoption of vocabularies and standards enables machine readability, integration, aggregation, tracking of data across datasets and derived products, on which the quality of data services depends. The recording and reporting of how data is analyzed computationally is also important for reproducibility of the workflow and assessment of final product quality, opening the issue of software availability.

A virtuous feedback loop between data providers, intermediaries, and users is established allowing a coproduction system devoted to convert ocean data into useful information for decision-making and capable to innovate and advance ocean knowledge at each iteration. Other derived advantages are the possibility to acknowledge all operators that contributed to the downstream product and monitor the products' uptake. The user-engagement (Mackenzie et al., 2019; Iwamoto et al., 2019) in a codesign and codevelopment process is finally the way to satisfy user needs and get maximum value from data in downstream products, applications, and services.

Product quality assessment

Nowadays decision-making by both public and private sectors is largely data-driven, namely based on data gathering, analysis, and derived information (i.e., from in-situ satellite and reanalysis). Zeng et al. (2019) state that also within the climate community it has become essential to assess the technical and scientific quality of the provided data and information products, including their value to users, to establish the relation of trust between providers of data and information and various downstream users. In order to avoid decisions based upon incorrect or misleading information, decision-making should include the conclusion of a logical process of quality assessment that provides a measure of the suitability of a product for a particular application (Batini et al., 2017; Bushnell et al., 2019; Zeng et al., 2019). A quantitative quality assessment methodology necessitates a common conceptual representation in the data domain or thematic, which allows to document it properly and completely. This conceptual representation is related to reference standards describing the full data life cycle and value chain. A utility assessment process is also essential for decision-making to translate user needs in conceptual requirements. In fact, while low-level user requirements could be satisfied only by subjective judgment (i.e., product visualization), high-level requirements related to decision-making process necessitate an accurate objective methodology.

Quality is defined by ISO9000 as the "degree to which a set of inherent characteristics fulfils requirements." A product of quality is obtained by defining and controlling its characteristics on the basis of what the potential customer intends to do. In short, quality means ability to achieve established objectives (effectiveness) and making the best use of available resources (efficiency). While quality assessment necessitates the handling of rich and meaningful metadata, the satisfaction of the "fitness for use" requisite deals also with the involvement of technical arrangements that ensure interoperability. Fichtinger et al. (2011) defines interoperability as "the possibility of

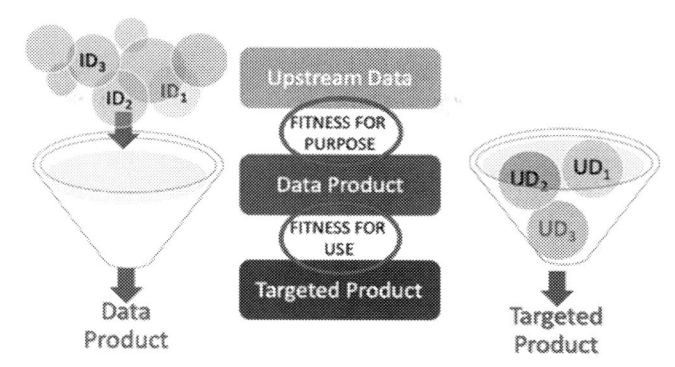

Figure 4.21 Schematic of some key concepts in the quality assessment process elaborated within the framework of the MedSea Checkpoint Project by Pinardi et al. (2018) (https://www.emodnet.eu/en/checkpoint/medsea). ID stands for Input Dataset and UD for Upstream Dataset.

datasets to be combined and for services to interact without repetitive manual intervention in such a way that the result is coherent and the added value of the datasets and services is enhanced." In fact, Toth et al. (2012) included interoperability as a fundamental user requirement in the elements characterizing data quality.

Fig. 4.21 summarizes some key concepts and terms related to the quality assessment process elaborated by Pinardi et al. (2018) within the framework of the MedSea Checkpoint Project (2013−2016). When designing a data product or application, it is necessary to define which input datasets (ID) are needed according to the demanded data product specifications and to select among all the available input dataset the ones that best fit the intended use. The selection of upstream datasets (UD) on the base of the producer data specification permits to generate the targeted data product. The agreement of the realized targeted product specifications with the theoretical product specifications determines its fitness for use and the fitness for purpose of its upstream datasets. Most complex or multidisciplinary products or applications need a larger number and different types of upstream datasets (in-situ, satellite, model-based forecasts, reanalyses, projections) demanding a wide community effort to define and adopt standard metadata documentation and quality information and finally implement a comprehensive product quality assessment framework (Zeng et al., 2019).

The conceptual product quality assessment framework developed starting from the EU INSPIRE Directive (https://inspire.ec.europa.eu/inspire-directive/2) and derived from the harmonization issue of spatial data and metadata (Fichtinger et al., 2011). The introduction of data specifications for data themes, containing data models and other requirements (e.g., encoding rules and spatial reference systems to use), created the possibility to combine data from heterogeneous upstream sources into integrated consistent information products in a way that is of no concern to the end-user. The first phase of data harmonization process consists of the definition of the target, e.g., data product specification (comprising a conceptual schema and other data characteristics) through the following steps:

- development and description of existing use cases;
- identification of user requirements;
- as–is analysis (analysis of existing data) including data harmonization issues;
- gap analysis;
- development of data product specification including application schema;
- implementation, test, and validation.

The transformation of data legacy to comply with product specification (effectiveness) needs efficient, cost-effective, and user-friendly tools that are nowadays spreading among the ocean science community and main marine data infrastructures thanks to FAIR data services (Tanhua et al., 2019).

Following Toth et al. (2012), the data product specification translates user requirements into data content and technical characteristics as defined by EN ISO 19131:2008 (data model, identifiers of the datasets, applicable coordinate reference and projection systems, encoding, portrayal, etc.). The data specification phase contributes to the metadata collection by specifying metadata for:

- discovery (identifiers of the dataset, producer, validity, etc.);
- evaluation and use (data content, scale, resolution, Coordinate Reference System-CRS, spatial representation, etc.);
- data quality and lineage.

The measure of quality consists in evaluating how far a product is from the ideal one, estimated in terms of "conformance," as defined by ISO 19105:2005, the fulfillment of specified requirements (Toth et al., 2012). The specified requirements (expressed in data content, technical characteristics, and data quality requirements) are compared with result values achieved in the production phase, and can be published as metadata elements expressed in terms of quality indicators.

Data quality dimensions

The quality of a product depends on the quality of the upstream data and methods used for it. The upstream data can be in-situ observations, remotely sensed data, or even model outputs. Data quality (DQ) is a multi-dimensional concept in which the data dimensions represent measurement attributes that can be assessed and interpreted (Satler, 2009; Wang and Strong, 1996; Malaki et al., 2012). Wand and Wang (1996) identified a certain number of data quality dimensions (Table 4.2) from the analysis of the representation mapping from states of the real-world system to states of the information system. Details on the quality dimensions in remote sensing were provided by Batini et al. (2017), who discussed the problems of approaching heterogeneous data sources and of providing a universally valid taxonomy of data quality dimensions.

There is not a consensus yet on a set of data quality dimensions in ocean science data. For example, Pinardi et al. (2018) in the framework of the EMODnet MedSea Checkpoint (https://www.emodnet.eu/en/checkpoint/medsea) used 17 partially different dimensions, and concepts were also adapted from those theorized by Wand and Wang (1996).

The development of INSPIRE implementing rules lead to more exact definition of dimensions on the base of ISO standards, such as (inter alia) ISO9000 family for product quality. ISO 19131:2007 specifies requirements for the specification of geographic data products, based upon the concepts of other ISO 19100 International Standards. It also provides help in the creation of data product specifications, so that they are easily understood and fit for their intended purpose.

Table 4.2 Data quality dimensions from Wand and Wang (1996).

Dimension		
Accuracy	Format	Comparability
Reliability	Interpretability	Conciseness
Timeliness	Content	Freedom from bias
Relevance	Efficiency	Informativeness
Completeness	Importance	Level of detail
Currency	Sufficiency	Quantitativeness
Consistency	Usableness	Scope
Flexibility	Usefulness	Understandability
Precision	Clarity	

Finally, ISO/IEC 25012 (https://iso25000.com/index.php/en/iso-25000-standards/iso-25012) provides a Data Quality (DQ) model composed of 15 characteristics (Table 4.3) that must be taken into account for the data quality assessment, understood as the degree to which they satisfy the requirements. This DQ model is divided in two parts:

Table 4.3 Data quality characteristics ISO/IEC 25012.

Inherent DQ	Accuracy	Syntactic accuracy	closeness of the data values to a set of values defined in a domain considered syntactically correct
		Semantic accuracy	closeness of the data values to a set of values defined in a domain considered semantically correct
	Completeness	The degree to which subject data associated with an entity has values for all expected attributes and related entity instances in a specific context of use	
	Consistency	The degree to which data has attributes that are free from contradiction and are coherent with other data in a specific context of use. It can be either or both among data regarding one entity and across similar data for comparable entities	
	Credibility	The degree to which data has attributes that are regarded as true and believable by users in a specific context of use	
	Currentness	The degree to which data has attributes that are of the right age in a specific context of use	
Inherent and system-dependent DQ	Accessibility	The degree to which data can be accessed in a specific context of use (including people who need supporting technology or special configuration because of some disability)	
	Compliance	The degree to which data has attributes that adhere to standards, conventions or regulations in force and similar rules relating to data quality in a specific context of use	

Table 4.3 Data quality characteristics ISO/IEC 25012.—cont'd

	Confidentiality	The degree to which data has attributes that ensure that it is only accessible and interpretable by authorized users in a specific context of use (ISO/IEC 13335-1:2004)
	Efficiency	The degree to which data has attributes that can be processed and provide the expected levels of performance by using the appropriate amounts and types of resources in a specific context of use
	Precision	The degree to which data has attributes that are exact or that provide discrimination in a specific context of use
	Traceability	The degree to which data has attributes that provide an audit trail of access to the data and of any changes made to the data in a specific context of use
	Understandability	The degree to which data has attributes that enable it to be read and interpreted by users, and are expressed in appropriate languages, symbols, and units in a specific context of use
System-dependent DQ	Availability	The degree to which data has attributes that enable it to be retrieved by authorized users and/or applications in a specific context of use
	Portability	The degree to which data has attributes that enable it to be installed, replaced, or moved from one system to another preserving the existing quality in a specific context of use
	Recoverability	The degree to which data has attributes that enable it to maintain and preserve a specified level of operations and quality, even in the event of failure, in a specific context of use

- Inherent data quality that refers to the degree to which quality character-istics have the intrinsic potential to satisfy stated and implied needs when data is used under specified conditions;
- System-dependent data quality that refers to the degree to which data quality is reached and preserved within a computer system when data is used under specified conditions;

A literature review on data quality, specific for the big data era has been provided by Cai and Zhu (2015), who formulated a hierarchical data quality framework and assessment process characterized by a dynamic feedback mechanism. While in the past data originators were both producers and users, now the users are less and less frequently the data providers, who keep the responsibility to provide all the metadata from start to end of the data life cycle. It implies, from a user's perspective, the rapid analysis of big data from various sources must be based on standardized and certified high quality data, avoiding further unaffordable integration and quality control procedures. Timeliness is in fact an important quality element of availability, that Bushnell et al. (2019) include among the five main quality dimensions. The others are usability, reliability, relevance, and presentation quality. Each dimension is divided in typical elements with its own corresponding quality indicator. Even though Cai and Zhu (2015) detail a quality assessment process, they wish for a practical assessment tool with weight coefficients for the assessment indicators and outputting of results.

The EMODnet checkpoints

The conceptual quality assessment process has been applied for the first time in the framework of the EMODnet Sea-basin Checkpoints between 2013 and 2018 (https://emodnet.eu/en/checkpoints) that were asked to assess the adequacy of the observing system at basin-scale of the EU marginal seas, according to specific end-user challenges applications related to marine conservation, climate, and the emerging blue growth (Martín Míguez et al., 2019; Pearlman et al., 2019; Buck et al., 2019). The proposed stress tests were conducted by experts' teams that developed targeted products using upstream datasets selected among the many potential input datasets in a limited time. The checkpoint teams developed different assessment methodologies, but common metrics or indicators have been used to evaluate the availability, accessibility, and appropriateness of EU marine data, providing an overview of the main gaps of the EU observing monitoring capacity and recommendations for its improvement.

All checkpoints developed a management system that could be turned into an efficient checkpoint service to guarantee a sustained and prioritized observing system that supports research, environmental and climate monitoring, forecasting activities, and society, as advocated also by Sloyan et al. (2019) at a global level.

The adequacy of data has been evaluated comparing data characteristics with user requirements, using a qualitative approach (expert opinion) or implementing an objective quantitative assessment methodology using ISO principles and following INSPIRE rules, as pioneered by the MedSea Checkpoint (Pinardi et al., 2018) and further consolidated in the Atlantic and the Black Sea. A checkpoints synthesis report (available at https://emodnet.eu/en/synthesis-emodnet-sea-basin-checkpoints-eoos-conference) lists all the assessment indicators considered by the experts and the seven core indicators obtained from a shared best practice. The construction of a dedicated meta-database for both data and products specifications allowed the computation of indicators, providing a first example of a checkpoint service to be developed to improve users' confidence and increase data and products uptake. Such a service could fit the European Ocean Observing System strategy (Mackenzie et al., 2019) to boost its codesign and coproduction potential.

Toward an ocean service based on essential ocean variables

The achievement of Sustainable Development Goal 14 (life below water) depends on the capability to maximize the ocean data and information flow along with the data value chain, to link efficiently the multitude of ocean data to the global reporting effort about the state of the ocean. Buck et al. (2019) and Sloyan et al. (2019) indicate the Essential Ocean Variables (EOVs) as the elements on which to build such a system through the most advanced web technology. However, the lack of a standardized quality assessment framework is slowing down this target, also due to data interoperability and integrity issues which still characterize the global landscape. In order to reach a digital ecosystem in which the stakeholders (humans or machines) can shop for ready-to-use data from different sources and easily combine them for rapid use, FAIR principles should be first widely adopted. This would enable the construction of a quality assessment system that exploits the available metadata documentation to derive the product quality information. Such information will guide the stakeholders to select and properly use the products and obtain reliable results. It will also preserve data provenance that has a twofold advantage; on one side it permits to acknowledge all the actors that played a role in the production and delivery workflow, on the other it makes feasible a feedback loop on the products and services quality.

EOVs are key priority physical variables to determine key ocean phenomena acting at many different space-time scales. The definition and continuous update of EOVs requirements are essential to answer scientific and societal challenges, allowing a synergic and optimal evolution of the multiscale Global Ocean Observing System (GOOS) and the relative technological solutions for data gathering, access, and management. The application and adaptation of the presented conceptual quality assessment model at global scale throughout the ocean data pipeline will enable an objective and practical assessment process as previously exemplified by the EMODnet checkpoints.

The climate community already moved in this direction and the European Copernicus Climate Change Service (C3S) is an example of service that is implementing a quality assurance strategy to deliver product quality indication (Zeng et al., 2019). Through the Evaluation and Quality Control (EQC) function, a component of the C3S architecture, it is possible to evaluate if the data are "climate compliant" or not, and to analyze to what extent the climate service meets the specific users' needs (or fitness for purpose). Indeed, C3S targets the downstream use of climate data and derived information products to flourish the market for climate services. For this purpose, the delivered Essential Climate Variable (ECV) data and related products, whether observations (satellite and/or in-situ) or reanalysis, shall be fully traceable, adequately documented, uncertainty quantified, and providing sufficient guidance for users to address their specific needs. Zeng et al. (2019) discusses how to deliver timely assessments of the quality and usability of single products, multiproducts, and thematic products.

Within the C3S Climate Data Store (CDS, https://cds.climate. copernicus.eu), namely the Copernicus climate data records archive, each available ECV (and corresponding dataset) is provided along with a single product (one variable-one dataset) and, when possible, multiproduct (one variable-multiple products) quality brief report. Based on an independent evaluation, these reports are conceived to provide high-level user guidance on basic characteristics, strengths, weaknesses, and limitations of the ECV product under analysis. Typically, a set of metrics (such as climatology, variability, and trends) are estimated in order to assess the suitability of the product for climate applications. Additional information, such as product metadata, accessibility, documentation, and validation traceability are also provided via the so-called System Maturity Matrix (SMM). SMM is a powerful tool developed to evaluate the maturity of a given product for climate service users in terms of the categories mentioned above (see, e.g., Yang et al., 2021). The multiproduct quality brief is based on an intercomparison of different products that provide the same ECV. This report shows the consistency and/or discrepancies among the analyzed products, providing a guidance for selecting the product that might be better suited for users' applications. A set of use cases can also be presented via the so-called Application Performance Matrix (APM), such as extreme events detection, marine resources (e.g., fishery) sustainability evaluation, etc. APM is the readiness indicator that facilitates the comparison of data characteristics and results of validation versus the user requirements for a certain application. Furthermore, thanks to the availability of a multiproduct inventory and of information about the maturity of the products, a thematic assessment is dedicated to documenting the physical consistency of multiple ECVs to close the budgets of energy, carbon, and water cycles. Finally, the usability assessment encompasses the whole value chain, and it requires uncertainty analysis details from each step to be able to estimate the overall product and how it might affect the decision-making.

The assessment of quality and usability of ECV products makes the climate information authoritative, establishing a relation of trust between information providers and various downstream users. This system, developed starting from remote sensing data products, has still limitations for applicability to in-situ and reanalysis products. The other main limitation relates to metrology traceability and the propagation of uncertainties along the value chain.

Conclusions and recommendations

An intense excursion on ocean data life cycle and value chain has been provided with the objective to touch the many aspects and actors that come into play and have to be in synergy for an effective advancement in ocean knowledge and sustainable development, as in the UN Ocean Decade scope.

The starting point of any scientific result is the availability of high-quality and fit-for-purpose data, which translates at large scale into the necessity of a sustained Global Ocean Observing System (GOOS) focused on Essential Ocean Variables (EOVs) relevant to operational ocean services, ocean health, and climate themes and capable to resolve the multiple spatial and temporal scales of ocean processes. Expert panels maintain and disseminate EOV specification sheets and recommendations, including what measurements are to be made, various observing options, and data management practices for data providers. At the same time the observing capacity is constantly reviewed thanks to the users' engagement process, which ensures the coupling of the observing and monitoring efforts to the various and up-to-date societal needs.

The management of observation-based data and associated metadata information according to the FAIR principles, from multiple but interconnected and interoperable infrastructures, shall enable the rapid generation of derived information for societal benefit. Timely and high-quality information enables (1) the rapid assessment of the ocean state; (2) the improvement of predictive skills for physical and biogeochemical ocean forecasting; (3) an effective decision-making process through the implementation of early warning and decision support systems.

The long-term monitoring of marine environmental conditions through the aggregation and harmonization of most recent and historical data, the subsequent derivation of ocean gridded climatologies and reanalysis products is instead important for reporting and communicating to various stakeholders the evolution of the marine environment, facilitating the progressive implementation of informed policies, adaptation solutions to the changing environment, mitigation strategies of multihazards, and applications for anticipatory planning.

The parallel progress of harmonization and standardization solutions within complementary disciplines and the involvement of the private sector has already encouraged the development of downstream services and applications in all blue economy sectors, promoting conscious use of natural resources and limiting human impact on ocean health.

The societal benefit derived from a "Decadal" blue revolution will be enormous, since it reduces costs, targets investments on the real needs, maximizes the information and knowledge at service of policy making and sustainable development that constantly preserve a good environmental status of our oceans. Ocean science is the engine of this process and shall continue toward openness and transparency of data, products, and processing steps along with the data life cycle and value chain. In a big data era, the increasing amount of data from new instrumentation demands full machine readability of metadata documentation, including quality metrics which enable users to select the right data or products for the intended use.

Data providers and information producers are thus charged with the responsibility of prior planning their data products and workflows, assisting them until their publication in public repositories, to enable their virtuous reuse. The corresponding additional workload shall be supplied by data and information managers that become key roles within collaborative scientific teams and guarantee the full compliance with the open science principles.

The multitude and increasing complexity of data, products, and applications are progressively detaching the end-users from the originators; thus, it is important to secure any associated provenance and lineage information. The obtained effects are traceability and transparency of the production process that, on one side, increases the product reliability; on the other side, it acknowledges originators, producers, and data managers of their work. The acknowledgment can derive from persistent identifiers like DOI and ORCID that can trace the impact of data, individuals, institutes, and projects in a circular generation of scientific knowledge.

The engagement of stakeholders is in the end crucial in upstream, midstream, and downstream services to maximize efficiency. Here communication and education come into play, to train users to a proper usage of data and products, and grow up a new generation to dedicate to ocean science and blue jobs. Ocean literacy will finally increase awareness of society on the key role of the oceans for human well-being.

To conclude, we report hereafter some recommendations for the Ocean Decade that has just started.

- Monitoring gaps and omissions in some specific ocean regions might be filled through peaceful cooperation among countries, capacity development, and technological transfer toward less developed countries.
- The historical data/metadata rescue and ingestion should be continued for their preservation and use. There are still many omissions in present data infrastructures of data used in scientific publications, which should somehow be retrieved before their permanent loss.
- The connection of more data sources and providers to the main data infrastructures should be promoted and strengthened through the implementation of specific data policies, such as the request of detailed data management plans for funded projects that include the final ingestion of data into public databases, the actual data availability for accepted scientific publications in compliance to common standards, etc.
- The data ingestion process should be facilitated thanks to tools and services (e.g., EMODnet Ingestion, https://www.emodnet-ingestion.eu/)

especially in academic and private sectors which might not recruit data managers.

- The promotion and recognition of data manager professionals start from their inclusion in academic curricula, their enrollment in research institutes but also in public and private bodies.
- Persistent identifiers for scientists, technicians, data managers, data, and products should be a common practice to acknowledge all the players within the marine data value chain and ocean services. It would allow the impact assessment of institutes and projects that produce, manage, and share data and data products usage, tracing the whole data cycle process.
- The speed-up of the data and production cycle can be obtained implementing automatic solutions and tools for coproduction, such as Virtual Research Environments (VREs). Such collaborative environments will facilitate the rapid and efficient reporting on the ocean state within specific programs and services. VREs will also be optimal means for the training of new generations of ocean professionals, with ready-to-use data, tools, and workflows. The rapid access to massive high-quality and certified data, thanks to cloud solutions in conjunction with VRE, will promote innovation and competition.
- Collaborative environments can optimize and make flexible QA/QC processes, keeping in tight connection data providers, information producers, and users in order to continuously monitor and upgrade the quality of the database content and of the derived products. Full metadata description of data enables a virtuous QA/QC loop with continuous feedback on data anomalies from intermediaries and users toward data providers.
- Adoption of big data and artificial intelligence approaches in ocean science will reduce human intervention during the QA/QC process, advance data assimilation techniques in prediction and reanalysis applications, and generate innovative multidisciplinary products and applications.
- The progressive implementation of a data quality assessment framework based on expected data product specifications, which evaluates the realized product characteristics according to users' requirements, will facilitate decision-making process in many public and private sectors. Moreover, it will raise the correct usage of data and products. Data quality assessment relies on the assignment of measurements uncertainty and its further propagation along the production chain, up to the final decision-making. The so-called metrological traceability needs to be improved to support reliable information for societal benefit.

References

Allison, L.C., Roberts, C.D., Palmer, M.D., Killick, R., Hermanson, L., Rayner, N.A., 2019. Towards quantifying uncertainty in ocean heat content changes using synthetic profiles. Environ. Res. Lett. 14, 8. https://doi.org/10.1088/1748-9326/ab2b0b.

Bakker, D.C.E., Pfeil, B., Landa, C.S., Metzl, N., O'Brien, K.M., Olsen, A., Smith, K., Cosca, C., Harasawa, S., Jones, S.D., Nakaoka, S., Nojiri, Y., Schuster, U., Steinhoff, T., Sweeney, C., Takahashi, T., Tilbrook, B., Wada, C., Wanninkhof, R., Alin, S.R., Balestrini, C.F., Barbero, L., Bates, N.R., Bianchi, A.A., Bonou, F., Boutin, J., Bozec, Y., Burger, E.F., Cai, W.-J., Castle, R.D., Chen, L., Chierici, M., Currie, K., Evans, W., Featherstone, C., Feely, R.A., Fransson, A., Goyet, C., Greenwood, N., Gregor, L., Hankin, S., Hardman-Mountford, N.J., Harlay, J., Hauck, J., Hoppema, M., Humphreys, M.P., Hunt, C.W., Huss, B., Ibánhez, J.S.P., Johannessen, T., Keeling, R., Kitidis, V., Körtzinger, A., Kozyr, A., Krasakopoulou, E., Kuwata, A., Landschützer, P., Lauvset, S.K., Lefèvre, N., Lo Monaco, C., Manke, A., Mathis, J.T., Merlivat, L., Millero, F.J., Monteiro, P.M.S., Munro, D.R., Murata, A., Newberger, T., Omar, A.M., Ono, T., Paterson, K., Pearce, D., Pierrot, D., Robbins, L.L., Saito, S., Salisbury, J., Schlitzer, R., Schneider, B., Schweitzer, R., Sieger, R., Skjelvan, I., Sullivan, K.F., Sutherland, S.C., Sutton, A.J., Tadokoro, K., Telszewski, M., Tuma, M., Van Heuven, S.M.A.C., Vandemark, D., Ward, B., Watson, A.J., Xu, S., 2016. A multi-decade record of high quality fCO_2 data in version 3 of the Surface Ocean CO_2 Atlas (SOCAT). Earth Syst. Sci. Data 8, 383−413. https://doi.org/10.5194/essd8-383-2016.

Balmaseda, M.A., Dee, D., Vidard, A., Anderson, D.L., 2007. A multivariate treatment of bias for sequential data assimilation: application to the tropical oceans. Q. J. R. Meteorol. Soc. 133, 167−179. https://doi.org/10.1002/qj.12.

Balmaseda, M.A., Hernandez, F., Storto, A., Palmer, M.D., Alves, O., Shi, L., Smith, G.C., Toyoda, T., Valdivieso, M., Barnier, B., Behringer, D., Boyer, T., Chang, Y.-S., Chepurin, G.A., Ferry, N., Forget, G., Fujii, Y., Good, S., Guinehut, S., Haines, K., Ishikawa, Y., Keeley, S., Köhl, A., Lee, T., Martin, M.J., Masina, S., Masuda, S., Meyssignac, B., Mogensen, K., Parent, L., Peterson, K.A., Tang, Y.M., Yin, Y., Vernieres, G., Wang, X., Waters, J., Wedd, R., Wang, O., Xue, Y., Chevallier, M., Lemieux, J.-F., Dupont, F., Kuragano, T., Kamachi, M., Awaji, T., Caltabiano, A., Wilmer-Becker, K., Gaillard, F., 2015. the Ocean reanalyses intercomparison project (ORA-IP). J. Oper. Oceanogr. 8 (Suppl. 1), s80−s97. https://doi.org/10.1080/1755876X.2015.1022329.

Barnes, S., 1964. A technique for maximizing details in numerical map analysis. J. Appl. Meteorol. 3, 395−409. https://doi.org/10.1175/1520-0450(1964)003<0396:ATFM-DI>2.0.CO;2.

Barth, A., Beckers, J.M., Troupin, C., Alvera-Azcárate, A., Vandenbulcke, L., 2014. divand-1.0: n-dimensional variational data analysis for ocean observations. Geosci. Model Dev. 7, 225−241. https://doi.org/10.5194/gmd-7-225-2014. http://www.geosci-model-dev.net/7/225/2014/.

Barth, A., Troupin, C., Reyes, E., Alvera-Azcárate, A., Beckers and, J.M., Tintoré, J., 2021. Variational interpolation of high-frequency radar surface currents using DIVAnd. Ocean Dynam. https://doi.org/10.1007/s10236-020-01432-x.

Batini, C., Blaschke, T., Lang, S., Albrecht, F., Abdulmutalib, M., Barsi, Á., Szabó, G., Kugler, Z., 2017. DATA QUALITY IN REMOTE SENSING. Int. Arch. Photogramm. Remote Sens. Spatial Inf. Sci., XLII-2/W7, pp. 447−453. https://doi.org/10.5194/isprs-archives-XLII-2-W7-447-2017.

Beckers, J.M., Barth, A., Troupin, C., Alvera-Azcárate, A., 2014. Approximate and efficient methods to assess error fields in spatial gridding with data interpolating variational analysis (DIVA). J. Atmos. Ocean. Technol. 31, 515—530. https://doi.org/10.1175/JTECH-D-13-00130.1.

Bell, M.J., Schiller, A., Le Traon, P.Y., Smith, N.R., Dombrowsky, E., Wilmer-Becker, K., 2015. An introduction to GODAE OceanView. J. Oper. Oceanogr. 8 (Suppl. 1), s2—s11. https://doi.org/10.1080/1755876X.2015.1022041.

Bensoussan, N., Chiggiato, J., Buongiorno Nardelli, B., Pisano, A., Garrabou, J., 2019. *Insights on 2017 marine heat Waves in the Mediterranean Sea.* Copernicus marine service ocean state report, issue 3. J. Oper. Oceanogr. 12, S101—S108. https://doi.org/10.1080/1755876X.2019.1633075.

Berto, D., Formalewicz, M., Giorgi, G., Rampazzo, F., Gion, C., Trabucco, B., Giani, M., Lipizer, M., Matijevic, S., Kaberi, H., Zeri, C., Bajt, O., Mikac, N., Joksimovic, D., Aravantinou, A.F., Poje, M., Cara, M., Manfra, L., 2020. Challenges in harmonized assessment of heavy metals in the adriatic and ionian seas. Front. Mar. Sci. 7, 717. https://doi.org/10.3389/fmars.2020.00717.

Borja, A., Elliott, M., Carstensen, J., Heiskanen, A.-S., van de Bund, W., 2010. Marine management — towards an integrated implementation of the European marine strategy framework and the water framework directives. Mar. Pollut. Bull. 60, 2175—2186.

Boyer, T.P., Baranova, O.K., Coleman, C., Garcia, H.E., Grodsky, A., Locarnini, R.A., Mishonov, A.V., Paver, C.R., Reagan, J.R., Seidov, D., Smolyar, I.V., Weathers, K., Zweng, M.M., 2018. World ocean database 2018. In: Mishonov Technical, A. (Ed.), NOAA Atlas NESDIS 87.

Boyer, T., Levitus, S., 1994. Quality Control and Processing of Historical Oceanographic Temperature, Salinity and Oxygen Data. NOAA Technical Report NESDIS81.

Brankart, J.M., Brasseur, P., 1996. Optimal analysis of in situ data in the Western Mediterranean using statistics and cross-validation. J. Atmos. Ocean. Technol. 13, 477—491. https://doi.org/10.1175/1520-0426(1996)013<0477:OAOISD>2.0.CO;2.

Bretherton, F.P., Davis, R.E., Fandry, C.B., 1976. A technique for objective analysis and design of oceanographic experiment applied to MODE-73. Deep Sea Res. 23, 559—582. https://doi.org/10.1016/0011-7471(76)90001-2.

Buck, J.J.H., Bainbridge, S.J., Burger, E.F., Kraberg, A.C., Casari, M., Casey, K.S., Darroch, L., Rio, J.D., Metfies, K., Delory, E., Fischer, P.F., Gardner, T., Heffernan, R., Jirka, S., Kokkinaki, A., Loebl, M., Buttigieg, P.L., Pearlman, J.S., Schewe, I., 2019. Ocean data product integration through innovation-the next level of data interoperability. Front. Mar. Sci. 6, 32. https://doi.org/10.3389/fmars.2019.00032.

Buongiorno Nardelli, B., Tronconi, C., Pisano, A., Santoleri, R., 2013. High and ultra-high resolution processing of satellite sea surface temperature data over southern European seas in the framework of MyOcean project. Rem. Sens. Environ. 129, 1—16. https://doi.org/10.1016/j.rse.2012.10.012.

Bushnell, M., Waldmann, C., Seitz, S., Buckley, E., Tamburri, M., Hermes, J., Henslop, E., Lara-Lopez, A., 2019. Quality assurance of oceanographic observations: standards and guidance adopted by an international partnership. Front. Mar. Sci. 6, 706. https://doi.org/10.3389/fmars.2019.00706.

Butenschön, M., Lovato, T., Masina, S., Caserini, S., Grosso, M., 2021. Alkalinization scenarios in the mediterranean sea for efficient removal of atmospheric CO_2 and the mitigation of ocean acidification. Front. Clim. https://doi.org/10.3389/fclim.2021.614537.

Cai, L., Zhu, Y., 2015. The challenges of data quality and data quality assessment in the big data era. Data Sci. J. 14, 2. https://doi.org/10.5334/dsj-2015-002.

Canter, M., Barth, A., Beckers, J.-M., 2016. Correcting circulation biases in a lower-resolution global general circulation model with data assimilation. Ocean Dynam. 67, 1—18. https://doi.org/10.1007/s10236-016-1022-3.

Carter, E.F., Robinson, A.R., 1987. Analysis models for the estimation of oceanic fields. J. Atmos. Ocean. Technol. 4, 49—74. https://doi.org/10.1175/1520-0426(1987) 004<0049:AMFTEO>2.0.CO;2.

Castelao, G.P., 2020. A framework to quality control oceanographic data. J. Open Source Softw. 5 (48), 2063. https://doi.org/10.21105/joss.02063.

Chelton, D.B., Wentz, F.J., 2005. Global microwave satellite observations of sea surface temperature for numerical weather prediction and climate research. Bull. Am. Meteorol. Soc. 86 (8), 1097—1116.

Commission of the European Community (CEC) and Intergovernmental Oceanographic Commission, 1993. Manual of Quality Control Procedures for Validation of Oceanographic Data. UNESCO, Paris, France, 436pp. (Intergovernmental Oceanographic Commission Manuals and Guides; 26). http://hdl.handle.net/11329/167.

Cowley, R., Killick, R.E., Boyer, T., Gouretski, V., Reseghetti, F., Kizu, S., Palmer, M.D., Cheng, L., Storto, A., Le Menn, M., Simoncelli, S., Macdonald, A.M., Domingues, C.M., 2021. International quality-controlled ocean database (IQuOD) v0.1: the temperature uncertainty specification. Front. Mar. Sci. https://doi.org/10.3389/fmars.2021.689695. In press.

Cressman, G., 1959. An operational objective analysis system. Mon. Weather Rev. 88, 327—342. https://doi.org/10.1175/1520-0493(1959)087<0367:AOOAS>2.0.CO;2.

Daley, R., 1993. Estimating observation error statistics for atmospheric data assimilation. Ann. Geophys. Atmos. Hydro. Space Sci. 11, 634—647.

Deser, C., Alexander, M.A., Xie, S.P., Phillips, A.S., 2010. Sea surface temperature variability: patterns and mechanisms. Ann. Rev. Mar. Sci. 2, 115—143.

Dierssen, H.M., Theberge, A.E., 2014. Bathymetry: assessment. In: Wang, Y., Pielke, R.A. (Eds.), Encyclopedia of Natural Resources - Water and Air - Vol II (1st ed.). CRC Press, Boca Raton. 2014. https://doi.org/10.1201/9780203757611.

Donlon, C., Berruti, B., Buongiorno, A., Ferreira, M.H., Féménias, P., Frerick, J., Goryl, P., Klein, U., Laur, H., Mavrocordatos, C., Nieke, J., Rebhan, H., Seitz, B., Stroede, J., Sciarra, R., 2012. The global monitoring for environment and security (GMES) sentinel-3 mission. Rem. Sens. Environ. 120, 37—57.

Embury, O., Merchant, C.J., Corlett, G.K., 2012. A reprocessing for climate of sea surface temperature from the along-track scanning radiometers: initial validation, accounting for skin and diurnal variability effects. Rem. Sens. Environ. 116, 62—78.

Emery, W.J., Thomson, R.E., 2001. Data Analysis Methods in Physical Oceanography, second ed. Elsevier, p. 654.

European Commission, 2000. Directive 2000/60/EC of the European Parliament and of the Council of 23 October 2000 establishing a framework for community action in the field of water policy. Off. J. Eur. Union 327, 1—73.

European Commission, 2002. Recommendation of the European Parliament and of the Council of 30 may 2002 concerning the implementation of integrated coastal zone management in Europe. Off J Eur Union 148, 24—27.

European Commission, 2008. Directive 2008/56/EC of the European Parliament and of the Council of 17 June 2008 establishing a framework for community action in the field of marine environmental policy (Marine Strategy Framework Directive). Off. J. Eur. Union 164, 19—40.

European Commission, 2010. Communication from the Commission to the European Parliament and the Council. Marine Knowledge 2020: Marine Data and Observation for Smart and Sustainable Growth. http://eur-lex.europa.eu/LexUriServ/LexUriServ. do?uri=CELEX:52010DC0461:EN:NOT. (Accessed 1 April 2021).

European Commission, 2013. Directive 2013/39/EU of the European Parliament and of the Council Amending Directives 2000/60/EC and 2008/105/EC as Regards Priority Substances in the Field of Water Policy. Brussels: EU.

European Commission, 2014. Directive 2014/89/EU of the European Parliament and of the Council of 23 July 2014 establishing a framework for maritime spatial planning. Off. J. Eur. Union 257, 135—145.

Fasham, M.J.R., Balino, B.M., Bowles, M.C., 2001. A new vision of ocean biogeochemistry after a decade of the Joint Global Ocean Flux Study (JGOFS). Ambio Spec. Rep. 3—31.

Fichaut, M., Garcia, M.J., Giorgetti, A., Iona, A., Kuznetsov, A., Rixen, M., MedAtlas Group, 2003. MEDAR/MEDATLAS 2002: A mediterranean and black sea database for operational oceanography. Elsevier Oceanogr. Ser. 69, 645—648.

Fichaut, M., Schaap, D., Maudire, G., 2013. SeaDataNet - Second phase achievements and technical highlights. International Conference on Marine Data and Information Systems 23-25 September, 2013 - Lucca (Italy). Bollettino di Geofisica teorica ed applicata, 54 supplement, pp. 15—16.

Fichtinger, A., Rix, J., Schäffler, U., Michi, I., Gone, M., Reitz, T., 2011. Data harmonisation put into practice by the HUMBOLDT project. Int. J. Spat. Data Infrastruct. Res. 6, 234—260.

Ganachaut, A., Wunsch, C., 2003. Large-scale ocean heat and freshwater transports during the World Ocean circulation experiment. J. Clim. 16, 696—705.

Gandin, L.S., 1965. Objective Analysis of Meteorological Fields. Israel Program for Scientific Translation, Jerusalem, 242 pp.

Gardner, W.D., Mishonov, A.V., Richardson, M.J., 2020. Global Transmissometer Database V3. https://doi.org/10.13140/RG.2.2.36105.26724. https://odv.awi.de/data/ocean/global-transmissometer-database/.

Gentemann, C.L., Meissner, T., Wentz, F.J., 2009. Accuracy of satellite sea surface temperatures at 7 and 11 GHz. IEEE Trans. Geosci. Rem. Sens. 48 (3), 1009—1018.

Giani, M., Kralj, M., Molina Jack, M.E., Lipizer, M., Giorgi, G., Rotini, A., Ivankovic, D., Ujevic, I., Matijević, S., Kaberi, H., Hatzianestis, J., Zeri, C., Iona, S., Fafandjel, M., Mikac, N., Cermelj, B., Bajt, O., 2020. Methodological Proposal for Data Quality Control Procedures WPT1 HarmoNIA Deliverable T1.1.3. https://doi.org/10.6092/db6877bc-9ebd-4d7dbc15-fd5997e4fc63.

Giese, B.S., Seidel, H.F., Compo, G.P., Sardeshmukh, P.D., 2016. An ensemble of ocean reanalyses for 1815—2013 with sparse observational input. J. Geophys. Res. Ocean 121, 6891—6910. https://doi.org/10.1002/2016JC012079.

Giorgetti, A., Lipizer, M., Molina Jack, M.E., Holdsworth, N., Jensen, H.M., Buga, L., Sarbu, G., Iona, A., Gatti, J., Larsen, M., Fyrberg, L., Østrem, A.K., Schlitzer, R., 2020. Aggregated and validated datasets for the European seas: the contribution of EMODnet chemistry. Front. Mar. Sci. 7, 583657. https://doi.org/10.3389/fmars.2020.583657.

Good, S.A., Martin, M.J., Rayner, N.A., 2013. EN4: quality controlled ocean temperature and salinity profiles and monthly objective analyses with uncertainty estimates. J. Geophys. Res. Ocean 118, 6704—6716. https://doi.org/10.1002/2013JC009067.

Good, S., Fiedler, E., Mao, C., Martin, M.J., Maycock, A., Reid, R., Roberts-Jones, J., Searle, T., Waters, J., While, J., Worsfold, M., 2020. The current configuration of the OSTIA system for operational production of foundation sea surface temperature and ice concentration analyses. Rem. Sens. 12, 720. https://doi.org/10.3390/rs12040720.

GOSUD, 2016. GOSUD Project-Global Ocean Surface Underway Data. SEANOE (SEA scieNtific Open data Edition). https://doi.org/10.17882/47403.

Griffin, P.C., Khadake, J., LeMay, K.S., et al., 2018. *Best practice data life cycle approaches for the life sciences* [version 2; peer review: 2 approved]. F1000Research 6, 1618. https://doi.org/10.12688/f1000research.12344.2.

Halliwell, G.R., Mehari, M., Shay, L.K., Kourafalou, V.H., Kang, H., Kim, H.-S., Dong, J., Atlas, R., 2017. OSSE quantitative assessment of rapid-response prestorm ocean surveys to improve coupled tropical cyclone prediction. J. Geophys. Res. Ocean 122, 5729—5748. https://doi.org/10.1002/2017JC012760.

Huang, B., Liu, C., Banzon, V., Freeman, E., Graham, G., Hankins, B., Smith, T., Zhang, H., 2021. Improvements of the daily optimum interpolation sea surface temperature (DOISST) version 2.1. J. Clim. 34 (8), 2923—2939. https://doi.org/10.1175/JCLI-D-20-0166.1.

IHO, 2008. IHO Standards for Hydrographic Surveys, fifth ed. Special Publication S44, Monaco.

Intergovernmental Oceanographic Commission, 2013. Ocean data standards volume 3. Recommendation for a quality flag scheme for the exchange of oceanographic and marine meteorological data. Paris, France, UNESCO-IOC, 5pp. & Annexes. Intergov. Oceanogra. Comm. Man. Guides 54 (3). https://doi.org/10.25607/OBP-6 (IOC/2013/MG/54-3).

Intergovernmental Oceanographic Commission of UNESCO, 2019. Ocean Data Standards: Technology for SeaDataNet Controlled Vocabularies for describing Marine and Oceanographic Datasets - A joint Proposal by SeaDataNet and ODIP projects. In IOC Manuals and Guides, n. 54, 4. IODE/UNESCO, Ostend. https://doi.org/10.25607/OBP-566.

Iona, A., Theodorou, A., Sofianos, S., Watelet, S., Troupin, C., Beckers, J.-M., 2018. Mediterranean Sea climatic indices: monitoring long-term variability and climate changes. Earth Syst. Sci. Data 10, 1829—1842. https://doi.org/10.5194/essd-10-1829-2018.

Iwamoto, M.M., Dorton, J., Newton, J., Yerta, M., Gibeaut, J., Shyka, T., Kirkpatrick, B., Currier, R., 2019. Meeting regional, coastal and ocean user needs with tailored data products: a stakeholder-driven process. Front. Mar. Sci. 6, 290. https://doi.org/10.3389/fmars.2019.00290.

Janowiak, J.E., Bauer, P., Wang, W., Arkin, P.A., Gottschalck, J., 2010. An evaluation of precipitation forecasts from operational models and reanalyses including precipitation variations associated with MJO activity. Mon. Weather Rev. 138, 4542—4560. https://doi.org/10.1175/2010MWR3436.1.

Kilic, L., Prigent, C., Aires, F., Boutin, J., Heygster, G., Tonboe, R.T., Roquet, H., Jimenez, C., Donlon, C., 2018. Expected performances of the Copernicus Imaging Microwave Radiometer (CIMR) for an all-weather and high spatial resolution estimation of ocean and sea ice parameters. J. Geophys. Res. Ocean 123 (10), 7564—7580.

Lellouche, J.-M., Greiner, E., Le Galloudec, O., Garric, G., Regnier, C., Drevillon, M., Benkiran, M., Testut, C.E., Bourdalle-Badie, R., Gasparin, F., Hernandez, O., Levier, B., Drillet, Y., Remy, E., Le Traon, P.Y., 2018. Recent updates on the Copernicus marine service global ocean monitoring and forecasting real-time 1/12° high resolution system. Ocean Sci. 14, 1093—1126. https://doi.org/10.5194/os-14-1093-2018.

Levitus, S., 2012. The UNESCO-IOC-IODE "global oceanographic data archeology and rescue" (GODAR) project and "World Ocean database" project. Data Sci. J. 11, 46—71.

Le Traon, P.Y., Reppucci, A., Alvarez Fanjul, E., Aouf, L., Behrens, A., Belmonte, M., Bentamy, A., Bertino, L., Brando, V.E., Kreiner, M.B., Benkiran, M., Carval, T., Ciliberti, S.A., Claustre, H., Clementi, E., Coppini, G., Cossarini, G., De Alfonso Alonso-Muñoyerro, M., Delamarche, A., Dibarboure, G., Dinessen, F., Drevillon, M., Drillet, Y., Faugere, Y., Fernández, V., Fleming, A., Garcia-Hermosa, M.I., Sotillo, M.G., Garric, G., Gasparin, F., Giordan, C., Gehlen, M., Gregoire, M.L., Guinehut, S., Hamon, M., Harris, C., Hernandez, F., Hinkler, J.B., Hoyer, J., Karvonen, J., Kay, S., King, R., Lavergne, T., Lemieux-Dudon, B., Lima, L., Mao, C., Martin, M.J., Masina, S., Melet, A., Buongiorno Nardelli, B., Nolan, G., Pascual, A., Pistoia, J., Palazov, A., Piolle, J.F., Pujol, M.I., Pequignet, A.C., Peneva, E., Pérez Gómez, B., Petit de la Villeon, L., Pinardi, N., Pisano, A., Pouliquen, S., Reid, R., Remy, E., Santoleri, R., Siddorn, J., She, J., Staneva, J., Stoffelen, A., Tonani, M., Vandenbulcke, L., von Schuckmann, K., Volpe, G., Wettre, C., Zacharioudaki, A., 2019. From observation to information and users: the Copernicus marine service perspective. Front. Mar. Sci. 6, 234. https://doi.org/10.3389/fmars.2019.00234.

Lipizer, M., Molina Jack, M.E., Giorgetti, A., Buga, L., Fyberg, L., Wesslander, K., Østrem, A.K., Iona, A., Larsen, M., Schlitzer, R., 2020. EMODnet Chemistry Guidelines for QC of Contaminants. https://doi.org/10.6092/54712172-641d-4c1c-8170-38d3e507b7a6, 12/05/2020, 18 pp.

Lipizer, M., Molina Jack, M., E. Lorenzon, S., Giorgi, G., Manfra, L., Trabucco, B., Cara, M., Čermelj, B., Fafandjel, M., Ivanković, D., Joksimović, D., Veliconja, M., Zeri, C., 2021. Harmonization requirements for MSFD and EcAp (contaminants) in the ADRION region: from sampling to data visualization. In: The Handbook of Environmental Chemistry. Springer, Berlin, Heidelberg. https://doi.org/10.1007/698_2020_719.

Liubartseva, S., Coppini, G., Lecci, R., Clementi, E., 2018. Tracking plastics in the Mediterranean: 2D Lagrangian model. Mar. Pollut. Bull. 129 (1), 151—162.

Locarnini, R.A., Mishonov, A.V., Baranova, O.K., Boyer, T.P., Zweng, M.M., Garcia, H.E., Reagan, J.R., Seidov, D., Weathers, K., Paver, C.R., Smolyar, I., 2018. In: Mishonov Technical, A. (Ed.), World Ocean Atlas 2018, volume 1: temperature. NOAA Atlas NESDIS 81, 52 pp.

Loew, A., Bell, W., Brocca, L., Bulgin, C.E., Calbet, X., Donner, R.V., Ghent, D., Gruber, A., Kaminski, T., Kinzel, J., Klepp, C., Lambert, J.C., Schaepman-Strub, G., Schoder, M., Verhoelst, T., 2017. Validation practices for satellite-based Earth observation data across communities. Rev. Geophys. 55, 779—817. https://doi.org/10.1002/2017RG000562.

Lonsdale, J.A., Nicholson, R., Judd, A., Elliott, M., Clarke, C., 2020. A novel approach for cumulative impacts assessment for marine spatial planning. Environ. Sci. Pol. 106, 125—135.

Mackenzie, B., Celliers, L., Assad, L.P.dF., Heymans, J.J., Rome, N., Thomas, J., Anderson, C., Behrens, J., Calverley, M., Desai, K., DiGiacomo, P.M., Djavidnia, S., dos Santos, F., Eparkhina, D., Ferrari, J., Hanly, C., Houtman, B., Jeans, G., Landau, L., Larkin, K., Legler, D., Le Traon, P.-Y., Lindstrom, E., Loosley, D., Nolan, G., Petihakis, G., Pellegrini, J., Roberts, Z., Siddorn, J.R., Smail, E., Sousa-Pinto, I., Terrill, E., 2019. The role of stakeholders in creating societal value from coastal and ocean observations. Front. Mar. Sci. 6, 137. https://doi.org/10.3389/fmars.2019.00137.

Malaki, A., Bertossi, L., Rizzolo, F., 2012. Multidimensional contexts for data quality assessment. In: 6th Alberto Mendelzon International Workshop on Foundations of Data Management. AMW. http://www.cs.toronto.edu/~flavio/amw2012.pdf.

Martín Míguez, B., Novellino, A., Vinci, M., Claus, S., Calewaert, J.-B., Vallius, H., Schmitt, T., Pititto, A., Giorgetti, A., Askew, N., Iona, S., Schaap, D., Pinardi, N., Harpham, Q., Kater, B.J., Populus, J., She, J., Palazov, A.V., McMeel, O., Oset, P., Lear, D., Manzella, G.M.R., Gorringe, P., Simoncelli, S., Larkin, K., Holdsworth, N., Arvanitidis, C.D., Molina Jack, M.E., Chaves Montero, M.M., Herman, P.M.J., Hernandez, F., 2019. The European marine observation and data network (EMODnet): visions and roles of the gateway to marine data in Europe. Front. Mar. Sci. 6, 313. https://doi.org/10.3389/fmars.2019.00313.

Merchant, C.J., Embury, O., Bulgin, C.E., Block, T., Corlett, G.K., Fiedler, E., Good, S.A., Mittaz, J., Rayner, N.A., Berry, D., Eastwood, S., Taylor, M., Tsushima, Y., Waterfall, A., Wilson, R., Donlon, C., 2019. Satellite-based time-series of sea-surface temperature since 1981 for climate applications. Sci. Data 6, 223. https://doi.org/10.1038/s41597-019-0236-x.

Mieruch, S., Demirel, S., Simoncelli, S., Schlitzer, R., Seitz, S., 2021. SalaciaML: a deep learning approach for supporting ocean data quality control. Front. Mar. Sci. 8, 611742. https://doi.org/10.3389/fmars.2021.611742.

Minnett, P.J., Alvera-Azcárate, A., Chin, T.M., Corlett, G.K., Gentemann, C.L., Karagali, I., Marullo, S., Maturi, E., Santoleri, R., Saux Picard, S., Steele, M., Vazquez-Cuervo, J., 2019. Half a century of satellite remote sensing of sea-surface temperature. Rem. Sens. Environ. 233, 111366.

Molina Jack, M.E., Kralj, M., Partescano, E., Giorgi, G., Rotini, A., Ivankovic, D., Iona, A., Čermelj, B., Fafandel, M., Castelli, A., Georgopoulou, C., Velikonja, M., Bakiu, R., Lipizer, M., 2020. Heavy Metals in the Adriatic-Ionian Seas: A Case Study to Illustrate the Challenges in Data Management when Dealing with Regional Datasets. Frontiers in Marine Science Special Issue: The Environmental Hazards of Toxic Metals Pollution Theme: Heavy Metals in Environmental Compartments. https://doi.org/10.3389/fmars.2020.571365.

Mora, D., 2014. Quality Assurance Monitoring Plan. Long-Term Marine Waters Monitoring, Mooring Program. Department of ecology, State of Washington. https://fortress.wa.gov/ecy/publications/publications/1403103.pdf. (Accessed 1 April 2021).

Moray, R., Hook, R., 1667. Directions for the observations and experiments to be made by masters of ships, pilots, and other fit persons in their sea-voyages. Phil. Trans. (2), 1666−1667. https://royalsocietypublishing.org/doi/10.1098/rstl.1666.0009.

Myroshnychenko, V., Simoncelli, S., 2020. SeaDataCloud Temperature and Salinity Climatology for the Black Sea (Version 2). Product Information Document (PIDoc). SeaDataNet. https://doi.org/10.13155/61812.

Nativi, S., Craglia, M., Pearlman, J., 2012. The brokering approach for multidisciplinary interoperability: a position paper. Int. J. Spat. Data Infrastruct. Res. 7, 1−15.

Noble, M.M., Harasti, D., Pittock, J., Doran, B., 2019. Linking the social to the ecological using GIS methods in marine spatial planning and management to support resilience: a review. Mar. Pol. 108, 103657.

Olsen, A., Lange, N., Key, R.M., Tanhua, T., Bittig, H.C., Kozyr, A., Àlvarez, M., Azetsu-Scott, K., Becker, S., Brown, P.J., Carter, B.R., Cotrim da Cunha, L., Feely, R.A., van Heuven, S., Hoppema, M., Ishii, M., Jeansson, E., Jutterström, S., Landa, C.S., Lauvset, S.K., Michaelis, P., Murata, A., Pérez, F.F., Pfeil, B., Schirnick, C., Steinfeldt, R., Suzuki, T., Tilbrook, B., Velo, A., Wanninkhof, R., Woosley, R.J., 2020. GLODAPv2.2020 − the Second Update of GLODAPv2. https://doi.org/10.5194/essd-2020-165.

Palmer, M.D., Roberts, C.D., Balmaseda, M., et al., 2017. Ocean heat content variability and change in an ensemble of ocean reanalyses. Clim. Dynam. 49, 909−930. https://doi.org/10.1007/s00382-015-2801-0.

Palmer, M.D., Boyer, T., Cowley, R., Kizu, S., Reseghetti, F., Suzuki, T., Thresher, A., 2018. An algorithm for classifying unknown expendable bathythermograph (XBT) instruments based on existing metadata. J. Atmos. Ocean. Technol. 35 (3), 429−440.

Pearlman, J., Bushnell, M., Coppola, L., Karstensen, J., Buttigieg, P.L., Pearlman, F., Simpson, P., Barbier, M., Muller-Karger, F.E., Munoz-Mas, C., Pissierssens, P., Chandler, C., Hermes, J., Heslop, E., Jenkyns, R., Achterberg, E.P., Bensi, M., Bittig, H.C., Blandin, J., Bosch, J., Bourles, B., Bozzano, R., Buck, J.J.H., Burger, E.F., Cano, D., Cardin, V., Llorens, M.C., Cianca, A., Chen, H., Cusack, C., Delory, E., Garello, R., Giovanetti, G., Harscoat, V., Hartman, S., Heitsenrether, R., Jirka, S., Lara-Lopez, A., Lantéri, N., Leadbetter, A., Manzella, G., Maso, J., McCurdy, A., Moussat, E., Ntoumas, M., Pensieri, S., Petihakis, G., Pinardi, N., Pouliquen, S., Przeslawski, R., Roden, N.P., Silke, J., Tamburri, M.N., Tang, H., Tanhua, T., Telszewski, M., Testor, P., Thomas, J., Waldmann, C., Whoriskey, F., 2019. Evolving and sustaining Ocean Best practices and standards for the next decade. Front. Mar. Sci. 6, 277. https://doi.org/10.3389/fmars.2019.00277.

Pinardi, N., Manzella, G., Simoncelli, S., Clementi, E., Moussat, E., Quimbert, E., Blanc, F., Valladeau, G., Galanis, G., Kallos, G., Patlakas, P., Reizopoulou, S., Kyriakidou, C., Katara, I., Kouvarda, D., Skoulikidis, N., Gomez-Pujol, L., Vallespir, J., March, D., Tintore, J., Fabi, G., Scarcella, G., Tassetti, A., Raicich, F., Cruzado, A., Bahamon, N., Falcini, F., Filipot, J., Duarte, R., Lecci, R., Bonaduce, A., Lyubartsev, V., Cesarini, C., Zodiatis, G., Stylianou, S., Calewart, J.B., Martin Miguez, B., 2018. Stress testing the EU monitoring capacity for the Blue economy. In: Buch, E., Fernández, V., Eparkhina, D., Gorringe, P., Nolan, G. (Eds.), Operational Oceanography Serving Sustainable Marine Development. Proceedings of the Eight EuroGOOS International Conference. 3-5 October 2017, Bergen, Norway, pp. 415—422. EuroGOOS. Brussels, Belgium. D/2018/14.040/1 ISBN 978-2-9601883-3-2. https://archimer.ifremer.fr/doc/00450/56156/.

Pisano, A., Marullo, S., Artale, V., Falcini, F., Yang, C., Leonelli, F., Santoleri, R., Buongiorno Nardelli, B., 2020. New evidence of mediterranean climate change and variability from sea surface temperature observations. Rem. Sens. 12 https://doi.org/10.3390/rs12010132.

Poncelet, C., Billant, G., Corre, M.-P., 2020. GLOBE (Global Oceanographic Bathymetry Explorer) Software. https://doi.org/10.17882/70460. https://wwz.ifremer.fr/flotte_en/Facilities/Shipboard-software/Analyse-et-traitement-de-l-information/GLOBE.

Prevenios, M., Zeri, C., Tsangaris, C., Liubartseva, S., Fakiris, E., Papatheodorou, G., 2018. Beach litter dynamics on Mediterranean coasts: distinguishing sources and pathways. Mar. Pollut. Bull. 129 (2), 448—457.

Redfield, A.C., 1934. On the Proportion of Organic Derivatives in Sea Water and Their Relation to the Composition of Plankton. James Johnstone Memorial Volume. University Press of Liverpool, pp. 176—192.

Rayner, N., Parker, D.E., Horton, E., Folland, C.K., Alexander, L.V., Rowell, D., Kent, E., Kaplan, A., 2003. Global analyses of sea surface temperature, sea ice, and night marine air temperature since the late nineteenth century. J. Geophys. Res. Atmos. 108 https://doi.org/10.1029/2002JD002670.

Rew, R.K., Davis, G.P., 1990. NetCDF: an interface for scientific data access. IEEE Comput. Gr.Appl. 10 (4), 76—82.

Robinson, I.S., 2004. Measuring the Oceans from Space: The Principles and Methods of Satellite Oceanography. Springer Science & Business Media.

Roemmich, D., Alford, M.H., Claustre, H., Johnson, K., King, B., Moum. J., Oke, P., Owens, W.B., Pouliquen, S., Purkey, S., Scanderbeg, M., Suga, T., Wijffels, S., Zilberman, N., Bakker, D., Baringer, M., Belbeoch, M., Bittig, H.C., Boss, E., Calil, P., Carse, F., Carval, T., Chai, F., Conchubhair, D.Ó., d'Ortenzio, F., Dall'Olmo, G., Desbruyeres, D., Fennel, K., Fer, I., Ferrari, R., Forget, G., Freeland, H., Fujiki, T., Gehlen, M., Greenan, B., Hallberg, R., Hibiya, T., Hosoda, S., Jayne, S., Jochum, M., Johnson, G.C., Kang, K., Kolodziejczyk, N., Körtzinger, A., Le Traon, P.-Y., Lenn, Y.-D., Maze, G., Mork, K.A., Morris, T., Nagai, T., Nash, J., Naveira Garabato, A., Olsen, A., Pattabhi, R.R., Prakash, S., Riser, S., Schmechtig, C., Schmid, C., Shroyer, E., Sterl, A., Sutton, P., Talley, L., Tanhua, T., Thierry, V., Thomalla, S., Toole, J., Troisi, A., Trull, T.W., Turton, J., Velez-Belchi, P.J., Walczowski, W., Wang, H., Wanninkhof, R., Waterhouse, A.F., Waterman, S., Watson, A., Wilson, C., Wong, A.P.S., Xu, J., Yasuda, I., 2019. On the future of Argo: a global, full-depth, multi-disciplinary array. Front. Mar. Sci. 6, 439. https://doi.org/10.3389/fmars.2019.00439.

Ryabinin, V., Barbière, J., Haugan, P., Kullenberg, G., Smith, N., McLean, C., Troisi, A., Fische, r A., Aricò, S., Aarup, T., Pissierssens, P., Visbeck, M., Eneveldsen, H.O., Rigaud, J., 2019. The UN decade of ocean science for sustainable development. Front. Mar. Sci. 6, 470. https://doi.org/10.3389/fmars.2019.00470.

Sammartino, M., Buongiorno Nardelli, B., Marullo, S., Santoleri, R., 2020. An artificial neural network to infer the mediterranean 3D chlorophyll-a and temperature fields from remote sensing observations. Rem. Sens. 12 (24), 4123.

Sattler, K.U., 2009. Data quality dimensions. In: Liu, L., Özsu, M.T. (Eds.), Encyclopedia of Database Systems. Springer, Boston, MA. https://doi.org/10.1007/978-0-387-39940-9_108.

Schlitzer, R., 2002. Interactive analysis and visualization of geoscience data with ocean data view. Comput. Geosci. 28 (10), 1211−1218.

Sepp Neves, A.A., Pinardi, N., Martins, F., 2016. IT-OSRA: applying ensemble simulations to estimate the oil spill risk associated to operational and accidental oil spills. Ocean Dynam. 66, 939−954. https://doi.org/10.1007/s10236-016-0960-0.

Sepp Neves, A.A., Pinardi, N., Navarra, A., Trotta, F., 2020. A general methodology for beached oil spill hazard mapping. Front. Mar. Sci. 7, 65. https://doi.org/10.3389/fmars.2020.00065.

Shepherd, I., 2018. European efforts to make marine data more accessible. Ethics Sci. Environ. Polit. 18, 75−81. https://doi.org/10.3354/esep00181.

Simoncelli, S., Coatanoan, C., Myroshnychenko, V., Bäck, O., Sagen, H., Scory, S, Oliveri, P., Shahzadi, K., Pinardi, N., Barth, A., Troupin, C., Schlitzer, R., Fichaut, M, Schaap, D., 2021. SeaDataCloud Data Products for the European marginal seas and the Global Ocean. Proceedings of the 9th International EuroGOOS Conference — Advances in Operational Oceanography: Expanding Europe's ocean observing and forecasting capacity. 3-5 May 2021, Online. EuroGOOS. https://eurogoos-conference.ifremer.fr/.

Simoncelli, S., Fichaut, M., Schaap, D., Schlitzer, R., Alexander, B., Fratianni, C., 2019. Marine open data: a way to stimulate ocean science through EMODnet and SeaDataNet initiatives. INGV Workshop on Marine Environment. In: Sagnotti, L., Beranzoli, L., Caruso, C., Guardato, S., Simoncelli, S. (Eds.), In: Miscellanea INGV, 51. Istituto Nazionale di Geofisica e Vulcanologia, Rome, pp. 99−101. https://doi.org/10.13127/misc/51.

Simoncelli, S., Oliveri, P., Mattia, G., 2020a. SeaDataCloud Mediterranean Sea - V2 Temperature and Salinity Climatology. https://doi.org/10.12770/3f8eaace-9f9b-4b1b-a7a4-9c55270e205a.

Simoncelli, S., Oliveri, P., Mattia, G., Myroshnychenko, V., Barth, A., Troupin, A., 2020b. SeaDataCloud Temperature and Salinity Climatology for the Mediterranean Sea (Version 2). Product Information Document (PIDoc). https://doi.org/10.13155/77514.

Sloyan, B.M., Wilkin, J., Hill, K.L., Chidichimo, M.P., Cronin, M.F., Johannessen, J.A., Karstensen, J., Krug, M., Lee, T., Oka, E., Palmer, M.D., Rabe, B., Speich, S., von Schuckmann, K., Weller, R.A., Yu, W., 2019. Evolving the physical global ocean observing system for research and application services through international coordination. Front. Mar. Sci. 6, 449. https://doi.org/10.3389/fmars.2019.00449.

Sonogashira, M., Shonai, M., Iiyama, M., 2020. High-resolution bathymetry by deep learning-based image super resolution. PLoS One 15 (7), e0235487. https://doi.org/10.1371/journal. pone.0235487.

Stammer, D., Balmaseda, M., Heimbach, P., Köhl, A., Weaver, A., 2016. ocean data assimilation in support of climate applications: status and perspectives. Annu. Rev. Mar. Sci. 8 (1), 491−518. https://doi.org/10.1146/annurev-marine-122414-034113.

Storto, A., Dobricic, S., Masina, S., Di Pietro, P., 2011. Assimilating along-track altimetric observations through local hydrostatic adjustment in a global ocean variational assimilation system. Mon. Weather Rev. 139, 738−754. https://doi.org/10.1175/2010MWR3350.1.

Storto, A., Masina, S., 2016. C-GLORSv5: an improved multipurpose global ocean eddy-permitting physical reanalysis. Earth Syst. Sci. Data 8, 679—696. https://doi.org/10.5194/essd-8-679-2016.

Storto, A., Masina, S., Balmaseda, M., Guinehut, S., Xue, Y., Szekely, T., et al., 2017. Steric sea level variability (1993—2010) in an ensemble of ocean reanalyses and objective analyses. Clim. Dynam. 49 (3), 709—729. https://doi.org/10.1007/s00382-015-2554-9.

Storto, A., Masina, S., Simoncelli, S., et al., 2019. The added value of the multi-system spread information for ocean heat content and steric sea level investigations in the CMEMS GREP ensemble reanalysis product. Clim. Dynam. 53, 287—312. https://doi.org/10.1007/s00382-018-4585-5.

Tanhua, T., Pouliquen, S., Hausman, J., O'Brien, K., Bricher, P., de Bruin, T., Buck, J.J.H., Burger, E.F., Carval, T., Casey, K.S., Diggs, S., Giorgetti, A., Glaves, H., Harscoat, V., Kinkade, D., Muelbert, J.H., Novellino, A., Pfeil, B., Pulsifer, P.L., Van de Putte, A., Robinson, E., Schaap, D., Smirnov, A., Smith, N., Snowden, D., Spears, T., Stall, S., Tacoma, M., Thijsse, P., Tronstad, S., Vandenberghe, T., Wengren, M., Wyborn, L., Zhao, Z., 2019. Ocean FAIR data services. Front. Mar. Sci. 6, 440. https://doi.org/10.3389/fmars.2019.00440.

Thiebaux, J., 1986. Anisotropic correlation functions for objective analysis. Mon. Weather Rev. 104, 994—1002. https://doi.org/10.1175/1520-0493(1976)104<0994:ACFFOA>2.0.CO;2.

Toth, K., Portele, C., Illert, A., Lutz, M., Nunes de Lima, M., 2012. A Conceptual Model for Developing Interoperability Specifications in Spatial Data Infrastructures. Publication office of the European Union, Luxembourg. https://doi.org/10.2788/21003. EUR — Scientific and Technical Research Series, ISSN1831-9424.

Treasure, A.M., Roquet, F., Ansorge, I.J., Bester, M.N., Boehme, L., Bornemann, H., Charrassin, J.-B., Chevallier, D., Costa, D.P., Fedak, M.A., Guinet, C., Hammill, M.O., Harcourt, R.G., Hindell, M.A., Kovacs, K.M., Lea, M.-A., Lovell, P., Lowther, A.D., Lydersen, C., McIntyre, T., McMahon, C.R., Muelbert, M.M.C., Nicholls, K., Picard, B., Reverdin, G., Trites, A.W., Williams, G.D., de Bruyn, P.J.N., 2017. Marine Mammals exploring the oceans Pole to Pole: a review of the MEOP consortium. Oceanography 30 (2), 132—138. https://doi.org/10.5670/oceanog.2017.234.

Troupin, C., Machín, F., Ouberdous, M., Sirjacobs, D., Barth, A., Beckers, J.-M., 2010. High-resolution climatology of the northeast atlantic using data-interpolating variational analysis (Diva). J. Geophys. Res. 115, C08005. https://doi.org/10.1029/2009JC005512.

Troupin, C., Barth, A., Sirjacobs, D., Ouberdous, M., Brankart, J.M., Brasseur, P., Rixen, M., Alvera-Azcárate, A., Belounis, M., Capet, A., Lenartz, F., Toussaint, M.E., Beckers, J.M., 2012. Generation of analysis and consistent error fields using the Data Interpolating Variational Analysis (DIVA). Ocean Model. 52 (53) https://doi.org/10.1016/j.ocemod.2012.05.002, 90—10.

UN General Assembly, October 21, 2015. Transforming Our World : The 2030 Agenda for Sustainable Development. A/RES/70/1, Available at: https://www.refworld.org/docid/57b6e3e44.html. (Accessed 19 February 2021).

von Schuckmann, K., Storto, A., Simoncelli, S., Raj, R.P., Samuelsen, A., de Pascual Collar, A., et al., 2018. Ocean heat content. In: von Schuckmann, K., Le Traon, P.-Y., Smith, N., Pascual, A., Brasseur, P., Fennel, K., et al. (Eds.), Copernicus Marine Service Ocean State Report, Issue 2, Journal of Operational Oceanography, vol. 11. Ditcot: Taylor & Francis, p. s41. https://doi.org/10.1080/1755876X.2018.1489208.

Wackernagel, H., 2003. Multivariate Geostatistics: An Introduction with Applications, third ed. Springer-Verlag.

Wahba, G., Wendelberger, J., 1980. Some new mathematical methods for variational objective analysis using splines and cross validation. Mon. Weather Rev. 108, 1122−1143. https://doi.org/10.1175/1520-0493(1980)108<1122:SNMMFV>2.0.CO;2.

Wang, R.Y., Strong, D.M., 1996. Beyond accuracy: what data quality means to data consumers. J. Manag. Inf. Syst. 12 (4), 5−33. http://mitiq.mit.edu/Documents/Publications/TDQMpub/14_Beyond_Accuracy.pdf.

Wand, Y., Wang, R.Y., 1996. Anchoring data quality dimensions in ontological foundations. Commun. ACM 39, 11. http://web.mit.edu/tdqm/www/tdqmpub/WandWangCACMNov96.pdf.

Widmann, M., Goosse, H., van der Schrier, G., Schnur, R., Barkmeijer, J., 2010. Using data assimilation to study extratropical Northern Hemisphere climate over the last millennium. Clim. Past 6, 627−644. https://doi.org/10.5194/cp-6-627-2010.

Wilkinson, M.D., 2016. The FAIR Guiding Principles for scientific data management and stewardship. Sci. Data 3. https://doi.org/10.1038/sdata.2016.18.

Wilson, J.R., 1998. Global Temperature-Salinity Profile Program (GTSPP) - Overview and Future. UNESCO, Paris, France, 33pp. (Intergovernmental Oceanographic Commission Technical Series; 49). http://hdl.handle.net/11329/159.

Wong, A.P.S., Wijffels, S.E., Riser, S.C., Pouliquen, S., Hosoda, S., Roemmich, D., Gilson, J., Johnson, G.C., Martini, K., Murphy, D.J., Scanderbeg, M., Bhaskar, T.V.S.U., Buck, J.J.H., Merceur, F., Carval, T., Maze, G., Cabanes, C., André, X., Poffa, N., Yashayaev, I., Barker, P.M., Guinehut, S., Belbéoch, M., Ignaszewski, M., Baringer, M.O., Schmid, C., Lyman, J.M., McTaggart, K.E., Purkey, S.G., Zilberman, N., Alkire, M.B., Swift, D., Owens, W.B., Jayne, S.R., Hersh, C., Robbins, P., West-Mack, D., Bahr, F., Yoshida, S., Sutton, P.J.H., Cancouët, R., Coatanoan, C., Dobbler, D., Juan, A.G., Gourrion, J., Kolodziejczyk, N., Bernard, V., Bourlès, B., Claustre, H., D'Ortenzio, F., Le Reste, S., Le Traon, P.-Y., Rannou, J.-P., Saout-Grit, C., Speich, S., Thierry, V., Verbrugge, N., Angel-Benavides, I.M., Klein, B., Notarstefano, G., Poulain, P.-M., Vélez-Belchí, P., Suga, T., Ando, K., Iwasaska, N., Kobayashi, T., Masuda, S., Oka, E., Sato, K., Nakamura, T., Sato, K., Takatsuki, Y., Yoshida, T., Cowley, R., Lovell, J.L., Oke, P.R., van Wijk, E.M., Carse, F., Donnelly, M., Gould, W.J., Gowers, K., King, B.A., Loch, S.G., Mowat, M., Turton, J., Rama Rao, E.P., Ravichandran, M., Freeland, H.J., Gaboury, I., Gilbert, D., Greenan, B.J.W., Ouellet, M., Ross, T., Tran, A., Dong, M., Liu, Z., Xu, J., Kang, K., Jo, H., Kim, S.-D., Park, H.-M., 2020. Argo data 1999−2019: two million temperature-salinity profiles and subsurface velocity observations from a global array of profiling floats. Front. Mar. Sci. 7, 700. https://doi.org/10.3389/fmars.2020.00700.

Wunsch, C., 2005. Towards the World Ocean Circulation Experiment and a Bit of Aftermath. http://ocean.mit.edu/~cwunsch/papersonline/wocehistory.pdf?origin=publicationDetail. (Accessed 1 April 2021).

Yang, C., Leonelli, F.E., Marullo, S., Artale, V., Beggs, H., Buongiorno Nardelli, B., Chin, T.M., De Toma, V., Good, S., Huang, B., Merchant, C.J., Sakurai, T., Santoleri, R., Vazquez-Cuervo, J., Zhang, H.-M., Pisano, A., 2021. Sea surface temperature intercomparison in the framework of the Copernicus climate change service (C3S). J. Clim. 1−102. https://doi.org/10.1175/JCLI-D-20-0793.1. Retrieved May 14, 2021, from https://journals.ametsoc.org/view/journals/clim/aop/JCLI-D-20-0793.1/JCLI-D-20-0793.1.xml.

Yang, C., Masina, S., Storto, A., 2017. Historical ocean reanalyses (1900−2010) using different data assimilation strategies. Q. J. R. Meteorol. Soc. 143, 479−493. https://doi.org/10.1002/qj.2936.

Zeng, Y., Su, Z., Barmpadimos, I., Perrels, A., Poli, P., Boersma, K.F., Frey, A., Ma, X., de Bruin, K., Goosen, H., John, V.O., Roebeling, R., Schulz, J., Timmermans, W., 2019. Towards a traceable climate service: assessment of quality and usability of essential climate variables. Rem. Sens. 11, 1186. https://doi.org/10.3390/rs11101186.

Zweng, M.M., Reagan, J.R., Seidov, D., Boyer, T.P., Locarnini, R.A., Garcia, H.E., Mishonov, A.V., Baranova, O.K., Weathers, K., Paver, C.R., Smolyar, I., 2018. World Ocean Atlas 2018, volume 2: salinity. A. Mishonov technical. In: NOAA Atlas NESDIS 82, 50 pp.

PART FOUR

Education

CHAPTER FIVE

Connecting marine data to society

Kate E. Larkin[1,2], Andrée-Anne Marsan[1,2], Nathalie Tonné[1,2], Nathalie Van Isacker[1,2], Tim Collart[1,2], Conor Delaney[1,2], Mickaël Vasquez[3], Eleonora Manca[4], Helen Lillis[4], Jan-Bart Calewaert[1,2]

[1]Seascape Belgium bvba, Brussels, Belgium
[2]European Marine Observation and Data Network (EMODnet) Secretariat, Ostend, Belgium
[3]Ifremer, Brest, France
[4]Joint Nature Conservation Committee (JNCC), Peterborough, United Kingdom

Introduction

High-quality, societally relevant ocean data and information are crucial for society, not least to underpin economic activities at sea, enable knowledge creation to further understand the ocean's complex role in the planetary system, understand human interactions and impacts, and empower decision-makers with information for evidence-based decision-making for sustainable management of the Ocean and adaptation to climate change. Such marine data are also a crucial component of the knowledge base to meet international policies including the UN2030 Agenda and its 17 Sustainable Development Goals, together with the Paris agreement treaty on climate change.

In Europe, well-integrated, accessible, and timely data information on the marine environment are already crucial for member states to assess the marine environmental status of their national waters, provide national marine spatial plans, and more, as required by Directives under the European Integrated Maritime Policy (Shepherd, 2018). In addition, marine data are increasingly important to meet the ambitious targets set out in the European Green Deal, and its related Climate Pact, pledging for Europe to be climate neutral by 2050 and the need for understanding our inextricable link with the ocean and its crucial role in shaping our natural and human-centric world (https://ec.europa.eu/info/sites/info/files/european-green-deal-communication_en.pdf). In turn, marine data will also present opportunities and solutions for a greener transition of the blue economy and for society as a whole.

Ocean Science Data
ISBN: 978-0-12-823427-3
https://doi.org/10.1016/B978-0-12-823427-3.00003-7

© 2022 Elsevier Inc.
All rights reserved.

Moreover, the COVID-19 global pandemic, which began in 2019, has emphasized still further the central and crucial role of open science and open data to enhance public access to research results and research data, and drive data interoperability, data combination, and reuse to enable scientific research and innovation. This is in line with the eight ambitions of the EU Open Science Policy, including the overarching goal to move toward a shared research knowledge system by 2030 (https://ec.europa.eu/research/openscience/pdf/ec_rtd_ospp-final-report.pdf).

Yet, despite the clear societal relevance and demand for open and free access to marine data, the achievement of standardized, harmonized, and integrated data from diverse sources is relatively recent. In 2012 when the European Commission Marine Knowledge 2020 Green Paper was published (European Commission, 2012a), much of Europe's marine data remained scattered in unconnected databases and repositories. Even when data were available, they were often not compatible, making the sharing of the information and data-aggregation impossible. The EC Marine Knowledge 2020 vision was therefore simple: *To bring together diverse and disparate marine data, harmonize and integrate these, and deliver high resolution maps of our European seas. Maps of seafloor topography, seabed geology, seabed habitats, all accompanied by collated, harmonized and integrated diverse marine data; from biodiversity to maritime human activities, providing also timely information on the physical, chemical and biological state of the water column.*

The solution to this was the European Marine Observation and Data Network (EMODnet), a long-term, marine data initiative funded by the European Maritime and Fisheries Fund. Over the past decade, EMODnet has formed and consolidated a network of over 120 organizations that deliver data management and technical expertise to enable free and open access to integrated, standardized, and harmonized marine data and data products for European seas and beyond (European Commission, 2012b; Calewaert et al., 2016). These organizations together have made EMODnet what it is today: A marine domain leader in Europe providing essential marine knowledge brokerage to access the most comprehensive in-situ marine datasets—and related human activity data at sea that are Findable, Accessible, Interoperable, and Reusable (FAIR) (Wilkinson et al., 2016; ECO Magazine, 2020b).

FAIR data are also crucial to achieving fully open data and enabling open science. The European Open Science Cloud (EOSC), initiated by the European Commission in 2015, and building on the EC Open Data Policy (European Commission, 2003) has further driven data providers and data services toward offering findable, accessible, interoperable, and reusable (FAIR) open data, as a contribution to open science and information for societal benefit (see Chapters 2 and 3). In fact, the EOSC initiative—a contribution to the European Cloud Initiative and EU digital single market (https://ec.europa.eu/digital-single-market/en/european-cloud-initiative)— has the vision to be an open, trusted, and scalable data space for science, research, and innovation, with the aim to support more than 1.7 million researchers and boosting interdisciplinary research in Europe.

At international levels there is a drive to build a global consensus on Open Science (UNESCO, 2019), and this will lead to concrete measures on Open Access and Open Data that bring citizens closer to science, and commitments for a better distribution and production of science in the world. An article in 2020 in *Nature* by Brett et al. advocated for communities to open up, share, and network information so that marine stewardship can mitigate climate change, overfishing, and pollution. Writing about the rapid growth in ocean information over the past decade and the metaphorical "data tsunami," they noted how sharing must be established as a new default, unless there are compelling constraints (see also Chapter 4).

The UN Ocean Decade on Ocean Science Sustainable Development is an overarching framework to build on existing capability and fully realize open data and open science to drive a transparent and accessible ocean, together with underpinning all the thematic areas of the UN Decade, including facilitating Ocean Literacy through democratizing access to data, information, and knowledge.

EMODnet: a marine knowledge broker for society

The EMODnet is a long-term, marine data initiative funded by the European Commission. Initiated in 2009, EMODnet is a pan-European marine knowledge broker, providing a gateway to marine data and data products. A central philosophy of EMODnet is to "collect [data] once and use many times" and to connect across the marine knowledge value chain with a central aim to increase productivity in all tasks involving marine data, to promote innovation, and to reduce uncertainty about the behavior of the sea (see the EMODnet Vision Document published by EMB-EuroGOOS, 2008).

The European Commission's vision for EMODnet was clear from its inception, with 2020 as the landmark year to deliver maps of our European seas covering topography, geology, habitats, and ecosystems, at the highest resolution possible in areas that have been surveyed. These were to be accompanied by collated, harmonized, and integrated marine data spanning from ocean surface to seafloor, providing timely information on the physical, chemical, and biological state of the overlying water column (European Commission, 2010; 2012a,b). A core principle throughout has been to enable the free and open access to marine data, both from scientific research and from wider stakeholders that collect marine data, e.g., from the public authority marine monitoring programs for policy requirements or private sector ocean observing platforms, providing datasets for blue economy operational purposes.

Following a decade of progress and achievements (https://www.emodnet.eu/en/ten-years-emodnet-ten-minutes), EMODnet is now a mature data initiative and marine knowledge broker, offering open and free access to data and data products in European seas and beyond, spanning seven broad disciplinary themes: bathymetry, geology, physics, chemistry, biology, seabed habitats, and human activities related to the sea (Calewaert et al., 2016; Martín Míguez et al., 2019; Larkin and Calewaert, 2020). These integrated, standardized, and harmonized data and data products are made possible by a large network of over 120 organizations and many more experts supported by the EU's Integrated Maritime Policy and wider European and national efforts, who work together to observe the sea, process the data according to international standards, and make that information freely available as interoperable data layers and data products. This large network and "ecosystem" of people, organizations, and networks include thousands of diverse data producers, e.g., national capability infrastructures, European Research Infrastructures, the European Global Ocean Observing System (EuroGOOS), government monitoring agencies, hydrographic organizations, civil society, and the private sector, taking ocean observations and collecting marine data at sea. To this end, EMODnet has also played a pivotal role in connecting across European ocean observing and marine monitoring communities, building on existing efforts (see Chapters 2, 3 and 4), (European Marine Board, 2013) and leading the organization of an international conference in 2018 on the topic of enhancing dialogue and connection across the existing European Ocean Observing capabilities, attended by over 250 key stakeholders representing the public and private sectors spanning research, industry, civil society, and policy (EOOS Conference Report, 2018 - Larkin et al., 2019).

EMODnet also works with many diverse entities involved on data quality control, assurance, curation, and management, initiated in most cases at the organizational and national levels, e.g., through national data centers, regional efforts, e.g., through the Regional Sea Convention data integration for sea-basin assessments, and to the European level through data infrastructures such as SeaDataNet and SeaDataCloud. Research infrastructures and e-infrastructures have also paved the way for the standardized, harmonized, and integrated multidisciplinary marine data. And, building on this capability, projects such as the Horizon 2020 Blue-Cloud pilot project (https://www.blue-cloud.org/)—involving direct partnership with such key data services—have further driven the dialogue and technical advancements toward web-based science and a marine domain contribution to EOSC.

EMODnet does not only aggregate data but it has been a key adopter of standards for the marine domain, both within and across disciplines. This is crucial for enabling interoperability of data allowing greater discovery and access by users and wider society. Working with data management experts at national, regional, and European level (e.g., from SeaDataNet), EMODnet standardizes and harmonizes the data to meet European and International standards such as the INfrastructure for SPatial InfoRmation in Europe (INSPIRE) geospatial data standards (European Commission, 2007a) which require rich metadata descriptions alongside datasets and products, clear information on the provenance of the data, meaning where the data comes from and how it has been treated—producer and provider and the quality and uncertainties associated. This is further explained in Chapters 2 and 3.

EMODnet overcomes many of the bottlenecks that previously prevented marine data delivering their full benefits, by connecting the huge diversity of data collection efforts resulting in data being scattered across different institutions in the EU, including hydrographic offices, geological surveys, local authorities, environmental agencies, research institutes, and universities (Shepherd, 2018; European Commission, 2012a,b). In delivering seamless access to comprehensive marine datasets this thereby reduces the risks (and costs) associated with private and public investments in the blue economy, and facilitates more effective protection of the marine environment. EMODnet is also a dynamic system and continuously enhances the number and type of platforms in the system by unlocking and providing high-quality data from a growing network of providers (Manzella et al., 2018), which include the wider research community including marine and wider environmental research infrastructures and the private sector, e.g., blue economy businesses that are increasingly submitting data to the EMODnet Data Ingestion service (https://www.emodnet-ingestion.eu/).

EMODnet works in a diverse and complex landscape. It continues to develop a number of key partnerships, including a close collaboration with other long-term European marine data initiatives such as Copernicus Marine Environment Monitoring Service (CMEMS) of the Copernicus Space Program which primarily delivers satellite-derived marine environmental datasets with global coverage, funded by the European Commission Directorate General for Defense, Industry, and Space (DG Defis). This continuing and expanding collaboration has led to cross-validation of satellite and in-situ datasets, user-driven data products, and complementary developments to further add value to the user-experience. Collaborations between EMODnet and CMEMS include operational data flow between the EMODnet thematics (particularly physics and chemistry) with the CMEMS in-situ Thematic Assembly Center (INSTAC), joint development of catalogs, e.g., for MSFD and MSP (in progress) and coordination, communication, and capacity building activities, e.g., joint workshops with industry sectors, hackathons, etc. Through this collaboration and complementarity, EMODnet and CMEMS, together with the Data Collection Framework for fisheries, collectively implement the EU's Marine Knowledge 2020 strategy.

EMODnet also increasingly operates in a global context. Activities span global data coverage for some themes, e.g., physics and biology, international dialogue in data sharing, including the EMOD-PACE (EMODnet PArtnership for China and Europe) project, and contribution of EMODnet Biology to the international Ocean Biogeographic Information System (OBIS), EMODnet Bathymetry's European contribution to the international Seabed 2030 initiative. EMODnet is also a European focal point for the developing IOC Ocean Information Hub, an initiative of IOC-IODE (EMODnet, 2020). All of these are made possible by standardization and interoperability of data, toward making it available to global users and wider society.

The benefits of working together are clearly illustrated by the ability to produce seamlessly integrated map layers of data collected from thousands of diverse sources as shown with some examples of reference maps in Fig. 5.1. The visualization of geospatial data is a powerful tool and useful resource, promoting the use of EMODnet data and data products by a greater range of users, beyond the classic technical research community to include other marine and maritime professionals in the fields of policy, civil society, and industry. An example is the particularly popular EMODnet Bathymetry high resolution Digital Terrain Model (DTM) data product (see Fig. 5.2 from

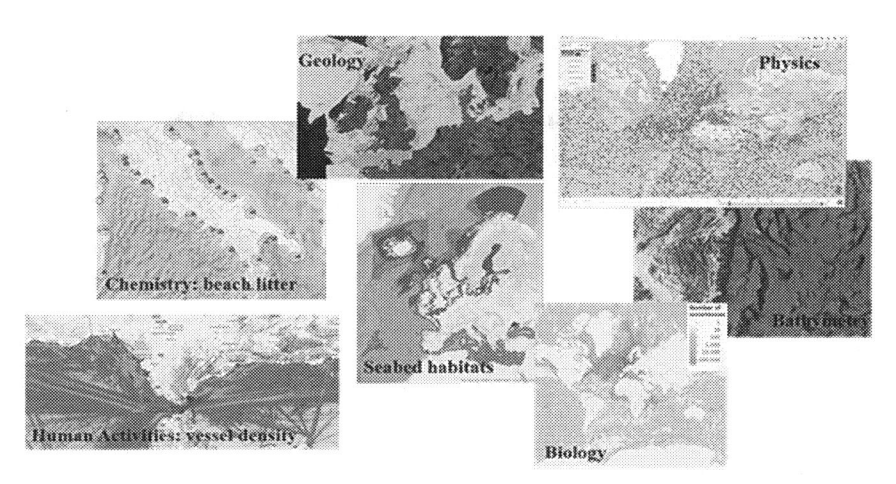

Figure 5.1 Examples of EMODnet map layers from each of the seven thematic areas of EMODnet bathymetry, biology, chemistry, geology, human activities, physics, and seabed habitats. Each thematic has many diverse datasets, available as integrated maps and for download as individual datasets, together with added value data products, for discovery at the EMODnet Central Portal www.emodnet.eu.

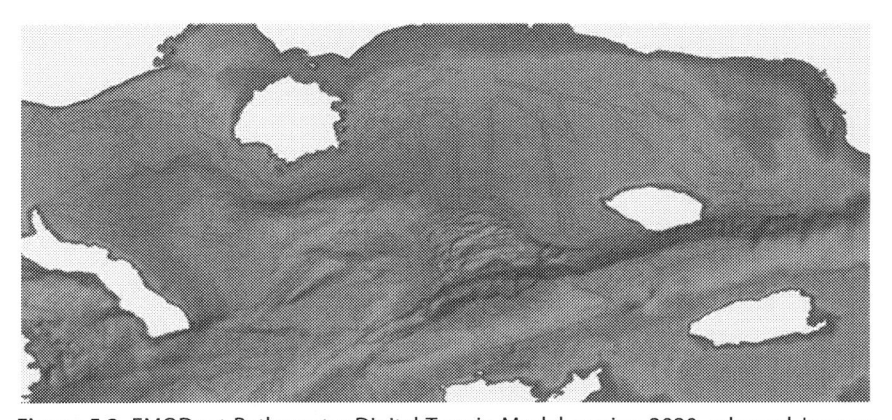

Figure 5.2 EMODnet Bathymetry Digital Terrain Model version 2020, released January 2021, displaying a region along the coast of Greece with new survey tracks highlighted since the previous release in 2018. The harmonized EMODnet DTM is complete for European sea regions (36W,15N; 43E,90N) based on selected bathymetric survey datasets, composite DTMs, Satellite Derived Bathymetry (SDB) data products, while gaps with no coverage are completed by integrating the GEBCO Global Bathymetry. The resolution of the EMODnet DTM (1/16 arc minutes) to indicate its added value over GEBCO (15 arc seconds/1/4 arc minutes). It is available free of charge for viewing and downloading, and sharing by OGC web services from the EMODnet Bathymetry portal. *EMODnet Bathymetry, for public use.*

the 2020 version – https://emodnet.eu/en/emodnet_bathymetry_digital_terrain_model_dtm_2020), which is a harmonized data product complete for European sea regions (36W,15N; 43E,90N) and based on selected bathymetric survey datasets, composite DTMs, Satellite-Derived Bathymetry (SDB) data products, with gaps in coverage being completed by integrating the GEBCO Global Bathymetry. In addition, in 2021 EMODnet Bathymetry released a World Base Layer Service (EBWBL), providing the highest resolution bathymetric worldwide layout (ECO Magazine, 2020a).

The number and diversity of private sector blue economy and ICT businesses that contribute to, and benefit from, EMODnet increase each year spanning dredging and offshore renewables (including those transitioning from the oil and gas sector), to maritime shipping and yachting, aquaculture, fisheries, and ocean sensor development companies (see https://emodnet.eu/en/use-cases for a full list of use cases).

EMODnet Seabed Habitats' EUSeaMap: an integrated, multidisciplinary data product for research, policy, and ecosystem-based management

EMODnet Seabed Habitats (https://www.emodnet-seabedhabitats.eu/) provides a single access point to European seabed habitat data and products by assembling individual point datasets, maps, and models from various sources and publishing them as interoperable data products for assessing the environmental state of ecosystems and sea basins. It provides access to a growing library of habitat maps and ground-truthing datasets from surveys across Europe, using multiple habitat classification systems, including Marine Strategy Framework Directive broad habitats and the European Nature Information System (EUNIS) classification. Using these many data sources, it has built and published several composite products showing the known extent of important habitats in Europe, including three of the Global Ocean Observing System's (GOOS) Essential Ocean Variables (EOVs), namely seagrass beds, macroalgal canopy, and cold-water coral reefs. Without EMODnet, habitat data and maps would otherwise be scattered in different portals, in several classification systems, if published at all. And the process of building a regional overview of the extent of specific important habitats would be much more difficult.

Furthermore, EMODnet Seabed Habitats has created, and continues to improve, its flagship product: the EMODnet broad-scale seabed habitat map for Europe, known as EUSeaMap. EUSeaMap is produced using a top-down modeling approach combining classified habitat descriptors to

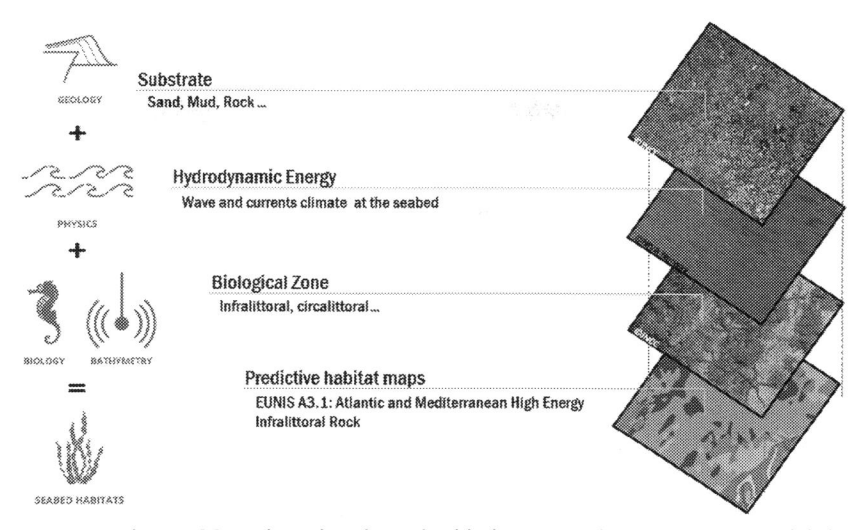

Figure 5.3 The EMODnet broad-scale seabed habitat map (EUSeaMap): A model data product with predictive capability made possible using EMODnet's diverse multithematic datasets, standardized to EU INSPIRE Directive geospatial data standards and integrated by EMODnet data experts, made available as open access and free of charge. Discover more at: https://www.emodnet-seabedhabitats.eu/.

describe a habitat type according to the EUNIS habitat classification and the MSFD broad habitat types. EUSeaMap covers all European Seas: the Mediterranean Sea, Black Sea, Baltic Sea, and areas of the north-east Atlantic extending from the Canary Islands in the south to the Barents Sea in the North. Habitat descriptors differ per region but include biological zone, energy class, oxygen regime, salinity regime, and seabed substrate (see Fig. 5.3).

EUSeaMap is a great example of the power of data integration, relying on the integration of data originating from at least four different EMODnet portals, namely bathymetry, biology, geology, and physics, to produce the map. This highlights what can be achieved through collaborations both within, and across, disciplines, to harmonize, standardize, and integrate data to create value-added products.

Application for policy and ecosystem-based management

Through its trusted service and powerful combination of integrated datasets, data visualization, and data products, EMODnet Seabed Habitats supports ecosystem-based management and evidence-based decision-making. For instance, EUSeaMap fulfills the EU member states' legal obligations under the Marine Strategy Framework Directive (MSFD): EU member states used broad-scale habitat maps, often in combination with survey data from the EMODnet library of habitat maps, in their first MSFD assessments in 2012, and most recently in the 2019 assessment. Regional broad-scale habitat maps are of particular use in cross-border assessments, which are required by the Marine Strategy Framework Directive (Andersen et al., 2018).

EUSeaMap was a key data source for defining ecosystem components in the Baltic Sea for computing cumulative pressure and impact indexes in the latest HELCOM holistic assessment of the ecosystem health of the Baltic Sea (HELCOM, 2018). EUSeaMap was also used for the assessment of cumulative impacts of human pressures on marine ecosystems in the north-east Atlantic. As part of the OSPAR Intermediate Assessment (2017) it was used in combination with fishing pressure data to produce maps of potential disturbance to benthic habitats due to fishing in the north-east Atlantic (OSPAR, 2017). Recently, the European Environment Agency performed a similar assessment at a Europe-wide level (Korpinen et al., 2019).

EMODnet Seabed Habitats supports Marine Protected Area (MPA) networks evaluations (including the Natura 2000 MPA network). EUSeaMap was key in evaluating whether the MPA network in the Western Mediterranean was achieving targets for benthic habitat coverage set by the Habitats Directive (Agnesi et al., 2017a). It was also applied in an assessment of the ecological coherence of the MPA network in the Baltic (HELCOM, 2016), and at European scale, for the assessment of Europe's MPA network by the European Environment Agency (Agnesi et al., 2017a).

EMODnet's broad-scale habitat maps have been identified as the best data source to represent marine benthic marine habitats in the Celtic Seas, with potential to support the implementation of a Marine Spatial Plan, as a result of the EU Marine Spatial Planning Directive 2014/89/EU (SIMCELT, 2016). EMODnet Seabed Habitats' products have already been used in Marine Spatial Planning in the United Kingdom (e.g., Defra, 2014a,b). Furthermore, EMODnet Seabed Habitats' datasets and products provide context for decision-makers involved in industry case work, hence supporting the implementation of the Habitats Directive and enforcing the protection of listed habitats.

Regarding the Birds Directive, the EUSeaMap mapping approach and EMODnet Seabed Habitats proposed Special Protection Areas designation. In Scotland, UK, for example, the data were used to identify the location and extent of some prey species and habitats that support inshore wintering waterfowl, and these will, in future, inform the management of Special Protection Areas.

Finally, an emerging use for EMODnet broad-scale habitat maps is the quantification of marine natural capital at a regional or national scale. EUSeaMap has been used in evaluating the capacity of ecosystems to provide services at a European level (Tempera et al., 2016) and initial marine natural capital accounts for the UK marine environment (Thornton et al., 2019). To find out more about these use cases and more, visit the EMODnet website (https://www.emodnet.eu/en/use-cases?field_portal_taxonomy_tid=29&field_case_type_tid=43).

Wider data visualization tools and applications

Data visualization is also in growing demand for use in complex and large-scale international research projects. Using open software such as GeoNode (Corti et al., 2019; https://geonode.org/) that can be tailored to specific functionalities and requirements enables geospatial data to be visualized during the lifetime of a project, with the possibility to set restrictions on data access if embargoes exist for scientific publication or commercial agreements. This offers researchers a community tool for sharing, visualizing, and conceptualizing data between project partners, even before the raw data are ready to be made open access. And as a tool for wider society, it serves as a "window" to promote a project's new data and research outputs to wider stakeholders, e.g., marine spatial planners, conservation managers, blue economy operators, etc. For example, the European Horizon 2020 project ATLAS (2015–20) provided the first coherent, integrated basin-scale assessment of Atlantic deep-water ecosystems and their Blue Growth potential. Building on, and expanding this concept, the H2020 iAtlantic project (started in 2019) is a multidisciplinary research program seeking to assess the health, e.g., the stability and vulnerability, of deep-sea and open-ocean ecosystems across the full span of the Atlantic Ocean.

Examples of the GeoNode data visualization tools for both projects are presented in Figs. 5.4 and 5.5. These examples demonstrate the potential for enhancing the exchange of (geospatial) data and information between project partners during the lifetime of the project, and with wider stakeholders, without requiring the technical expertise to access and plot the diverse datasets. In this way, using GIS platforms to visualize data improves the promotion, communication, and ultimately the transfer of research outputs to science, policy, and industry stakeholders. It also encourages wider data providers to share data layers, including from more diverse knowledge sources including industry, civil society, and even indigenous

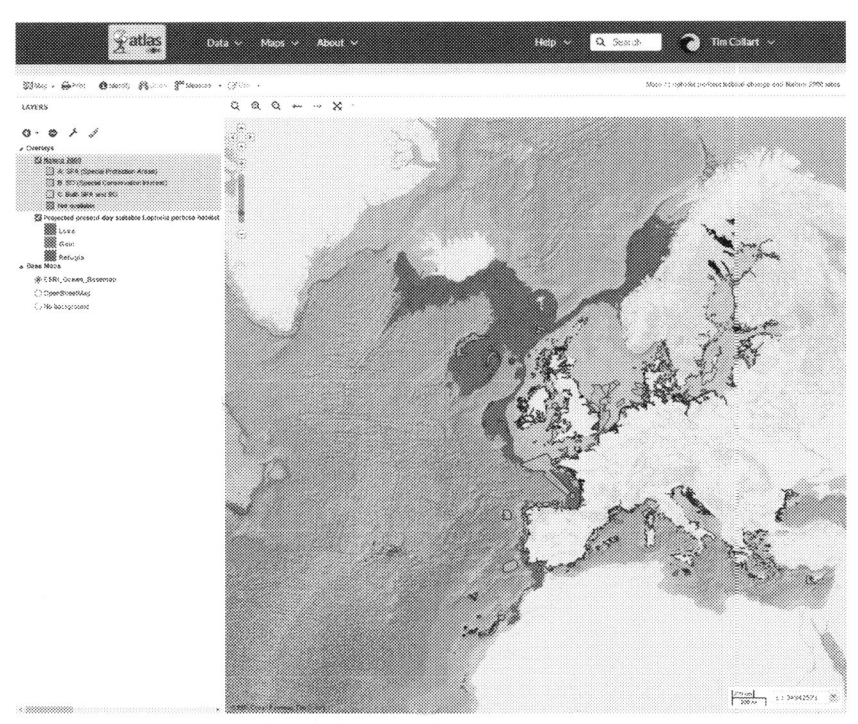

Figure 5.4 H2020 ATLAS GeoNode (www.atlas-horizon2020.eu) of the European H2020 ATLAS project (2016–20) http://www.eu-atlas.org/. The map shows the projected change in the habitat of the deep water reef building coral, *Lophelia pertusa*, as determined by the H2020 Atlas project, data citation: https://doi.org/10.1111/gcb.14996. This data layer is overlain with the outline of Natura 2000 marine protected areas provided by EMODnet Human Activities project, www.emodnet-humanactivities.eu, funded by the European Commission Directorate General for Maritime Affairs and Fisheries. The EMODnet dataset of Natura 2000 sites was created by Cogea for EMODnet in 2014 based on GIS Data from the European Environmental Agency (EEA), plus additional info, links, and selected EEA tabular data joined to the feature attributes, as well as a calculation by Cogea of marine and coastal location of features. Natura 2000 is an ecological network composed of sites designated under the Birds Directive (Special Protection Areas, SPAs) and the Habitats Directive (Sites of Community Importance, SCIs, and Special Areas of Conservation, SACs). The dataset covers the whole EU and the UK. The whole dataset is available for download on the EMODnet Human Activities portal.

peoples and local communities, further democratizing the access to data and the inclusivity of data sharing and codesign of research. Finally, these tools allow for easy transfer and ingestion of data layers to long-term open data systems, e.g., EMODnet for harmonization and integration into datasets that can be used to create data products for wider user uptake, beyond the lifetime of a project.

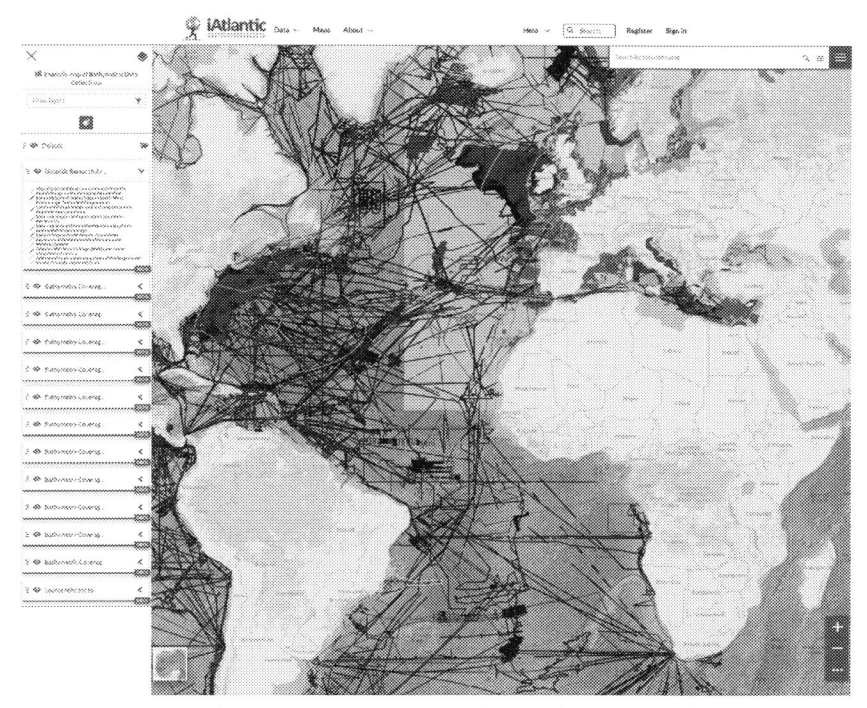

Figure 5.5 H2020 iAtlantic Composite map combining the outlines of existing bathymetry data from Data Center for Digital Bathymetry (DCDB), Helmholtz Center for Ocean Research Kiel (GEOMAR), bathymetric surveys used in the creation of the EMODnet European Sea Digital Terrain Model (linked as a remote layer) with the preliminary boundaries of the iAtlantic study areas (red [gray in printed version] outlined areas).

Marine data visualization for assessing blue economy human activity during COVID-19

Marine data visualization can also be used to provide socioeconomic information related to human activity at sea. For example, EMODnet vessel density maps offer composite maps tracking maritime shipping (trajectories, types of vessels, etc.) that are free and open access (https://www.emodnet.eu/en/new-insights-european-maritime-traffic-new-emodnet-vessel-density-maps) and European Maritime Safety Agency (EMSA) route density maps, also made available by EMODnet (http://www.emsa.europa.eu/newsroom/latest-news/item/3775-traffic-density-maps-for-a-better-understanding-of-maritime-traffic.html). When the COVID-19 pandemic first hit in 2020, data visualization was used in a number of ways to see the effect that COVID-19 was having on the blue economy and the marine environment. Since the early stages of the COVID-19 pandemic, the European Market Observatory for Fisheries and Aquaculture Products (EUMOFA) published a COVID-19 bulletin (https://www.eumofa.eu/COVID-19), highlighting how the fisheries sector activity at sea declined in European waters during the worst months of the pandemic (Fig. 5.6). In many regions of European seas, the fishing activity declined during 2020, with a reduction in EU sea basins, on average, of 18% compared to the same period in 2019, with greater reduction seen in areas of the Mediterranean Sea and the Celtic Sea.

Virtual nature

With the emergence of new digital tools and the wealth of marine data, it is now possible to create 3D visualizations of marine ecosystems by combining marine science and computer-generated imagery (CGI). But to make photorealistic visualizations and re-creations of the natural world, we need to base these virtual reality environments and virtual ecosystems (https://johanvandekoppel.nl/virtual-ecosystems/) on spatial ecology, including spatial self-organization models and accurate field data (Onrust et al., 2017). Such computer-aided digital simulations of the marine environment, e.g., in the Wadden Sea (see Fig. 5.7) make the ocean more tangible, inspiring citizens to engage in learning more about the ocean. Such tools can also really make a difference to decision-makers to visualize and conceptualize how ecosystems form, how they can be best managed, and predict how they may change. In the future, such virtual ecosystems open up all kinds of opportunities in terms of augmented and virtual reality.

The European Atlas of the Seas: an EU online communication tool for an increasingly blue, ocean literate society

To achieve an ocean literate society and connect people to our blue planet, it is crucial to communicate marine data and information in an attractive, easy-to-digest, and interactive way. This is the objective of the European Atlas of the Seas (ec.europa.eu/maritimeaffairs/atlas/maritime_atlas) (referred to here as "the Atlas"), launched in 2010 by the European Commission, Directorate-General for Maritime Affairs and Fisheries (DG MARE) (Barale et al., 2014). The European Atlas of the Seas is a free-to-use educational and interactive web mapping application for the general public, schools, and nonexpert professionals. It brings at-a-glance data in a comprehensive and fully visual way to a broad public, while at the same time serving as a support tool for marine policy and the blue economy (see Fig. 5.8).

April 2019

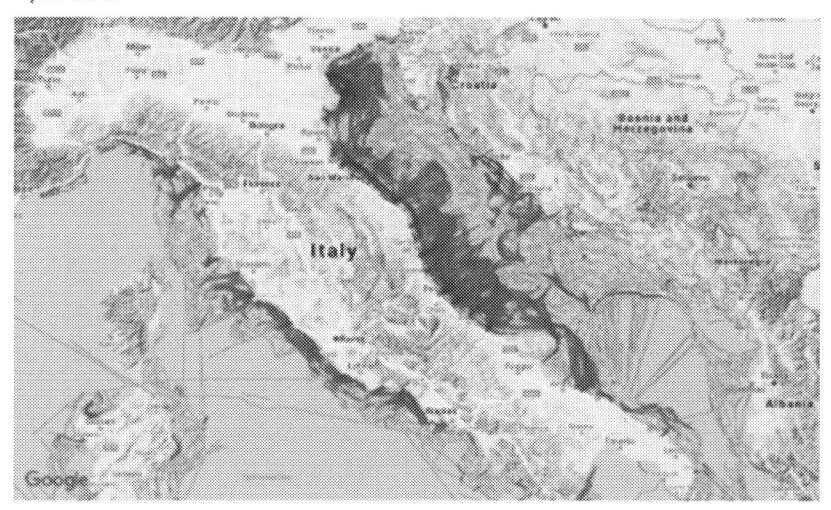

April 2020

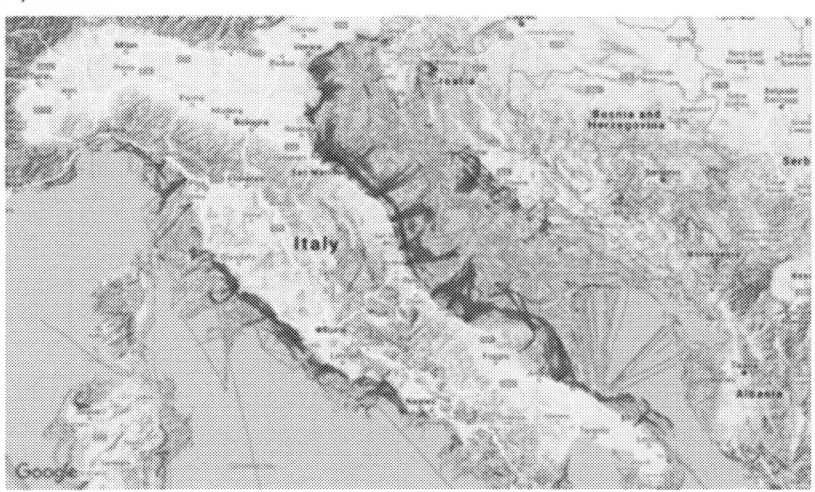

Figure 5.6 Fishing route density maps, derived from signals from ships' automatic identification systems (AIS) unique to the vessel and used for a number of maritime security purposes, including to avoid collisions. Similar conclusions can be drawn from the Vessel Monitoring System (VMS) used for fisheries control. Neither system includes the smaller artisanal vessels that fish nearer the coast, which make up a particularly high proportion of vessels in the Mediterranean. More information can be found here: https://www.emodnet-humanactivities.eu/blog/?p=1258.

Figure 5.7 A hound shark in Oosterschelde using computer-generated imagery (CGI) and virtual reality applications. *Jacco de Kok, Mo4Com visualisations/NIOZ. For further information, visit https://johanvandekoppel.nl/virtual-ecosystems/.*

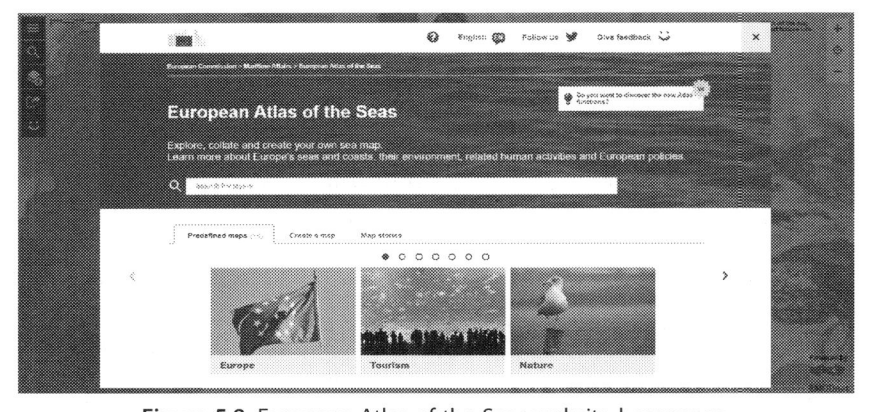

Figure 5.8 European Atlas of the Seas website homepage.

Marine and coastal maps for society, powered by open access data and science

The Atlas hosts a wealth of information about Europe's marine environment and related human activities, covering topics such as nature and sea life, environment, litter, tourism, security, energy, transport, sea bottom, sea surface temperature trends, sea level rise, coastal erosion, fish consumption and fishing activities, aquaculture, and much more. A mission of the Atlas is to raise awareness about Europe's seas and coasts in the context of the EU's Integrated Maritime Policy (https://ec.europa.eu/maritimeaffairs/policy_en). It also provides key data and information relevant to societal challenges addressed in European policies in the framework of the European Green Deal (https://ec.europa.eu/info/strategy/priorities-2019-2024/european-green-deal_en), including the European Climate Law and the European Climate Pact, the EU Biodiversity Strategy for 2030, the Farm to Fork strategy, the EU Strategy on offshore renewable energy, and the future zero pollution action plan.

The Atlas was set up as an interactive map viewer, providing nontechnical users with an easy interface to explore maps based on diverse source data and learn about Europe's marine environment and the ecosystem services it provides to society. It aims to encourage public engagement and promote learning about the ocean by giving the general public access to reliable information and by customizing the tool to their needs according to feedback provided by users and educators.

Users can explore an enriched catalog with almost 300 map layers that continues to grow with the data provider and further information marked clearly in the descriptive metadata. The Atlas is powered by EMODnet, with more than half of the datasets being provided by network's standardized and integrated datasets. EMODnet map layers are also automatically updated through the EMODnet web services, which means that the visitors of the Atlas have access to up-to-date information and data at all times.

The Atlas also includes a growing number of maps sourced from other open access datasets from diverse data collectors and providers across Europe. These include the European Commission, the European Environment Agency (EEA), the Joint Research Center (JRC), Copernicus, Eurostat, as well as Horizon 2020 research projects, and in the future others including iAtlantic. The contribution of these data providers is crucial for the success of the application and ensures a wide diversity of content.

For example, a collaboration between EMODnet and Copernicus Marine Service (CMEMS) has enabled a number of CMEMS open source data being made available on the Atlas. This includes the CMEMS Ocean Monitoring Indicators (OMI) layers, which are key variables used to track the vital health signs of the ocean and changes related to climate change such as sea surface temperature trends and sea level trends (see Fig. 5.9). Such maps are accompanied by a full abstract written for the lay-person, nonexpert audience to allow greater accessibility to users to understand these important datasets and the information they portray.

To further improve accessibility to all EU citizens, the EMODnet Secretariat launched, in September 2020, a fully revamped Atlas (version 6) which includes officially translated content in 24 EU languages and a number of new features. The launch of the new version was accompanied by a communication campaign including 24 ambassadors where a representative for each EU language introduces the Atlas and explains why the ocean is important for her or him. These developments have made the Atlas an even more inclusive tool and allowed to increase the visibility of the Atlas in countries across Europe. This has led to an increasing number and diversity of users with over 10,000 visits a month by the end of 2020 and with the growing audience coming from a more diverse range of (European) countries, underlining the added value of the high-quality translations now available in the Atlas.

To increase the collaboration with the users and respond to their needs, a feedback form is available on the Atlas where they are able to comment and ask questions in their own language. The Atlas, as a tool for the wider public, is therefore partly developed and improved by the users themselves, in order to feel part of this initiative more than just as users.

European Atlas of the Seas: Educational activities, ocean literacy, and possible applications

Already, the Atlas is an important tool for ocean literacy and education, as the web mapping application is being used by schools, public aquariums, and nonexpert professionals. A Teachers' Corner (https://webgate.ec. europa.eu/fpfis/wikis/display/AtlasOfSeas) is accessible since autumn 2018 to promote and facilitate the use of the Atlas by teachers and educators. It contains a number of exercises for different age groups based on the use of Atlas maps as well as communication tools specifically designed for teachers and students. The Teachers' Corner is an open platform where all users can find and share inspiration, information, educational material, and much more. Exercises are currently available in English and in French, and soon in Portuguese.

In order to make the Atlas even more aligned with the values, needs, and expectations of society, and more specifically of schools, two memorandums of understanding were signed in 2019 to promote and further develop the Atlas as a strong and must-have educational tool. A partnership was established between the EMODnet Secretariat and Nausicaá (www.nausicaa. fr), the biggest aquarium in Europe, with an important educational program. Furthermore, the Atlas has also partnered with the Directorate General for Maritime Policy (DGPM) (https://www.dgpm.mm.gov.pt/) in Portugal that is driving the project Escola Azul, a program with the goal to improve the level of ocean literacy in schools, leading to responsible and active generations that contribute for the ocean's sustainability.

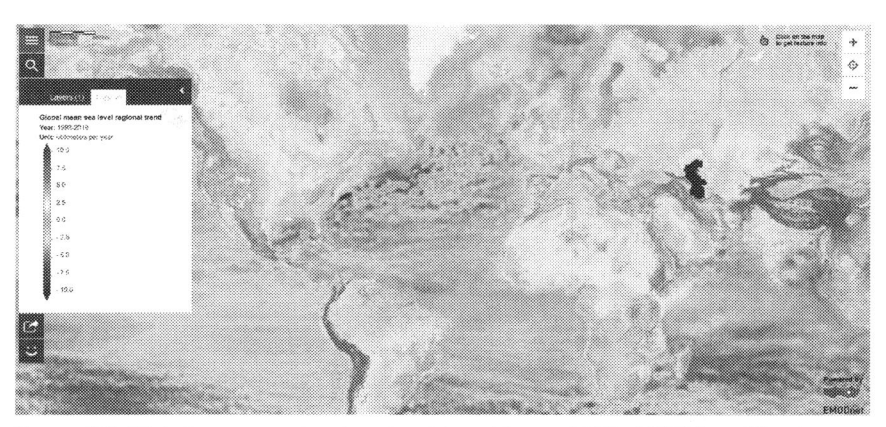

Figure 5.9 Global mean sea level regional trend (years 1993–2019) in millimeters per year from Copernicus Marine Environment Monitoring Service (CMEMS) data, made available on the European Commission communication tool, the European Atlas of the Seas, powered by EMODnet. https://ec.europa.eu/maritimeaffairs/atlas/maritime_ atlas/#lang=EN;p=w;bkgd=1;theme=128:0.8;c=617910.1422549915,6661522.512668 013;z=4; e = t Time-series images are available here: https://resources.marine. copernicus.eu/documents/IMG/GLOBAL_OMI_SL_area_averaged_anomalies-hq.png.

While the Atlas provides easy-to-use and reliable interactive marine information and maps, Nausicaá and the DGPM develop educational programs and tools for students and teachers, making the ocean a central focus of sustainable development and education of young people. The Atlas and these two partners are therefore complementary, with the Atlas being a content provider while Nausicaá and DGPM are experts in developing educational resources.

Nausicaá has already included the Atlas in their visitors' smartphone application, exhibition area, as well as in guides for teachers while they continue to provide expertise in ocean education. With DGPM, a series of four workshops with Escola Azul teachers is taking place to provide valuable feedback on how to use the Atlas in classrooms as well as to develop exercises for different age groups, using the Atlas.

These exciting collaborations as well as various upcoming workshops with schools and teachers are at the core of the development of the Atlas in order to make it a must-have resource in the context of ocean literacy. In this framework, the Atlas has been presented at the Hamburg Climate Week (https://www.klimawoche.de/) and to teachers across Europe through Scientix (http://www.scientix.eu/), a project managed by European Schoolnet that collects and promotes best practices in Science,

Technology, Engineering, and Mathematics (STEM) teaching and learning in Europe.

The Atlas can be used to teach many school subjects including, for example, geography, STEM, information and communication technology (ICT), economy, and languages. It enables students to gain new competences (e.g., reading and understanding maps and graphs, searching for information in a catalog of maps, working in different languages, etc.) and to extend their knowledge and stimulate their interest for a wide range of topics (e.g., climate change, environmental issues, nature conservation, blue economy, etc.).

As another way to connect with the public and raise awareness on relevant marine topics, a "Map of the week" is published every Friday on Twitter (@EuropeAtlasSeas), on the EMODnet website, and on the European Commission's Maritime forum (https://webgate.ec.europa.eu/maritimeforum/en/frontpage/1351). These short stories highlight an interesting map layer to explore based on current events, specific challenges, or general news.

The Atlas promotes a stronger collaboration between policymakers, scientists, and citizens. New maps are continuously added to the catalog, with the aim to further diversify the data sources, and to bring in maps with compelling information that can be described and used for educational purposes. More recently, map stories have been added in the Atlas that provide a more in-depth explanation of the marine phenomena and human activities portrayed in the maps in an engaging way (see Fig. 5.10).

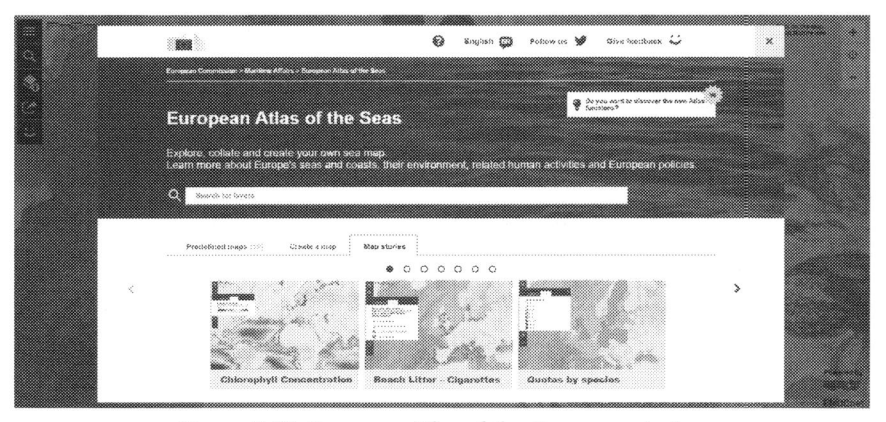

Figure 5.10 European Atlas of the Seas map stories.

Catalyzing and mobilizing citizens through ocean literacy

Digital tools for EU4Ocean

With the increased recognition across society that we live in a context of climate emergency, there is a need to raise societal awareness, engagement, and action on the role of the global ocean in regulating the wider planetary system. In connecting people to our Blue Planet and making them better understand the impact of their actions and the role they can play to better protect our ocean and environment, ocean literacy contributes to our well-being and that of the planet.

The European Ocean Coalition (EU4Ocean - https://webgate.ec. europa.eu/maritimeforum/en/frontpage/1482) was initiated in 2020 to connect diverse organizations, projects, and people that contribute to **ocean literacy** and the sustainable management of the ocean. Supported by the European Commission, this bottom-up inclusive initiative aims at uniting the voices of Europeans to make the ocean a concern of everyone (Fig. 5.11).

The Coalition has three key communities: Firstly, an EU4Ocean Platform as a focal point for a wide diversity of stakeholders spanning the areas

Figure 5.11 EU4Ocean coalition on the EC Maritime forum.

of marine research, science-policy, blue economy industry and the private sector, civil society, arts, education, youth, and media (Larkin et al., 2020). In addition, a Youth4Ocean community connects Europe's youth between the ages of 16−30, and a schools component aims to set up a European-wide Blue Schools network, building on the existing capability in some European countries (e.g., Portugal).

The EU4Ocean Coalition is an inclusive network and members codesign activities spanning three key thematic areas, namely Food from the Ocean; Climate and Ocean; and a Healthy and Clean Ocean. These activities build on and promote existing capabilities working across discipline and sector silos to communicate, engage, and inspire citizens to take action for our ocean and planet across multiple scales, from local and national organizations to regional sea basins and European initiatives.

The European Atlas of the Seas is a key digital tool for EU4Ocean and has been actively used in the design of a Teachers' Handbook for European Blue Schools and to visualize the membership of the growing EU4Ocean Platform network of more than 100 stakeholders (https://ec.europa.eu/maritimeaffairs/atlas/maritime_atlas/#lang=EN;p=w;bkgd=5;theme=354:0.8;c=926104.2403007429,6377788.263673515;z=5;e=t). It will be increasingly used in EU4Ocean educational workshops, and in the "Find the Blue" EU4Ocean campaign to get European schools involved in bringing ocean topics into school curricula and extracurricular activities (https://webgate.ec.europa.eu/maritimeforum/en/node/5494). Finally, the Atlas is used as input to the EU4Ocean Platform thematic Working Group activities and to the sea-basin workshops taking place in 2021 and 2022.

International context of ocean literacy

European efforts in ocean literacy are set in the context of international ocean literacy which is spearheaded by UNESCO and its Intergovernmental Oceanographic Commission. In 2017 *Ocean Literacy for All. A Toolkit* (https://unesdoc.unesco.org/ark:/48223/pf0000260721) was published, which together with an online platform (https://oceanliteracy.unesco.org) is galvanizing and promoting society as a whole to become engaged in ocean literacy. Data visualizations through images, and moving animations, and exploring the nexus between science, art, and music remain crucial ways to bring the ocean to life in educational and wider ocean literacy activities. An ocean literacy strategic plan and framework has been developed for the UN Decade of Ocean Science for Sustainable Development (https://www.oceandecade.org/). This has identified key interlinked elements to this framework, including societal outcomes, learning opportunities, contributions, and stakeholders.

Marine data visualization for science communication and storytelling

While a map can say more than 1000 words, a 3D or even 4D animation is a further tool pushing the potential of societal engagement even more. Data visualizations are already used widely in scientific research, and there is an increasing capability for taking these into spherical displays that powerfully communicate the global ocean and allow projections of geospatial data time series to show ocean parameters, e.g., currents, chlorophyll-a moving in time and space. Such visualizations are increasingly being used in science outreach centers, museums, and planetariums, with increasing collaborations between marine scientists and science engagement professionals to visualize ocean data for large dome displays.

In these visualizations, a range of data, e.g., geo-referenced bathymetry, biological parameters, and also subsea video, images, etc., can all be converted for dome shows that help to bring the information to life, combining powerful visuals and storytelling. Fig. 5.12 shows static images taken from a dome show and permanent exhibition at Dynamic Earth, Edinburgh. This uses "Digistar6" planetarium software to visualize open source data available through the EMODnet data service, e.g., bathymetry and new data; e.g., larval particle tracks produced by the North Atlantic international research project ATLAS, funded by European research and innovation program Horizon 2020 (H2020).

Projecting high-resolution model outputs to offer further information on ocean past, present, and future, for society, is also being piloted, e.g., through new research collaborations through the international iAtlantic EU H2020 project. These visualizations are also crucial to communicate just how large and interconnected the global ocean is, and for citizens to understand and connect with the high-seas—the waters beyond territorial waters and exclusive economic zones (EEZs)—which are largely unexplored and yet represent 66% of the world's ocean. These emerging "planet earth and ocean show" visualizations have great potential for engaging millions of citizens, both locally due to the accessible nature of many citizens to science centers, and the considerable reach, and worldwide since the software, e.g., Digistar, can be linked to a global network of planetariums. A further example is the Puffer Fish displays (https://pufferfishdisplays.com/solutions/displays/) used in the Royal Society Christmas Lecture 2020 Planet Earth: A user's guide (https://www.rigb.org/christmas-lectures/2020-planet-earth-a-users-guide). Such resources and science centers and museums have a local, yet considerably large reach, with the ability to

Figure 5.12 Images from the ATLAS visualization project, showcasing ATLAS consortium research for general audiences in a full-dome format, taken from within the planetarium at Dynamic Earth. As of early 2021, this project is still evolving. (A) An example bathymetry view from the Azores region, rendered for viewing in real-time by Digistar 6 planetarium software (https://www.es.com/digistar/). Here, the base layer is sourced from EMODnet, the higher-resolution foreground region from IMAR and IH (Bathymetry data: The Portuguese Hydrographic Institute & Portuguese Navy [https://www. hidrografico.pt/]). (B) A still from a sequence detailing the progression of simulated larval particle tracks, in this case approximating the behavior for species *Lophelia pertusa* https://www.researchgate.net/publication/342554467_Larval_behaviour_dispersal_and_ population_connectivity_in_the_deep_sea. *Image credits: Alastair Bruce, Dynamic Earth, UK https://www.dynamicearth.co.uk/.*

communicate to tens of thousands—if not hundreds of thousands—of people each year, showing the potential of engaging with communities and citizens at local levels and connecting to science to society.

Hackathons: innovating marine data and solutions by society, for society

Hackathons are innovation events where teams compete over a short-term duration, e.g., 48–72 h, to develop user-applications and societal solutions using marine and other open-source data. In recent years, hackathons have become more and more popular as a tool for both innovation and

Figure 5.13 A selection of images from the Open Sea Lab I Hackathon, organized by EMODnet, in collaboration with the EC, DG MARE, Flanders Marine Institute (VLIZ), and IMEC https://www.emodnet.eu/conference/opensealab/.

communication among multinationals, SMEs, start-ups, governmental organizations, and NGOs. Examples of hackathons in the marine domain include Open Sea Lab (see Fig. 5.13; https://www.emodnet.eu/conference/opensealab/) and the Ocean Hackathon (https://www.campusmer.fr/Ocean-Hackathon_-Ocean-Hackathon_-2021-3567-1307-0-0.html).

These hackathons are typically a collaboration across marine and maritime data services and wider experts, with the goal to utilize the full availability of open access marine—and wider domain, e.g., socioeconomics—data available through remote web services. The teams themselves are also often multi- and/or transdisciplinary groups, and it is exactly this mix of participants which is key, as diversity in backgrounds and expertise areas generates more innovation (see also Chapter 6). Participants do not need to be limited to "technical" experts (although these are indispensable), but could also involve subject matter experts (e.g., offshore engineers, scientists, policy experts, etc.), creative minds, and entrepreneurs (such as start-ups in relevant sectors, designers, concept/product developers, architects, gaming experts, etc.). Through this diversity, unique solutions can be cocreated to make the vast resources of available marine data more accessible to users, e.g., as data visualizations, decision-support tools, etc. A hackathon generally pitches a number of challenges to stimulate brainstorming and ideation, and these can range from high-level topics including a sustainable blue economy and marine environment protection and management to more technical, specific challenges, depending on the participants.

Hackathons are being increasingly viewed as a great creative methodology for stimulating innovation through brainstorming, prototyping, and launching new ideas. To harness innovative ideas generated by younger generations, the European Commission DG MARE is developing and hosting a Hack4Oceans (https://hack4oceans.eu/) that will be primarily a youth innovation tool, with more than half of participants from higher education (undergraduates at university level), to cocreate opportunities for prosperity through ocean conservation and sustainable use of marine resources with marine and professional stakeholders.

In addition to the development of innovative applications, hackathons have a great communication value to promote the benefits of open data which in turn can further encourage data sharing by wider stakeholders beyond research, including the private sector. Participating, sponsoring, and mentoring such hackathons also gives the private sector access to innovative ideas that can also be taken forward to catalyze solutions and applications to meet societal challenges.

Toward a transparent, accessible, and digital ocean

2021 sees the start of the United Nations Decade of Ocean Science for Sustainable Development. This is set to be a crucial global connector and unifier for ocean scientists and wider marine and maritime professionals to connect with society to codesign, conduct, and communicate about the science we need for the ocean we want. Open-source marine data and connecting such data to society are an important foundation to create a transparent and accessible ocean, underpinning wider societal engagement and democratizing knowledge for all.

In Europe, open data and open science are also crucial to unlocking the potential of information and driving new knowledge to achieve decarbonization and the European Green Deal (https://ec.europa.eu/info/sites/info/files/european-green-deal-communication_en.pdf), in line with the EU's commitment to global climate action under the Paris Agreement. How will Europe adapt to the digital revolution, and how will it translate into job transformation, a digital economy, and society, with higher capability for science and innovation? The following section looks at the rise of big data, the challenges of this data deluge, and how to better connect (marine) environmental data to society in the increasingly digital age.

The rise of big data

We are already in a data-rich society, and as technology continues to improve the "data deluge" will only increase (European Marine Board et al., 2019; Alexander et al., 2021). This offers the potential for "big data," i.e., large volumes of high variety data collected at high velocity, to be utilized more routinely in marine science and other marine and maritime applications. Big data has been made possible and routine by the cloud. There are clear advantages in terms of using big data to conduct truly holistic ecosystem-based management which requires a huge volume of multiparametric data. Big data by its sheer number of data points can also further constrain and reduce the uncertainty in modeling outputs, e.g., for the ocean and ocean–climate interactions and for producing more accurate low uncertainty predictions and forecasts. However, the challenge for the marine community is to publish data in such a way as to encourage big data processing, the quickest way being to make these data open. Commercial companies like Amazon are offering no cost hosting of data, with minimal costs for computer processing. For example, with a relatively small investment, e.g., just less than 1000 Euros, cloud processing of big data—and by big we mean around 400 terabytes of data—is now being made possible, using 10 s of terabytes of RAM, thousands of CPUs, and using cloud computing and/or cloud-free mosaics. And, while some data lend themselves to big data (e.g., some physical oceanographic parameters with automated quality control, vessel density tracking data, etc.), other data are less easily suitable (e.g., marine biological datasets). For the marine community, a key challenge will be to ensure a stepwise approach to scaling up toward wider automation and near real-time delivery of big data, to ensure the quality and provenance of data (e.g., though metadata descriptions) is maintained and interoperability remains paramount. And despite the growing capability for near-real-time delivery, there would also remain a need for delayed mode, harmonized, and standardized, integrated datasets, for example, to validate models.

Marine and wider environmental data for society in the digital era

Bauer et al. (2021) report the European plans to create a digital twin, or digital replica, of the Earth, to support the green transition envisaged by the EU's Green Deal. This would enable a leap in information technology and systems and a new era for geospatial information for society, elevating the existing capability for open discovery and access of data and contributing to the GreenData4All and Destination Earth information systems. It would also produce capability for the highest precision model of the Earth, setting the stage for the highest quality, low uncertainty, modeling simulations with predictive capabilities to assess the past, present, and future planetary processes as they happen and to make informed decisions in management, business, and society. The European Commission has launched the Destination Earth initiative (DestinE) which will develop such digital twins. The first will be a Digital Twin on Climate Adaptation for generating analytical insights and testing of predictive scenarios in support of climate adaptation and mitigation policies at decadal timescales with unprecedented reliability at regional and national levels. The next will be a Digital Twin of the Ocean (DTO) which will build on existing capability to integrate a wide range of data sources and model simulations to transform data into knowledge and to connect, engage, and empower citizens, governments, and industries by providing them with the capacity to inform their decisions.

In 2019 the European Marine Board's Navigating the Future V (European Marine Board et al., 2019) included recommendations for Europe to build a common "ecosystem" of applications to exchange and utilize ocean data for societal and business usage, and as an action to deliver a common augmented or virtual reality platform where all information about the ocean can be uploaded and used to describe how the underwater world works, in real time. And there is extensive capability to build upon. In Europe, existing long-term marine data infrastructures and services, including the European Marine Observation and Data Network (EMODnet) and the Copernicus Marine Service of the Copernicus Space program, are already providing open and free access to integrated high resolution marine data from in situ and satellites, respectively, offering mature, long-term data services that will make the next step—change toward full interoperability of marine data, model outputs, and data products. These data infrastructures—paired with key e-infrastructures including D4Science, EUDAT, and DIAS (WEkEO— see Chapter 3)—represent the fundamental building blocks for delivering an interoperable, trusted ecosystem of marine data, data products, and applications, to achieve a European Digital Twin Ocean framework with a multitude of applications for society. Projects such as the EU H2020 Blue-Cloud (https://www.blue-cloud.org/) are piloting web-based open science through collaborations across existing data infrastructures, research infrastructures, and e-infrastructures in the marine domain toward this goal.

A Digital Twin for Europe will be a real opportunity for Europe to collectively build a next generation capability for transformative science delivering integrated information and knowledge for meeting the ambitious European Green Deal targets, support decision-making for global issues such as mitigating climate change, overfishing and pollution, and supporting smarter and greener operations in the blue economy (Alexander et al., 2021). It will also capitalize and link with the digitalization across much of the Blue Economy, including shipping industry that will also develop digital twins including virtual models of physical ships, producing valuable insights from data (https://safety4sea.com/cm-the-digital-twin-concept-explained/). Finally, a European Digital Twin Ocean framework with a more complete accessibility to integrated datasets would also allow to more easily identify gaps, to inform future ocean observing infrastructure and data collection efforts, leading to a more coordinated and efficient European ocean observing and monitoring framework.

Conclusions

There is clearly an increasing demand for open access marine data for society. Timely access to high-quality marine data is key to meet the operational needs, and the green transition, of the blue economy, in both traditional, e.g., shipping, and emerging, e.g., renewable energy, sectors. High-quality, open-source, and interoperable data and information are also vital to enable a transparent and accessible ocean, with information available to all, to drive a knowledge-based society and evidence-based decision-making, e.g., for climate adaptation, marine planning and management, and to meet the wider ambitious targets of the European Green Deal (European Commission, 2019a), and international agreements, e.g., the UN 2030 Agenda and the Paris agreement on climate change. In Europe, marine data brokerage services such as EMODnet, CMEMS, and societal information tools, e.g., the European Atlas of the Seas offer integrated data and data products that can be accessed and used by marine and maritime professionals and wider society (ECO Magazine, 2021). In addition, there is a high potential to expand the use of data visualization, animation, and storytelling, to engage and connect the ocean to citizens. Community hackathons are also increasingly used as a mechanism for harnessing the power of integrated, open, and free marine data for creating societal applications through innovation and codesign.

Future technology advancements in the domains of information technology and communications, artificial intelligence, and cloud computing will no doubt drive future innovations in online web-based open data, including for digital twin innovations and related simulations and scenario-setting. However, simulations and models are only as good as the data that are available. A key bottleneck to the full potential of open marine data use remains the ability for the international ocean community to come together to agree—and collectively apply common standards for data and metadata to drive forward truly FAIR marine data that is fully interoperable and that has a clear data provenance. It is these data that will be the cornerstone of all future marine knowledge services, digital frameworks, and decision support tools that will be a cross-cutting resource for the UN Decade of Ocean Science for sustainable development. Achieving truly FAIR data is also essential for producing truly big data that is transdisciplinary and heterogeneous through interoperability with other digital twins from different domains and for tackling transdomain issues, e.g., in coastal areas at the land-sea interface, and at the ocean-climate nexus. In addition, the scale-up toward automated data collection, processing, and analysis needs to be done in a stepwise, structured process to ensure data quality so that big data are fit-for-purpose.

Today we are presented with an important opportunity to direct the wealth of marine knowledge toward broader public to create a more ocean literate society. The challenge is to connect professionals spanning marine science, wider maritime, education, and engagement to develop powerful tools, resources, and outputs to really integrate knowledge and make marine data come alive for the benefit of all citizens, which will in turn ensure better policymaking.

The ocean is increasingly visible to society in political agendas, e.g., G7, in the mass media, e.g., Blue Planet II (https://www.bbcearth.com/blueplanet2/), and in Ocean Literacy and related educational activities, e.g., EU4Ocean. In addition, public funding streams for ocean-related research and innovation and also bringing diverse sectors together to tackle ocean-related issues, e.g., the upcoming European Mission Starfish 2030: Restore our Ocean and waters by 2030 (https://ec.europa.eu/info/publications/mission-starfish-2030-restore-our-ocean-and-waters_en). This coming decade is a huge opportunity to build on this momentum and existing mature capability for open marine data, and connect across disciplines, linking marine and maritime professionals with education specialists, ICT and marine data users, to develop powerful tools that can be used by citizens to further connect to the ocean and related environment. Only then will society—en masse—start connecting with the ocean at an emotional level, and move toward changing attitudes and behaviors toward a more green and sustainable society.

With an ambitious European Green Deal and with the United Nations 2030 Agenda and Sustainable Development Goals, the UN Decade of Ocean Science for Sustainable Development and UN Decade for ecosystem restoration, marine data, and knowledge are certain to be the foundation, glue, and connector for all knowledge production. And as we further connect, coordinate, and integrate marine data into interoperable, transparent ways for societal use, the data and information to support societal change will be so compelling and the evidence so strong that it can no longer be ignored.

Acknowledgments

EMODnet is a truly network-driven pan-European collaborative initiative. As with all EMODnet developments and achievements, this paper would not have been possible without the efforts and contributions of the entire EMODnet community. The authors recognize funding from the European Union European Maritime and Fisheries Fund (EMFF) for the EMODnet Secretariat, Central Portal, seven thematics and data ingestion. Data/Information used in this book chapter was made available by the EMODnet thematic lots, all available through www.emodnet.eu in particular the EMODnet Human Activities project, www.emodnet-humanactivities.eu, EMODnet Bathymetry, https://www. emodnet-bathymetry.eu/and EMODnet Seabed Habitats project https://www.emodnet. eu/en/seabed-habitats, all funded by the European Commission Directorate General for Maritime Affairs and Fisheries.
The authors also acknowledge funding from the European Union's Horizon 2020 research and innovation program under grant agreement No. 678760 (ATLAS), grant agreement No. 818123 (iAtlantic), and grant agreement No. 862409 (Blue-Cloud), for work outlined and referenced in this article. This article reflects only the author's view and the European Union cannot be held responsible for any use that may be made of the information contained therein. The authors acknowledge the inputs of Alastair Bruce, Hermione Cockburn (Dynamic Earth), and Murray Roberts (University of Edinburgh) related to H2020 ATLAS and iAtlantic work and Joana Gafeira (BGS) and Pascal Derycke (formerly SSBE) related to the H2020 ATLAS GeoNode.

References

Agnesi, S., Annunziatellis, A., Mo, G., Reker, J., 2017a. Applying Modelled Broad Scale Habitat Maps in MPA Network Evaluations - the Western Mediterranean Sea Case Study — in Program and Abstracts, 2017 GeoHab Conference, Dartmouth, Nova Scotia. Available at: https://www.emodnet-seabedhabitats.eu/resources/use-cases/policy-makers/#h08099a44f176485e98b88ea334e60461.
Alexander, B., Kellett, P., Coopman, J., Muñiz Piniella, Á., Heymans, J.J., 2021. 7th EMB Forum Proceedings. Big Data in Marine Science: Supporting the European Green Deal, EU Biodiversity Strategy, and a Digital Twin Ocean. https://doi.org/10.5281/zenodo.4541335.
Andersen, J., Manca, E., Agnesi, S., Al-Hamdani, Z., Lillis, H., Mo, G., Populus, J., Reker, J., Tunesi, L., Vasquez, M., 2018. European broad-scale seabed habitat maps support implementation of ecosystem-based management. Open J. Ecol. 8, 86—103. https://doi.org/10.4236/oje.2018.82007.

Barale, V., Assouline, M., Dusart, J., Gaffuri, J., 2014. The European atlas of the seas: relating natural and socio-economic elements of coastal and marine environments in the European union. Mar. Geodes. 38 (1), 79—88. https://doi.org/10.1080/01490419.2014.909373.

Bauer, P., Stevens, B., Hazeleger, W., 2021. A digital twin of Earth for the green transition. Nat. Clim. Change. https://doi.org/10.1038/s41558-021-00986-y.

Brett, A., Leape, J., Abbott, M., et al., 2020. Ocean data need a sea change to help navigate the warming world. Nat. Comm. 582.

Calewaert, J.B., Weaver, P., Gunn, V., Gorringe, P., Novellino, A., 2016. The European marine data and observation network (EMODnet): your gateway to European marine and coastal data. In: Quantitative Monitoring of the Underwater Environment (2016). Volume 6 of the Series Ocean Engineering & Oceanography, 31. Springer, pp. 31—46. https://doi.org/10.1007/978-3-319-32107-3_4.

Corti, P., Bartoli, F., Fabiani, A., Giovando, C., Kralidis, A.T., Tzotsos, A., 2019. GeoNode: an open source framework to build spatial data infrastructures. PeerJ Preprints 7, e27534v1. https://doi.org/10.7287/peerj.preprints.27534v1.

Defra, 2014a. East Inshore and East Offshore Marine Plans. Annex 1: Supporting Information on the Production of Maps, 23 pp.

Defra, 2014b. South Inshore and South Offshore Marine Plan Areas: South Plans Analytical Report (SPAR) June 2014, 170 pp.

ECO Magazine, 2020a. EMODnet Bathymetry Now Offers the Highest Resolved Bathymetric Worldwide Layout. https://www.ecomagazine.com/news/industry/emodnet-bathymetry-now-offers-highest-resolved-bathymetric-worldwide-layout.

ECO Magazine, 2020b. Breaking Borders in Ocean Information: A European Success Story. https://www.ecomagazine.com/in-depth/featured-stories/breaking-borders-in-ocean-information-a-european-success-story.

ECO Magazine, 2021. EMODNet: Euorpe's free and open access marine knowledge broker http://digital.ecomagazine.com/publication/frame.php?i=707374&p=1&pn=&ver=html5.

EMB-EuroGOOS, 2008. The European Marine Observation and Data Network (EMODnet) Vision Document. https://www.marineboard.eu/sites/marineboard.eu/files/public/publication/EMODNET-7.pdf.

EMODnet Secretariat, 2020. Create Your Own Marine and Coastal Maps with the New European Atlas of the Seas. http://www.emodnet.eu/create-your-own-marine-and-coastal-maps-new-european-atlas-seas.

EOOS conference 2018 report and call to action. In: Larkin, K., Marsan, A.-A., Tonné, N., Calewaert, J.-B. (Eds.), 2019. Connecting European Ocean Observing Communities for End-To-End Solutions. The Egg, Brussels, 21—23 November 2018. ISBN: 9789492043719.

European Commission, 2003. Open Data Policy. https://ec.europa.eu/digital-single-market/en/open-data.

European Commission, 2007a. INSPIRE Directive. https://inspire.ec.europa.eu/inspire-directive/2.

European Commission, 2019a. European Green Deal. https://ec.europa.eu/info/sites/info/files/european-green-deal-communication_en.pdf.

European Commission, 2010. Marine Knowledge 2020 Marine Data and Observation for Smart and Sustainable Growth. Commission Communication COM (2010) 461, Publications Office of the European Union.

European Commission, 2012a. Green Paper: Marine Knowledge 2020. From Seabed Mapping to Ocean Forecasting. https://ec.europa.eu/maritimeaffairs/sites/maritimeaffairs/files/docs/body/marine-knowledge-2020-green-paper_en.pdf.

European Commission, 2012b. Roadmap for EMODnet. https://ec.europa.eu/maritimeaffairs/sites/maritimeaffairs/files/docs/body/roadmap_en.pdf.

European Marine Board, 2013. Navigating the Future IV. Position Paper 20 of the European Marine Board. Ostend, Belgium, ISBN:9789082093100. https://www.marineboard. eu/publication/navigating-future-iv.

European Marine Board, Guidi, L., Fernandez Guerra, A., Canchaya, C., Curry, E., Foglini, F., Irisson, J.-O., Malde, K., Marshall, C.T., Obst, M., Ribeiro, R.P., Tjiputra, J., Bakker, D.C.E., 2019. Big data in marine science. In: Alexander, B., Heymans, J.J., Muñiz Piniella, A., Kellett, P., Coopman, J. (Eds.), Future Science Brief 6 of the European Marine Board, Ostend, Belgium. ISSN, pp. 2593—5232. ISBN: 9789492043931. https://www. marineboard.eu/sites/marineboard.eu/files/public/publication/EMB_FSB6_BigData_ Web_v4.pdf.

HELCOM, 2016. Ecological coherence assessment of the marine protected area network in the Baltic. Balt. Sea Environ. Proc. No. 148.

HELCOM, 2018. State of the Baltic Sea — Second HELCOM Holistic Assessment 2011-2016. Baltic Sea Environment Proceedings, vol. 155. ISSN 0357-2994. Available at: www.helcom.fi/baltic-sea-trends/holistic-assessments/state-of-the-baltic-sea-2018/ reports-and-materials/.

Korpinen, S., Klančnik, K., Peterlin, M., Nurmi, M., Laamanen, L., Zupančič, G., Popit, A., Murray, C., Harvey, T., Andersen, J.H., Zenetos, A., Stein, U., Tunesi, L., Abhold, K., Piet, G., Kallenbach, E., Agnesi, S., Bolman, B., Vaughan, D., Reker, J., Royo Gelabert, E., 2019. Multiple Pressures and Their Combined Effects in Europe's Seas. ETC/ICM Technical Report 4/2019: European Topic Centre. https://www.eionet. europa.eu/etcs/etc-icm/products/etc-icm-report-4-2019-multiple-pressures-and-their-combined-effects-in-europes-seas.

Larkin, K., Marsan, A.-A., Calewaert, J.-B., September 2020. European Ocean Literacy EU4Ocean Stakeholder Platform Terms of Reference & Key Facts.

Larkin, K., Calewaert, J.-B. (Eds.), 2020. EMODnet Annual Report 2019.

Manzella, G.M.R., Novellino, A., Angelo, P.D., 2018. EMODnet physics: benefits from marine data sharing. Mod. Approaches Oceanogr. Petrochem. Sci. 1 (5).

Martín Míguez, B., Novellino, A., Vinci, M., Claus, S., Calewaert, J.-B., Vallius, H., Schmitt, T., Pititto, A., Giorgetti, A., Askew, N., Iona, S., Schaap, D., Pinardi, N., Harpham, Q., Kater, B.J., Populus, J., She, J., Palazov, A.V., McMeel, O., Oset, P., Lear, D., Manzella, G.M.R., Gorringe, P., Simoncelli, S., Larkin, K.E., Holdsworth, N., Arvanitidis, C.D., Molina Jack, M.E., del Mar Chaves Montero, M., Herman, P.M.J., Hernandez, F., 2019. The European Marine Observation and Data Network (EMODnet): Visions and Roles of the Gateway to Marine Data in Europe. Special Issue: OceanObs'19: An Ocean of Opportunity: Frontiers in Marine Science. https://doi.org/10.3389/fmars.2019.00313.

Onrust, B., Bidarra, R., Rooseboom, R., van de Koppel, J., 2017. Ecologically sound procedural generation of natural environments. Int. J. Comput. Games Technol. vol. 2017. Number 7057141 — 2017. https://graphics.tudelft.nl/Publications-new/2017/OBRK 17/OBRK17.pdf.

OSPAR, 2017. Extent of Physical Damage to Predominant and Special Habitats. Intermediate Assessment 2017. Available at: https://oap.ospar.org/en/ospar-assessments/ intermediate-assessment-2017/biodiversity-status/habitats/extent-physical-damage-predominant-and-special-habitats/.

Shepherd, I., 2018. European efforts to make marine data more accessible. Ethics Sci. Environ. Polit. 18, 75—81.

SIMCelt, 2016. SIMCelt MSP Evaluation Report. https://www.msp-platform.eu/practices/simcelt-msp-evaluation-report.

Tempera, F., Liquete, C., Cardoso, A.C., 2016. Spatial Distribution of Marine Ecosystem Service Capacity in the European Seas. EUR 27843. Publications Office of the European Union, Luxembourg. https://doi.org/10.2788/753996.

Thornton, A., Luisetti, T., Grilli, G., Donovan, D., Phillips, R., Hawker, J., 2019. Initial Natural Capital Accounts for the UK Marine and Coastal Environment. Final Report. Report Prepared for the Department for Environment Food and Rural Affairs.

UNESCO, 2019. Consolidated Roadmap for a Possible UNESCO Recommendation on Open Science. https://unesdoc.unesco.org/ark:/48223/pf0000369699.

Wilkinson, M., Dumontier, M., Aalbersberg, I., et al., March 15, 2016. The FAIR Guiding Principles for scientific data management and stewardship. Sci. Data. https://doi.org/10.1038/sdata.2016.18.

Further reading

Agnesi, S., Mo, G., Annunziatellis, A., Chaniotis, P., Korpinen, S., Snoj, L., Globevnik, L., Tunesi, L., Reker, J., 2017b. Assessing Europe's marine protected area networks — proposed methodologies and scenarios. In: Künitzer, A. (Ed.), ETC/ICM Technical Report 2/2017, Magdeburg: European Topic Centre on Inland, Coastal and Marine Waters, 72 pp Available at: https://www.eionet.europa.eu/etcs/etc-icm/products/etc-icm-reports/assessing-europes-marine-protected-area-networks-proposed-methodologies-and-scenarios.

Barale, V., 2018. A supporting marine information system for maritime spatial planning: the European Atlas of the Seas. Ocean Coast Manag. https://doi.org/10.1016/j.ocecoaman.2018.03.026.

Bayraktarov, E., Ehmke, G., O'Connor, J., Burns, E.L., Nguyen, H.A., McRae, L., Possingham, H.P., Lindenmayer, D.B., 2019. Do Big Unstructured Biodiversity Data Mean More Knowledge? Frontiers in Ecological Evolution. https://doi.org/10.3389/fevo.2018.00239.

Carroll, S.R., Garba, I., Figueroa-Rodríguez, O.L., Holbrook, J., Lovett, R., Materechera, S., Parsons, M., Raseroka, K., Rodriguez-Lonebear, D., Rowe, R., Sara, R., Walker, J.D., Anderson, J., Hudson, M., 2020. The CARE principles for indigenous data governance. Data Sci. J. 19 (1), 43. https://doi.org/10.5334/dsj-2020-043.

European Commission, 2007b. Integrated Maritime Policy. https://ec.europa.eu/maritimeaffairs/policy_en.

European Commission, 2014. The Marine Spatial Planning Directive. https://publications.europa.eu/en/publication-detail/-/publication/14d91dc4-4b1c-4284-b2fe-6446e407e938.

European Commission, 2018. The Open Science Policy Platform. https://ec.europa.eu/research/openscience/pdf/ec_rtd_ospp-final-report.pdf.

European Commission, 2019b. Towards a Sustainable Europe 2030. https://ec.europa.eu/info/publications/towards-sustainable-europe-2030_en.

European Commission, 2020a. Ocean Observation: Sharing Responsibility Inception Impact Assessment (and EMODnet Response). https://ec.europa.eu/info/law/better-regulation/have-your-say/initiatives/12539-Ocean-observation-sharing-responsibility.

European Commission, 2020b. 2020 Blue Economy Report. https://ec.europa.eu/commission/presscorner/detail/en/ip_20_986.

European Commission, 2020c. A European Strategy for Data. https://ec.europa.eu/digital-single-market/en/content/european-digital-strategy.

European Marine Board, 2019. Navigating the Future V: Marine Science for a Sustainable Future. Position Paper 24 of the European Marine Board, Ostend, Belgium. https://doi.org/10.5281/zenodo.2809392. ISBN: 9789492043757. ISSN: 0167-9309. https://www.marineboard.eu/navigating-future-v.

IOC-UNESCO, Valdés, et al. (Eds.), 2017. Global Ocean Science Report — the Current Status of Ocean Science Around the World. UNESCO Publishing, Paris.

IOC-UNESCO, 2020. In: Isensee, K. (Ed.), Global Ocean Science Report 2020 — Charting Capacity for Ocean Sustainability. UNESCO Publishing. UNESCO, Paris. https://unesdoc.unesco.org/ark:/48223/pf0000375147.

MacKenzie, B., Celliers, L., de Freitas Assad, L.P., Heymans, J.J., Rome, N., Thomas, J., Anderson, C., Behrens, J., Calverly, M., m Desais, K., DiGiacomo, P.M., Djavidnia, S., dos Santos, F., Eparkhina, E., Ferrari, J., Hanly, C., Houtman, B., Jeans, G., Landau, L., Larkin, K., Legler, D., Le Traon, P.-Y., Lindstrom, E., Loosley, D., Nolan, G., Petihakis, G., Pellegrini, J., Roberts, Z., Siddorn, J.R., Smail, E., Sousa-Pinto, I., Terrill, E., 2019. The Role of Stakeholders in Creating Societal Value from Coastal and Ocean Observations. Special Issue: OceanObs'19: An Ocean of Opportunity: Frontiers in Marine Science. https://doi.org/10.3389/fmars.2019.00137.

McMeel, O., Calewaert, J.B., 2016. Marine Data Portals and Repositories and Their Role in Knowledge Transfer to Support Blue Growth. Deliverable 4.2 H2020 COLUMBUS Project No.652690.

CHAPTER SIX

How can ocean science observations contribute to humanity?

Giuseppe M.R. Manzella[1,2], William Emery[3]
[1]The Historical Oceanography Society, La Spezia, Italy
[2]OceanHis SrL, Torino, Italy
[3]University of Colorado, Boulder, CO, United States

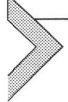

The importance of the ocean in the human environment

> The sea is the mother not only of all waters, of all springs, and of all rivers to which it supplies their existence by means of evaporation, but also of all that exists; and still at present it produces and feeds more species of animals than the land.
> Immanuel Kant, *Physical Geography*, 1750—90

Our world is the result of a marvelous exchange of ideas and cultures dictated by economical needs and an innate curiosity. It is a global environment where people are in contact across continents and countries. This concept was expressed by Fernand Braudel (1985) for the Mediterranean cultural area, but today it can be expanded to the entire world.

What characterizes this variety of countries is not their differences, but their many common elements. Commerce, population migrations, cultures, and religious struggles have directly influenced the world. Population movements create conditions for the development of unique local societies in a worldwide context and the creation of new cultural spaces containing a variety of viewpoints that mutually interact, overlap, intersect, and potentially conflict. The oceans and their adjacent seas have been and are connecting highways between these countries and continents.

Ocean Science Data
ISBN: 978-0-12-823427-3
https://doi.org/10.1016/B978-0-12-823427-3.00005-0

© 2022 Elsevier Inc.
All rights reserved.

World's oceans are a defining feature of our planet that cover 70% of the Earth's surface. They provide a variety of benefits to humans while they directly and indirectly impact our daily lives. In addition to their role in the global transportation of goods and material, they are a source of inspiration, recreation, and discovery. The ocean offers new opportunities for smart, sustainable, and inclusive growth while at the same time it is something that we take for granted and abuse by allowing all of our waste to eventually reside in it. To realize the ocean's potential, there is a need to develop a collaborative science-technology-society working together to best utilize the ocean while preserving its basic character.

In 2012, at Rio+20, the nations of the world agreed on Sustainable Development Goals integrating environmental and developmental indicators to set targets for the future. They also discussed other options and opportunities for environmental stewardship and equitable development. Rio+20 calls on science to provide the knowledge base to build a sustainable, equitable, and prosperous future for current and future generations. The Future We Want (United Nations, 2012) provides a clear statement in this direction, stressing the need to engage civil society and incorporate science into policy and recognizing the importance of voluntary commitments on sustainable development.

A knowledge society can be advanced by the so-called "Matthew effect" (the rich get richer): the more you know, the more you can learn. An important principle is that knowledge needs to be of value to people in its current form, not merely banked against future needs. From a social standpoint, the ability to connect discourse within and between communities opens new possibilities for barrier-crossing, mutual support, and cross-fertilization. A knowledge-based society implies not only that scientific and civil societies work together but that they also perceive a common need for a great deal of new information and are able to work to acquire it and analyze it, creating new results, evaluating them, and finding new uses for them.

Production of knowledge and the educational implications of this shift toward a knowledge-based society are only beginning to be worked out, and they present a formidable new challenge that contains *lifelong learning, flexibility, creativity, higher-order thinking skills, collaboration, distributed expertise, learning organizations, innovation,* and *technological and science literacy*.

One of today's challenges is trying to answer fundamental questions such as how and why the global environment is changing:

- What are the likely future changes (both immediate and long-term)?
- What are the risks and implications for human development and for the diversity of life on Earth in light of these future changes?

- How can societies adapt to the physical, social, and ecological consequences of a warming ocean, and what are the barriers, limits to this adaptation, and potential opportunities provided by it?
- How can the integrity, diversity, and functioning of ecological and evolutionary systems be sustained so as to sustain life in the oceans and its ecosystems to equitably enhance human needs and well-being?

Answers to these questions will stimulate long-lasting debates involving many social communities, and there will be a need to continuously acquire new information and extract new knowledge from large and complex datasets to better answer these questions.

This chapter presents specific experiences in preparing young researchers capable of responding to these knowledge needs related to the marine environment. The most important need is to educate students to work in an interdisciplinary environment and complex processes. Hence the title of the chapter, which is the elaboration of a sentence contained in Alexander von Humboldt's "*Cosmos*" (1845) about nature as unity in diversity. The goal set here for ocean data science education is interdisciplinary crossfertilization, improvement of critical thinking and skills, and ultimately knowledge-based skills. In the modern world the plethora of ocean science data, relayed to shore by satellites from autonomous drifting platforms and from the satellites themselves, dictates that the limiting factor in the analyses of these data is the number of scientists/researchers available to analyze these data.

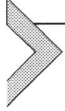

Ocean data science

Natural History seems at present to lie under some disgrace, upon account of the small Benefit that is presumed to arise from the Study of it. And indeed, if a Notion be formed of the whole thereof, from a few of its Parts which have been dryly treated by some Virtuosi, a rigid Philosopher might be apt to condemn it as trifling and almost useless to Mankind. A real philosophic Genius, bent upon its own Improvement, sees with Indignation, the Life of a Virtuosi spent among Shells and Insects; and wishes the same Time and Application said out to nobler Purposes. These Things however have their Use; and we ought not to despise all that we ourselves have no Relish for. There will, and 'tis necessary there should be Men of various Tastes; and 'tis happy for the World that no Part of Philosophy remains uncultivated. But if any Man

(Continued)

> has a despicable Opinion of Natural History in general, let him look upon it in that View wherein Mr. Boyle considered it; for here, as in everything else, our excellent Author has regarded 'Usefulness and the Benefit of Mankind. Natural History, as managed by him, has no superfluous Branches; Nothing that barely amuses the Mind, without gratifying and delighting it; Nothing that entertains the Fancy, without being serviceable in Life; and his History of the Air, his say of examining the Freshness of Water, &c. are eminent Instances of it. But Natural History, in its Extent, he found to be a very large Field, that required a great Number of Hands to be employed in it; and therefore, to render it as useful and complete as possible, he endeavoured to engage many of his Friends in the Study thereof.
>
> Peter Shaw, *The Philosophical Works of the Honourable Robert Boyle* Esq., 1728

Science (scientia in Latin) means knowledge. Regardless of the definition of "knowledge" (Plato in *Theaetetus*; Wittgenstein, 1975; Popper, 1984) science is "a continuous questioning of the foundations of our worldview to continually improve it" (Rovelli, 2014). Starting from the 17th century, scientific progress had two basic research paradigms: experimental and theoretical. The method of inquiry or the "Method of Making Experiments" (this book Chapter 1) consists of the formulation of hypotheses and collection of data through observation and experimentation to test these hypotheses. Simulation has added an essential third paradigm to this hypotheses testing (Bell et al., 2009) while it has greatly enabled methods to test theoretical ideas. Theoretical ideas can now be explored with numerical simulations in sufficient detail (spatial and temporal resolutions) so as to really simulate ocean behavior. At the same time, it is not adequate to simply simulate ocean behavior, and it is important to combine both numerical models and ocean observations.

The scope of data collection in marine science has been expanded from more traditional physical, geological, and biochemical observations to more interdisciplinary and complex ones including the many ecological patterns. In addition, observations have gone from measurements made from small groups of research scientists on dedicated research vessels to autonomous systems that collect vertical and horizontal data in the global ocean and relay their data through satellites. These platforms are often contributed by a large group of countries making the collection of these data a global enterprise. High volumes and high frequency of incoming data, as well as a large variety of information assets, require a "big data" approach, a revolution in information processing that has not yet been completely reached in the marine science domain. A key factor will be a close collaboration among data producers, data users, and information specialists (Chapter 3).

Data collection starts with scientific "inquiries" that include the investigation hypotheses possibly linked to societal needs. "Fitness for purpose" (or data producer needs) means that datasets or dataset series should be suitable for the intended purposes of the data producer (Pinardi et al., 2017). Its concept include logistics, sampling strategies, and quality assurance procedures, i.e., those sets of systematic actions necessary to satisfy data product requirements (both quantity and quality). Such systematic actions include the management of platforms, instruments, materials, products, and services in an end-to-end process (see Chapter 2). Preparation and operation of the entire data collection cycle requires the participation of scientists, engineers, and data managers.

Once created, data are valuable resources that can be used and reused for present and future environmental, scientific, management, and educational purposes. Sharing data avoids duplicate data collection, facilitates new scientific inquiry, and provides rich real-life resources for user-defined objectives (e.g., environmental protection as well as education and training). A revolution that has taken place in the past four to five decades is the transition from wanting to reserve newly collected data for the data collection scientist to analyze and publish before releasing it to the public to a modern view that all data should be shared with anyone who is interested in it. This is a consequence of the large volumes of data generated by automated systems and orbiting satellites. Today the limiting factor in the analysis of oceanographic data is the number of scientists interested in analyzing the data. Thus, oceanographic data have graduated from being the sole province of one or more investigators to a general resource available to all for any and all science applications. For this reason, it is critically important that those involved in the measurement processes themselves and the curation of these data make every effort possible to ensure the accuracy and precision of the data that are produced.

Data-driven methods have been developed to better utilize observations and simulations, as well as reanalyze results (e.g., Lguensat et al., 2019). Chapters 2, 3 and 4 demonstrate that data can be transformed into products and knowledge by using (inter alia) statistics and modeling. This is a natural consequence of the overabundance of data that are now available making it necessary to evaluate their utility by using statistics to express their fundamental characteristics. These considerations call again for an intersection of programming and computer networking skills, mathematics and statistical knowledge, working knowledge and skills as defined in different science domains (Press, 2013). Thus, oceanography has transitioned from a field where the scientist was personally familiar with each of his/her data points to one where statistics are employed to be able to assess the validity and utility of the data without addressing the value of each data point.

Data adequacy

> But in order to give Mr. Simplicius overabundant satisfaction and, if possible, remove him from the error, I say that in our century we have new and such accidents and observations, that I have no doubt that if Aristotle lived in our age, he would change his opinion.
> Galileo Galilei, *Dialogue on the Two Major Systems*, 1632

The following considerations must be considered when reusing data:
- Data may be a good fit for the intended use but not necessarily for all potential uses.
- Data are not perfect, and all users need to be aware of the limitations of the data by considering their accuracy and precision.
- Users of existing data should use other statistics to characterize the data such as means, medians, standard deviations, histograms, etc.

Products are generally derived from data. The issues are
- data integration and analyses to meet related needs,
- development of strategies to analyze multidisciplinary data,
- clear views on requirements for integration fitted to an application,
- propagation of data errors through the analytical process to create products.

The exchange and integration of heterogeneous data from different sources is key for many aspects of ocean data science. Analytical techniques (qualitative and quantitative) in data science are also important for product generation by data integration. The reuse of data is highly dependent on the optimal selection of the Universe of Discourse and its corresponding quality attributes. The Universe of Discourse is defined as "the view of the real world that includes everything of interest" (ISO 19101 - https://inspire.ec. europa.eu/glossary/UniverseOfDiscourse). It requires specification of products against which the quality of data is evaluated (e.g., Pinardi et al., 2017).

It is important to note that data science has an inherent and integral involvement in the selection of sources and origins of data, or, in other words, in the definition of data adequacy to generate quality products. Fichtinger et al. (2011) analyzed problems of combining and integrating data from different sources. The integration of data requires also their harmonization and the definition of a data management scheme that starts from data quality specifications through to product specifications. The complete scheme includes an evaluation of data assuring relevance, reliability, and fitness-for-purposes, adequacy, comparability, and compatibility (Chapter 4). In addition, the propagation of errors through the analysis procedure must be carefully considered to be able to define the uncertainty of the final product.

Datasets consist of both actual measurements and descriptive information, generally referred to as metadata. The generally accepted view is that metadata comprise location, time, units, accuracy, precision, methods of measurement, sampling and analytical methodologies, investigator, references to publications describing the dataset, and a description of the protocol used to process the data collected (Chapter 2 and 3).

Quality control of data is an essential component of oceanographic data management. Data quality information tells users of the data how it was gathered, how it was checked, processed, what algorithms have been used, what errors were found, and how the errors have been corrected or flagged. Without it, data from different sources cannot be combined or reused to gain the advantages of integration, synthesis, and the development of (e.g.,) long-time series or large spatial datasets. The quality control requirements may be different depending on sampling interval, latency, and use to which data are to be put.

Following internationally agreed upon standards, the quality of the data must be assessed "To ensure the data consistency within a single data set and within a collection of data sets and to ensure that the quality and errors of the data are apparent to the user who has sufficient information to assess its suitability for a task." (IOC/CEC, 1993).

Data has to be validated against "Confirmation by examination and provision of objective evidence that the particular requirements for a specific intended use are fulfilled" (ISO 9000). In other words, data validation will allow an objective assessment of whether the data are fit enough.

Adequate data (in light of the intended use) are used for the creation of specific products (e.g., environmental assessments, fish stock assessments, spatial planning, etc.). At the end of this process a verification of the results has to be accomplished for the "Confirmation by examination and through provision of objective evidence that specified requirements have been fulfilled" (ISO 9000).

Data adequacy was well-defined by Pinardi et al. (2017) as: "the fitness for use of the data for a particular user or for a variety of users. Since different applications require different properties of. the data itself, 'adequacy' should be defined objectively using standardized nomenclature and methods." Nomenclature and methods used by Pinardi et al. were derived from Sea-DataNet common vocabularies (https://www.seadatanet.org/Standards/Common-Vocabularies) and ISO standards.

The entire process from data selection to the creation of products and the verification of their quality requires the collaboration of data managers, statisticians, mathematicians, biologists, and other specialists. The Pinardi et al. (2017) "stress test" (data adequacy report) was an extensive research of data and information in data portals and publications, based on predetermined product specifications. The "fit for use" assessment was done with data quality elements presented in Chapter 4. The selected data were used to realize targeted products and the verification process consisted in the calculation of the "distance" between the targeted products (obtained from data) and the product specifications (predetermined by user requirements).

 ## Added value chain in ocean data science education

> The principal impulse by which I was directed, was the earnest endeavor to comprehend the phenomena of physical objects in their general connection, and to represent nature as one great whole, moved and animated by internal forces.
> Alexander von Humboldt, *Kosmos*, 1845

The UN General Assembly held in 2017 decided to proclaim the United Nations Decade of Ocean Science for Sustainable Development for the 10-year period beginning on January 1, 2021 (art. 292 UN resolution A/RES/72/73; https://undocs.org/A/RES/72/73). The main motivations of the decade are contained in very challenging general objectives such as the following (Ryablin et al., 2019):

• Generate scientific knowledge and the underpinning infrastructure

- Implement the partnerships necessary for sustainable development
- Provide ocean science data and information to be delivered to environmental managers and policy makers in support of all Sustainable Development Goals of the 2030 Agenda.

The last bullet is the main topic of this book.

Education in ocean data science is one of the building blocks of interdisciplinary research, in which various disciplines participate in revealing the intimate connections and relationships between the environmental phenomena that make life possible. Data science includes the ability to generate meaningful data, information, and answer questions in interdisciplinary and multistakeholder environments. Data creation consists of three phases: design, collection, and analysis.

Interdisciplinary science is one of the paradigms in universities providing education in Earth and marine sciences, ocean sciences, environmental sciences, ocean science and engineering, marine systems and policies, etc. University courses include such ocean disciplines as physics, chemistry, biology, geology, and geophysics. To effectively convey knowledge and make students active participants in building knowledge, schools are applying teaching methods containing

- Conceptual Learning,
- Interdisciplinary Research,
- Problem Solving,
- Science—Society Integration.
  Specific teaching methods include
- Formal Presentations and Demonstrations,
- Group Interaction,
- Hands-on Experience and Experimentation,
- Discovery, Inquiry, and Experimentation.

Experiences in postdoctoral courses and internships are also part of the education system. The purpose of the courses/internships is to prepare qualified personnel to work in solution-oriented projects in the future. The lessons are based on concepts that were developed experientially throughout the cycle of the lessons: the result is teaching oceanography as an evolving science and deriving mutual understanding and the enrichment of data to create information.

An evolving science

The definition of science as *a continuous questioning of the foundations of our world view* also has a corollary in the education, in particular in the demonstration that culture, science, and technology are strongly linked. The development of a rigorous methodology for conducting research at sea has been one of the major efforts since the first "proposition for diligent observations" (Chapter 1). The strong interaction between navigation and understanding of ocean phenomena during the 19th century began to shape the way oceanographic research was conducted.

Innovations in data collection technologies and methodologies have emerged in recent decades. One consequence was the need to coordinate observation activities internationally. Advances in understanding ocean processes have been achieved with the integrated efforts of specialists in various disciplines. These efforts have left behind information, knowledge, trained technicians, and young scientists. This legacy must be communicated to students and lessons in the history of oceanography can help them understand the paths toward knowledge and the need to continually improve working methods and concepts.

Important examples can be cited: the evolution of the concepts of heat and salt, obtained by applying the principles of thermodynamics, leads to a better analysis of conservation laws in the results of numerical models.

- The concept on "heat" evolved from in-situ temperature to conservative temperature (McDougal, 2003; see Chapter 1).
- The content of salt in sea water is calculated with a complex formula that will require further refinements in the future.
- Historically, specific gravity was measured instead of seawater density; today this is calculated from measurements of temperature and salinity.

Students need to understand what the consequences of formula changes/refinements are in terms of errors in specific applications (e.g., conservation of heat and mass). They also need to understand how these formulations are originally derived.

Mutual understanding

Instructions for students from different science backgrounds (biologists, engineers, physicists, geologists) can be challenging to formulate. Each of them has personal beliefs derived from his/her studies.

Before lecturing on data, it is important that all the students understand each other. Discipline-derived personal beliefs should be presented by students and debate encouraged. Discussions during the lessons highlight semantic differences and allow the definitions of terms with well-defined meanings and the adoption of common vocabularies, an important element in data management and interoperability.

The ultimate goal of this approach (which lasts for the entire course) is the transition from personal beliefs to a shared understanding and group perspective (Stahl, 2000). This is achieved by encouraging students to ask, find, or propose answers to the questions posed during the lessons.

Putting different ideas together without contradictions is the challenge. Students are asked to evaluate the quality of scientific information on the basis of the technologies and methodologies used to generate it. As stated earlier in this world of data abundance, researchers are forced to use descriptive statistics to judge the value of the data they are working with. Students may agree or disagree on the validity of one piece of information or theory over another, and it is hoped that this disagreement will lead to a discussion that evolves to a position of agreement. The debate to be held in the classroom must be guided to reach mutual understanding, extracting from each information, which is deemed correct and possibly leading to a theory that all students recognize as an improvement over past concept. If successful, the approach will get a shift from discipline-oriented practice to shared understanding and common best practices.

Requirements for commonly accepted best practices were defined by Pearlman et al. (2019) and include activities such as strategies for data collection and for use, generation of data and information products, ethical and governance aspects in data collection and management. Pearlman et al. underlined that the development of best practices across activities are an increase and *pressing concern of global science*.

Data enrichment

Discussions with students reveal their knowledge of major marine environmental phenomena and their different time and space scales. From here, lessons on sampling strategy, data collection devices, errors, quality assurance, and quality control procedures (e.g., Chapter 2 and 3) are carried out. These serve to stimulate a further debate on the minimum essential information that makes the data usable by various users.

Students are invited to define their information needs, based on the disciplinary knowledge that constituted their background. Furthermore, they are asked to include in their list of needs also information that they believed necessary for other disciplines. Surprisingly, in the experience carried out in some classes, it was possible to verify that many information needs were common to almost all the students and there were also complementary issues that, put together, gave a metadata model consistent with what is used in the best practices:

- Cruise information:
  - Sampling objective; sampling frequency
  - Principal investigator
  - Ship name and call sign; name of commander
  - Facilities of ship (e.g., laboratories)
  - Instrumentation used; information on sensor; calibration

Since the measurements are executed as "stations," students also defined additional mandatory information:

- Station information:
  - Station name or sampling identifier
  - Environmental characteristics (e.g., in front of a river)
  - Existing anthropogenic activities
  - Sea state
  - Meteorological data
  - Date, time, coordinates, depth
  - Initial and final date, time, position, depth

The data model selected by students was an alphanumeric comma separated file containing the following data: sampling depth, samples/parameters, error associated to sensors, quality flag, date, time, cruise name, name and number of operators.

Here, "sample" refers to any "object" in water, sediment, or biota.

It has been curious to see that students, having in mind the collection of multiparametric data and samples, were thinking of many operators with different roles, all worthy of mention. Asking, "why this inclusion in the data file?," the answer was that the quality of data also depends on the technician's skills.

The lessons continued by comparing the proposed data/metadata model with those adopted in the main global observation programs. Commonly used metadata and vocabularies were part of the final systematization of the data (see Chapter 3).

However, even in the case of very positive results, students must be asked other questions about data reuse:

- What was the original purpose of the data collection?
- What were the applied quality assurance/quality control protocols?
- How were the data originally used?
- Have the data been reused?
- How has the data been assessed for fitness of use?

The solution to these questions is the inclusion in the metadata of references to protocols and publications (Manzella, 2015; Manzella et al., 2017). In this way, students are obliged to carry out a bibliographic search that may eventually provide them with new knowledge.

From data to information/understanding

The critical part of the "evolution from data to information and understanding" is similar to the transition from academic education to solution-oriented learning, without losing the vision of more general knowledge. Data sufficiency analysis is an important component of this process, along with the ability to analyze problems, separate out the most relevant components, and make sense of the observational data and numerical model results. Students must be able to

- understand cause and effect in the solution-oriented specific process,
- identify data and information required for the solution-oriented process,
- carry out a synthesis of the data (e.g., statistics) to higher order products,
- represent this synthesis in an understandable way.

The development of these skills could be done with on-the-job experiences, formal apprenticeships, professional training, or mentoring programs. These educational approaches force students to think more about what they are trying to solve and to deepen their knowledge of the specific Universe of Discourse.

The goal of these programs is the acquisition of core competencies constituting mental habits guiding behavior in a competitive environment. In normal practice, the effect of "on-the-job experience/professional training" could be that students acquire domain-specific skills, i.e., both have an outcome contrary to the desired "Matthew effect."

However, the knowledge of the specific Universe of Discourse together with the task of improving understanding is a crucial component of solution-oriented educational approach when students are able to conceptualize, analyze, evaluate, synthesize information derived from observations and experiences from collaborative-interdisciplinary research programs.

A blend of mentoring program and professional/solution-oriented training that includes a wide range of investigation can effectively educate students by connecting theories and practices. This can be achieved by having the students work with real-world data, helping them to find solutions within the working group in which they have been inserted.

Students must be enabled to determine whether the findings can be applied to other sectors as well and defend them in public discussions as recommended by Hooke's experimental method (Chapter 1):

- After finishing the Experiment, to discourse, argue, defend, and further explain, such Circumstances and Effects in the ... Experiments, as may seem dubious' or difficult': And to propound what new Difficulties and Queries do occur, that require other Trials and Experiments to be made, in order to their clearing and answering: And farther, to raise such Axioms. and Propositions, as are thereby plainly' demonstrated and proved.
- To register ... the Objections and Objectors, the Explanation and Explainers, the Proposals and Propounded of new and farther Trials ...

Conclusions

The long introduction to this chapter was the premise for a definition of educational goals in which knowledge leads to "thinking skills"—critical and creative thinking—and therefore the ability to solve complex problems. In practice, a mentoring program that integrates data from many disciplines has been approached in a resourceful way, trying to make rational adjustments in the educational strategy every time that it was necessary.

The educational approach presented in this book is a blend of cultural/technical methods, their historical evolution, application, and impact on science and society.

As reported in Chapter 1, scientific interest in interrelationships between various natural phenomena began to develop in the second half of the 19th century. In particular, Alexander von Humboldt had a vision of nature as a dynamic unitary complex in which terrestrial phenomena, far from being isolated from each other, are governed by universal physical laws. In the preface of the book "*Cosmos*", Humboldt (2005) revealed his determination to "comprehend the phenomena of physical objects in their general connection, and to represent nature as one great whole, moved and animated by internal forces." In this context, it can be emphasized that the *Challenger* expedition (1872—76) was an effort to investigate "*everything about the sea*" (National Research Council, 1999).

However, Humboldt was aware of the difficulties in scientifically describing the *mutual dependence and connection existing between classes of phenomena*, and in the introduction of volume 1 added the following sentence: "It remains to consider whether, by operation of thought, we may hope to reduce the immense diversity of phenomena comprised by the Cosmos to the unity of a principle, and the evidence afforded by rational truths. In the present state of empirical knowledge, we can scarcely flatter ourselves with such hope. Experimental sciences, based on the observation of the external world, cannot aspire to completeness; the nature of things, and the imperfection of our organs, are alike opposed to it."

The history of ocean science (e.g., McConnel, 1982) clearly shows how the accumulation of knowledge underlies evolving scientific research and changes in mental habits. The important lesson from the history of ocean data science is that methodologies and technologies have been continually adapted and will continue to be adapted in the future, in an endless process of improving accuracy and complexity.

Teaching ocean data science requires knowledge of history, mathematics, physics, chemistry, and biology. There are many approaches to ocean data science in various universities, and this chapter does not propose specific rules but intends to contribute some ideas and tools to a way of teaching that helps students acquire critical thinking skills (Loseby, 2019).

The educational elements contained in this chapter are
- Detect problems and determine appropriate solutions
- Work with others in solution-oriented activities
- Think critical and constructive using existing information
- Cope with complexity

Interdisciplinary and technical skills are needed to transform raw ocean data into useful information to support (for example) sustainable goals and a sustainable blue economy. Reliable information, with its associated uncertainties, allows the creation of value-added products capable of supporting decision-making processes.

Acknowledgments

The lessons referred to in this chapter were carried out in the context of various regional projects. GMRM feels its duty to thank especially the students of the OPTIMA course (Operational oceanography and information technologies for maritime security) organized by the National Research Council (CNR). After them, other experiences have been carried out in projects carried out in the context of initiatives of the Liguria Region, the National Institute of Geophysics and Volcanology (INGV) and the Ligurian Cluster of Maritime Technologies (DLTM).

References

Bell, G., Hey, T., Szalay, A., 2009. Beyond the data deluge. Science 323, 1297—1298.

Braudel, F., 1985. La Mediterranée. Flammarion, Paris.

Fichtinger, A., Rix, J., Schäffler, U., Michi, I., Gone, M., Reitz, T., 2011. Data Harmonisation Put into Practice by the HUMBOLDT Project. Int. J. Spatial Data Infra. Res. 6, 234—260.

Humboldt (von), A., 2005. COSMOS: A Sketch of the Physical Description of the Universe, vol. 1 (The Project Gutenberg eBook, Translated by E.C. Otte).

IOC/CEC, 1993. Manual and Guides 26, Manual of Quality Control Procedures for Validation of Oceanographic Data. SC-93/WS-19. UNESCO.

Lguensat, R., Viet, P.H., Sun, M., Chen, G., Fenglin, T., Chapron, B., Fablet, R., 2019. Data-driven interpolation of sea level anomalies using analog data assimilation. Rem. Sens. 11, 858. https://doi.org/10.3390/rs11070858.

Loseby, D.L., 2019. Critical Thinking Skills. Behavioural Procurement Project, CIPS Knowledge. https://www.researchgate.net/publication/336058016_Critical_Thinking_ Skills.

Manzella, G.M.R., March 2015. Knowledge building and computer tools. In: Diviacco, P., Fox, P., Pshenichy, C., Leadbetter, A. (Eds.), Collaborative Knowledge in Scientific Research Networks. IGI Global. ISBN: 978-1-4666-6567-5.

Manzella, G.M.R., Bartolini, R., Bustaffa, F., D'Angelo, P., De Mattei, M., Frontini, F., Maltese, M., Medone, D., Monachini, M., Novellino, A., Spada, A., 2017. Semantic search engine for data management and sustainable development: marine planning service platform. In: Diviacco, P., Leadbetter, A., Glaves, H. (Eds.), Oceanographic and Marine Cross-Domain Data Management for Sustainable Development. IGI Global. https://doi.org/10.4018/978-1-5225-0700-0.ch006, 2016.

McConnell, A., 1982. No Sea Too Deep. Adam Hilger Ltd, Bristol.

McDougall, T.J., 2003. Potential enthalpy: a conservative oceanic variable for evaluating heat content and heat fluxes. J. Phys. Oceanogr. 33 (5), 945—963. https://doi.org/10.1175/ 1520-0485(2003)033<0945:PEACOV>2.0.CO;2.

National Research Council, 1999. Global Ocean Science: Toward an Integrated Approach. The National Academies Press, Washington, DC. https://doi.org/10.17226/6167.

Pearlman, J., Bushnell, M., Coppola, L., Karstensen, J., Buttigieg, P.L., Pearlman, F., Simpson, P., Barbier, M., Muller-Karger, F.E., Munoz-Mas, C., Pissierssens, P., Chandler, C., Hermes, J., Heslop, E., Jenkyns, R., Achterberg, E.P., Bensi, M., Bittig, H.C., Blandin, J., Bosch, J., Bourles, B., Bozzano, R., Buck, J.J.H., Burger, E.F., Cano, D., Cardin, V., Llorens, M.C., Cianca, A., Chen, H.. Cusack, C., Delory, E., Garello, R., Giovanetti, G., Harscoat, V., Hartman, S., Heitsenrether, R., Jirka, S., Lara-Lopez, A., Lantéri, N., Leadbetter, A., Manzella, G., Maso, J., McCurdy, A., Moussat, E., Ntoumas, M., Pensieri, S., Petihakis, G., Pinardi, N., Pouliquen, S., Przeslawski, R., Roden, N.P., Silke, J., Tamburri, M.N., Tang, H., Tanhua, T., Telszewski, M., Testor, P., Thomas, J., Waldmann, C., Whoriskey, F., 2019. Evolving and sustaining ocean best practices and standards for the next decade. Front. Mar. Sci. 6, 277. https://doi.org/10.3389/fmars.2019.00277.

Pinardi, N., Simoncelli, S., Clementi, E., Manzella, G., Moussat, E., Quimbert, E., et al., 2017. EMODnet MedSea CheckPoint Second Data Adequacy Report (Version 1). European Marine Observation and Data Network. https://doi.org/10.25423/cmcc/medsea_checkpoint_dar2. https://www.emodnet.eu/zh-hans/checkpoints/reports.

Popper, K., 1984. Realism and the Aim of Science from Postscript to the Logic of Scientific Discovery (It. Translation) Il Saggiatore Editor.

Press, G., 2013. A Very Short History of Data Science. Forbes. Retrieved from. https:// www.forbes.com/sites/gilpress/2013/05/28/a-very-short-history-of-data-science/? sh=3fc3d53355cf. (Accessed 14 December 2020).

Ryabinin, V., Barbière, J., Haugan, P., Kullenberg, G., Smith, N., McLean, C., Troisi, A., Fischer, A., Aricò, S., Aarup, T., Pissierssens, P., Visbeck, M., Enevoldsen, H.O., Rigaud, J., 2019. The UN decade of ocean science for sustainable development. Front. Mar. Sci. 6, 470. https://doi.org/10.3389/fmars.2019.00470.

Rovelli, C., 2014. Che cos'è la scienza. La rivoluzione di Anassimandro. Mondadori Editor, Milan.

Stahl, G., 2000. A model of collaborative knowledge-building. In: Fishman, B., O'Connor-Divelbiss, S. (Eds.), Fourth International Conference of the Learning Sciences. Erlbaum, Mahwah, NJ, pp. 70–77.

United Nations, 2012. Future We Want — Outcome Document. Retrieved from. https://sustainabledevelopment.un.org/futurewewant.html. (Accessed 14 December 2020).

Wittgenstein, L., 1975. On certainty. In: Anscombe, G.E.M., von Wright, G.H. (Eds.), Translated by Denis Paul and E.M. Anscombe Basil Blackwell, Oxford, 1969–1975.

PART FIVE

Appendix

CHAPTER SEVEN

Oceanography: a recent scientific discipline with ancient origins

Giuseppe M.R. Manzella[1,3], Antonio Novellino[2]
[1]The Historical Oceanography Society, La Spezia, Italy
[2]ETT SpA, Genova, Italy
[3]OceanHis SrL, Torino, Italy

This book has only three objectives: to tell how oceanographic knowledge developed over the centuries and to show how the access and integration of data and information are serving science and society and how marine infrastructures could become significant components of a global knowledge environment. It identifies the strategic fil rouge that is overcoming cultural distances between knowledge producer and users. "Public engagement with science" has become an almost obligatory passage point for science policy in most of the Earth countries, but its substantive forms and meanings still need development. The knowledge society is based on the possibility to have free and open access to data and information that users can transform into knowledge. Data, information, and knowledge must be shared not only among scientific and technological communities but must be made available to all societal groups.

The UN General Assembly Resolution 56/183 (December 21, 2001) endorsed the World Summit on the Information Society (WSIS) goal "to build a people-centered, inclusive and development-oriented Information Society, where everyone can create, access, utilize and share information and knowledge, enabling individuals, communities and peoples to achieve their full potential in promoting sustainable development and improving quality of life, premised on the purposes and principles of the Charter of the United Nations and respecting fully and upholding the Universal Declaration of Human Rights." This book is applying the concepts behind this statement: support an inclusive and development-oriented Information Society, where everyone can access, utilize, create, and share information and knowledge, enabling scientists, individuals, communities, and peoples to achieve not only sustainable development but also improve the quality of life.

Ocean Science Data
ISBN: 978-0-12-823427-3
https://doi.org/10.1016/B978-0-12-823427-3.00002-5

© 2022 Elsevier Inc.
All rights reserved.

Data in the Ocean Sciences are being used by a far broader community than ever before, and a multidisciplinary approach requires the "representation," "processing," and "communication" of information from different disciplines. Data are collected, analyzed, organized, and used by people, and their good use/re-use can be obtained with social practices, technological and physical arrangements intended to facilitate collaborative knowledge building, decision-making, and inference. This is what is called "knowledge environment" and requires a "knowledge building system" to address the needs of different communities to create knowledge. This includes provision of products and services, network connectivity, and management of intellectual capital.

Access to marine data and information is of vital importance for marine research and a key issue for various studies, from climate change prediction to offshore engineering. Giving access to and harmonizing marine data from different sources will help industry, public authorities, researchers, and students find the data and make more effective use of them to develop new products, services, and improve our understanding of how the seas behave.

Chapter 1 of this book provides a general view on technologies and methodologies for data collection and the consequences of the rapid evolution in knowledge and technology turn-over on working procedures and education.

Re-use of data requires the knowledge on objectives of original data collection, protocols in QA/QC, findings of data analysis, etc. It is not easy to assess "a priori" the re-use of data that could not be fitted for other general purposes. It is critical for any data collection system to provide additional information and documentation that can allow a correct use and re-use of the data also for purposes not envisaged during the data collection processes.

Chapters 2 and 3 underline opportunities to start a new roadmap for data management issues and open a plan for future collaboration based on knowledge management efforts. These focus on organizational objectives such as improved performance, competitive advantage, innovation, the sharing of lessons learned, integration, and continuous improvement of management organization.

Chapter 4 looks for case studies on strong collaborative frameworks among data producers, managers, and users. This is achieved in an effective way by preparing specific messages to targeted communities of users.

Chapters 5 and 6 contribute on transfer of research results to stakeholders from various interest groups including politics, business, and civil society, as well as concepts for the promotion of young scientists.

To complete the book, 56 slides have been prepared by authors of this book for presentations, with the idea that they can intrigue auditors and attract young researchers to a recent scientific discipline of ancient origins.

Oceanography:
a recent scientific discipline with ancient origins

Federico De Strobel[1],
William Emery[2]
Lavinio Gualdesi[1],
Giuseppe M.R. Manzella[1,4],
Antonio Novellino[3],

[1] The Historical Oceanography Society
[2] Colorado University
[3] ETT SpA
[4] OceanHis SrL

(references in Chapter 1 of this book, if not specified below figures)

16th sounding technology

Olaus Magnus (1490-1558) De Gentibus Septentrionalibus (1555)

• Important text on the populations of Northern Europe.

• It contains hints on the currents and depths of the Norwegian coasts assuming a similarity between the depths and the emerged lands.

• It illustrates the sounding technique with line and lead

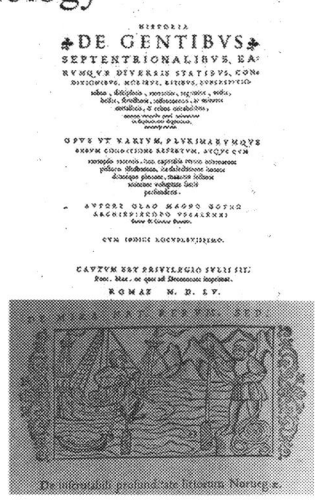

16th century sounding technology

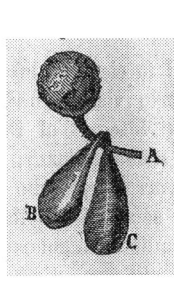

- Sounder without line conceived by Leon Battista Alberti (1404 – 1472) and reported by Bartoli in "Piacevolezze Matematiche (Mathematical Pleasures)" (1568).
- Float made to sink with a weight that is released by touching the bottom. The time, between launch and arrival on the surface, provides information on the depth.
- Sounder without line described in "Le Cosmolabe" (1567) by Jacques Besson (1540? – 1573)

Bartoli   Besson

 # Galileo thermoscope

- Galileo Galilei (1564 – 1642)
- *And so after three years of reading in Pisa, in September 26, 1592, he obtained the reading of mathematics in Padua for six years from the Ser.ma Republic of Venice … In these same times he found thermometers, that is, those glass instruments, with water and air, to distinguish the changes of heat and cold and the variety of heats of places* (Vincenzo Viviani: Racconto istorico della vita di Galileo, e-book, www.liberliber.it)

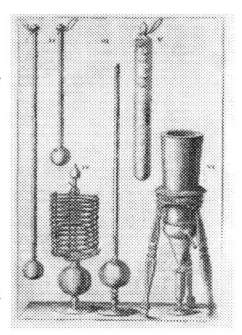

Thermoscopes from Magalotti L. (1667). Saggi di naturali esperienze fatte nell'Accademia del Cimento sotto la protezione del Serenissimo Principe Leopoldo di Toscana e descritte dal segretario di essa Accademia. Cocchini Giuseppe all'insegna della Stella, Firenze

Agrippa current meters 1595

Camillo Agrippa (1535 – after 1595)

 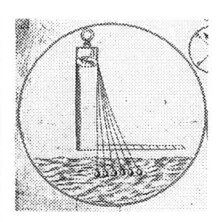

• Two current meters - inclination and still floating.

Intensity of the current provided by the inclination of a leaded cable or by the pull on the sail measured with a scale.

Direction referred to a compass through rotation on the axis of a tube connected to the side oriented by the rudder or sail

 # Crescentio water 'sampler'

Bartolomeo Crescentio (1595? - ?) 'Nautica Mediterranea' 1601

• A primitive system for taking up fresh water from depths

Kircher - Mundus Subterraneus

Athanasius Kircher (1602 – 1680)

• First map of the Gulf Stream

• Description of a tide gauge and a current meter

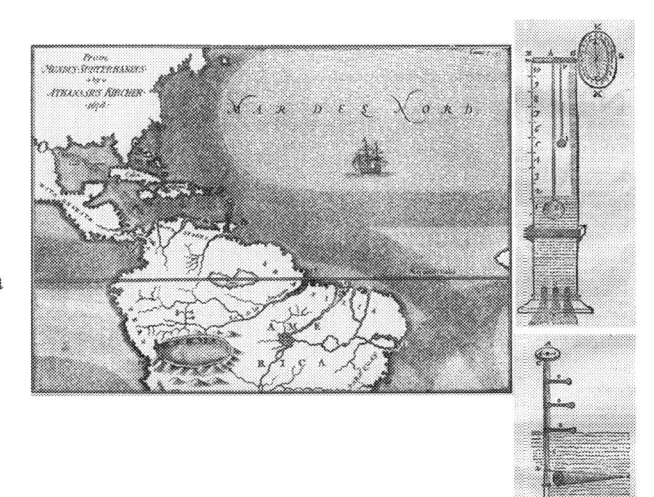

The rediscovery of the great scientists through the oceanographic vision

Robert Boyle (1627 – 1691)
• Studies on salinity and thermal characteristics of the sea (1671)

In the absence of efficient thermometers capable of measuring temperature at depth, he also used information provided by divers

TRACTS

WRITTEN

By the Honourable
ROBERT BOYLE,

The Cosmicall Qualities of things,
Cosmicall Suspicion,
The Temperature of the Subterraneal Regions,
The Temperature of the Submarine Regions,
The Bottom of the Sea.

To which is Prefixt,

An Introduction to the History of Particular
QVALITIES.

OXFORD,
Printed by W. H. for Ric. Davis,
M. DC. LXXI.

TRACTS
Consisting of
OBSERVATIONS
About the
SALTNESS of the SEA:
An Account of a
STATICAL HYGROSCOPE
And its USES:
Together with an APPENDIX
about the
FORCE of the AIR'S MOISTURE:
A FRAGMENT about the
NATURAL and PRETERNATURAL
STATE of BODIES.

By the Honourable ROBERT BOYLE.

To all which is prefix'd
A SCEPTICAL DIALOGUE
About the POSITIVE or PRIVATIVE
NATURE of COLD:
With some Experiments of Mr. Boyle's referr'd
to in that Discourse.

By a Member of the ROYAL SOCIETY.

London, Printed by E. Flesher for R. Davis Bookseller
in Oxford, M. DC. LXXIV.

Hooke's soundings and water sampler

Robert Hook (1635 — 1703). Sounding technique without lines

• Battista Alberti's type ball (1663) [top-right]

• 'Pressure sounder' (1691): depth inferred from the amount of water forced to enter the air cavity of the float [bottom-right]

• Propeller (1691): depth inferred from the number of turns of a propeller active in descent, blocked in ascent [bottom left]

• Water sampler composed by a 'Square Wooden Bucket' having two valves that remained open during the descent of the sampler and closed in the ascent [top left]

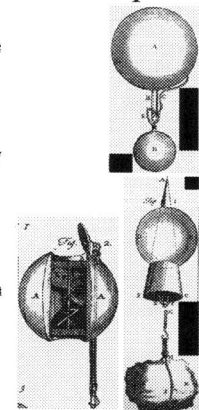

Hooke's explorators

• Explorator Profunditatis [left] Depth was inferred from the rotors inside the apparatus.

• Explorator Qualitatum [right] *I shall next shew how to fetch a Quantity of Water from the Bottom, or from any intermediate Space, or Distance from the Top.*

(Derham, 1726. Philosophical experiments and observations of the late eminent Dr. Robert Hooke)

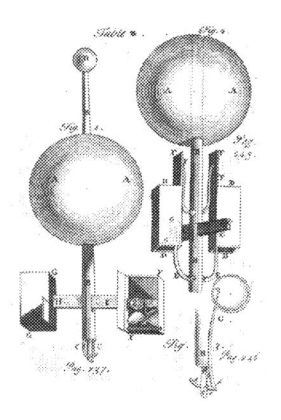

Hooke's improvements of scales

Weighting specific gravity (Derham, 1726. Philosophical experiments and observations of the late eminent Dr. Robert Hooke)

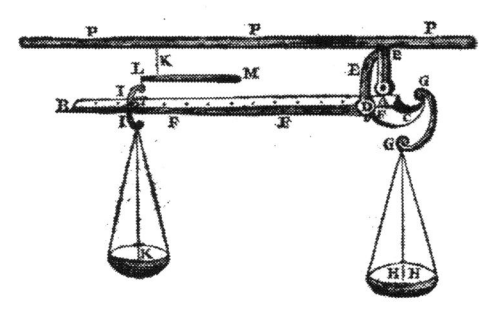

 Marsili and the Bosporus

Luigi Ferdinando Marsili (1658 – 1730)

• A precursor study of the two-layered movement of the Bosporus (Black Sea towards the Mediterranean on the surface and vice versa on the bottom) by obtaining information from fishermen and experimentally checking it from the configuration taken by a line lowered from a boat.

• He built an experimental model in a tank with waters of different densities to explain the reasons (1681)

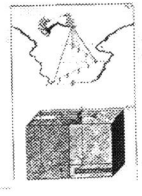

Marsili - Histoire Physique 1725

- First text of modern oceanography.
- Centered on studies in the Sea of Provence, it covers the various aspects of this new discipline:
  - ➤ Chemical-physical
  - ➤ biological
  - ➤ geological

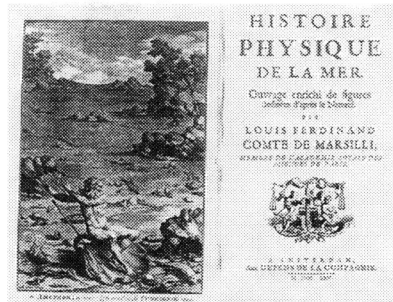

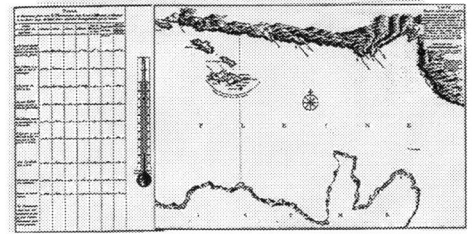

Cavendish thermometer 1757

- Henry Cavendish 1731 – 1810) invented many thermometers based on the dilatation of different 'liquors' (spirit of niter, oil of vitriol) including Spirit of Wine and Mercury.
- Thermometers initially were constructed for atmospheric applications and then adapted for use in the sea

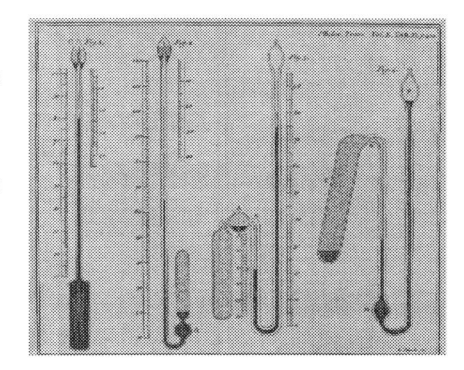

Six thermometer 1757

James Six (1731 – 1793) invented the minimum — maximum thermometer. The description of the instrument was published post mortem in 1794: *The construction and use of a thermometer for showing the extreme temperature in the atmosphere*

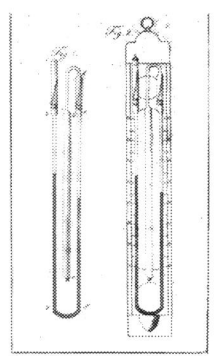

Six, J. (1794) *The construction and use of a thermometer*.
J. Blake, G. Wilkie and T. Wilkie, London

 ## Scoresby's Marine diver 1817

- Designed by Scoresby and used in whaling campaigns in Greenland

- Deep water sampler, for density and temperature measurements.

- Thermally insulated container, with closure activated by rapid ascent. Inserted two windows to facilitate the reading of the thermometer (Six protected max-min type).

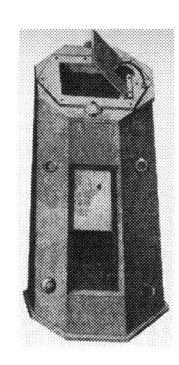

Ross Deep Sea Clam 1819

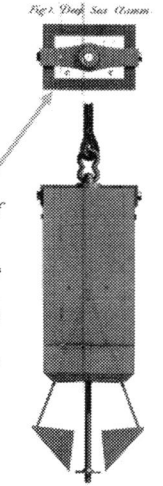

Fig 2. Deep Sea Clamm.

A. *hollow parallelogram of cast iron (1 cwt.), eighteen inches long, six by six, and three-quarter inches in the outside square, and in the inside four by five inches wide.*

C. *Is a view of the top, and a strap of iron across it, through which the spindle passes, and two inches below another strap of the same kind is placed.*

D. *Diagonal view of the forceps which are attached by a joint to the spindle, and which are kept extended by the joint bolt, No. 2. The cast-iron weight is, by the forceps being thus extended, kept up until the bolt touches the ground; the joint bolts No. 2. are then detached by No. 3., and the cast-iron weight slips down the spindle to which the rope is fixed, and shuts the forceps, which are by this time on the ground, by the power of the inclined plane enclosing and keeping fast the contents until taken out.*
JOHN ROSS, Captain,
H. M. S. Isabella.

Ross Hydraphorus 1819

The upper part, where the machinery is fixed, is square, having on one side a small aperture to admit water. This is covered by a circular plate in which another aperture is made to coincide with the former, ... a cover is fitted to protect this plate, the edge of which being divided into 800 equal parts, the aperture on the outside can be set to the required position. On the opposite side of the instrument there is a similar plate or wheel, which moves the former; and both are turned by the rotator as the Instrument descends, by the action of the water, the former in a proportion as one is to one hundred. The vanes of the rotator are made to fix in any position, ... At the top of the instrument there is a spring valve, for the double purpose of allowing the air to escape when the water enters, and to let the air enter when the water is drawn off by the stop cock at the bottom, and in the latter case the valve must be moved up by hand.

Fig 3. Hydrophorus.

Cup-lead dredge ~1819

Derived from the deep-sea clam for *moderate depths*. The conical iron cup was penetrating the mud or the sand and the was hauled up with a sample of the sea bottom. The *washer* n was pressed down during the ascent of the apparatus, retaining the sample (C. Wyville Thomson, 1873)

Fig. 37.—The "Cup lead."

 # Ball's Dredge 1847

In the practical use of the dredge, I have no experience, but hope, ere long, 'to be afloat' ... the two scrapers ABCD and ABCD are each twenty inches in length, by two inches in breadth; parallel with their lower edges CD and CD, about fourteen holes, equidistant from each other, are pierced to receive the laces of the bag, and these two plates are joined at their lower extremities, by means of two cross bars CC and DD ... the arms EF and EF are each sixteen inches in length by three-eighths in diameter, and play upon the cross bars by means of double swivel joints, as seen at EE and EE. Their anterior extremities at F are beaten flat ... and are pierced for the reception of the bolt H, which ... passes through the extremities of what may perhaps be termed the bridle ring G, to which the rope is affixed. (A. Hepburn, 1847; The Zoologist)

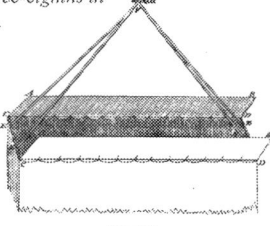

Ball's Dredge.

 # Brooke's apparatus ~1854

The instrument as devised by Brooke is very simple. A 64lb shot E is cast with a hole through it. An iron rod A has a chamber B at the lower end, and two moveable arms hinged to the upper end ... When the instrument strikes, the end of the rod is driven into the material of the bottom ... the two jointed arms fall down, the loops of the sling are relieved from the teeth, and the rod slips through the hole in the shot and comes up alone ... (C. Wyville Thomson, 1873)

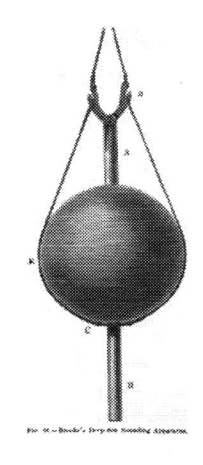 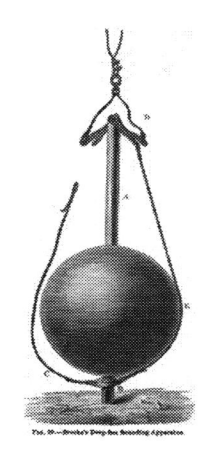

 # Bull-dog sounding machine 1860

An adaptation of the Ross' deep sea clam by Steil. *It was invented during the famous sounding voyage of H.M.S. 'Bulldog' in the year 1860* (C. Wyville Thomson, 1873)

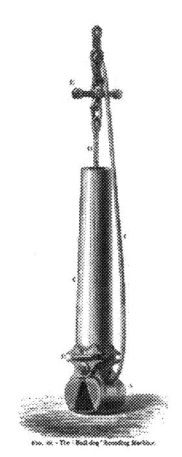

Miller - Casella thermometer 1868

Nearly a hundred years ago some one thought of employing Six's maximum and minimum thermometer for temperature observations in the sea, various modifications being introduced, until finally in 1868 it became quite serviceable as made by Casella under the direction of Dr. Miller. The Miller-Casella thermometer was the one principally used on board the "Challenger" and during other great expeditions. (Helland-Hansen, 1912)

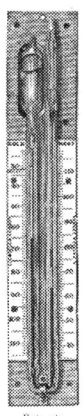

FIG. 154.
MILLER-CASELLA
THERMOMETER.

 # Buchanan stopcock water bottle ~1872

The water bottle designed by J.Y. Buchanan for the Challenger expedition. On the left the bottle during the descent, halfway when the stopcocks (taps - BB) are closed, on the right during the ascent. (Helland-Hansen, 1912)

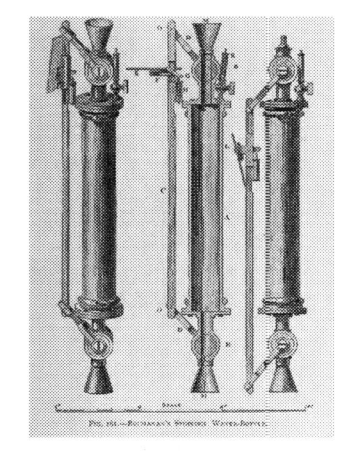

FIG. 161.—BUCHANAN'S STOPCOCK WATER-BOTTLE.

Baillie sounding machine ~1872

Baillie sounding machine used during the Challenger expedition.

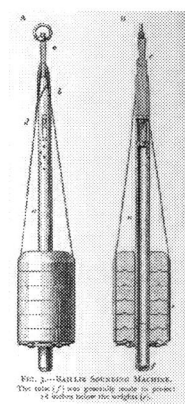

We see, therefore, that sounding in about 3000 fathoms took nearly an hour and a half, whereas for about 4500 fathoms two and a half hours were required, which must be considered very quick work. On the same line and with the same arrangement as for sounding, series of temperatures were taken and deepwater samples obtained. (Murray, 1912)

Challenger trawl ~1872

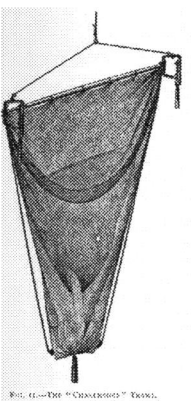

The " Challenger " used a trawl constructed like the ordinary beam-trawl, which was employed particularly by the fishermen in the shallow waters off the flat English coasts. The beams were of different lengths, 17, 13, and 10 feet, but the 10-feet length was found to be the best for deep water. (Murray, 1912)

Challenger tow-net ~1872

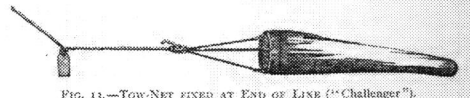

FIG. 13.—TOW-NET FIXED AT END OF LINE ("Challenger").

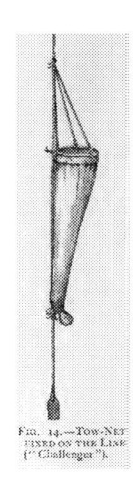

Tow-nets used during the Challenger expedition *for making pelagic captures* (Murray, 1912)

FIG. 14.—TOW-NET FIXED ON THE LINE ("Challenger").

Sigsbee dredge apparatus ~1877

During the winter of 1877-78 the United States Coast Survey steamer "Blake" undertook a cruise in the Gulf of Mexico, under the command of Captain Sigsbee and under the personal supervision of the late Alexander Agassiz. As it was proposed to carry out investigations with the dredge and trawl along the bottom, Agassiz suggested the use of a wire rope instead of hemp ropes. (Murray, 1912)

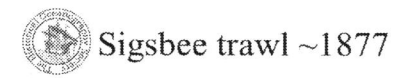 Sigsbee trawl ~1877

It was, however, difficult to tell, when the depth was at all great, whether the trawl had reached the bottom right side up, and whether it was open while being towed. Sigsbee solved this difficulty by having tripping lines on both sides. (Murray, 1912)

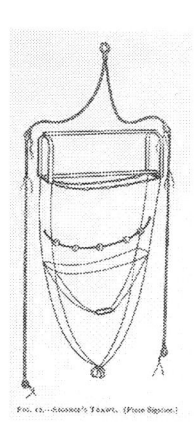

Fig. 12.—Sigsbee's Trawl. (From Sigsbee.)

Reversing thermometer 1878

In 1878 Negretti and Zambra of London constructed a reversing thermometer, which has played a prominent part in physical oceanography. In this form there is a narrowing of the tube just above the bulb; the mercury fills the tube above the narrowing to a greater or lesser extent according to the temperature, and when the thermometer is tipped over, the mercury breaks off at the narrowing, the portion which was above that point sinking down to the end of the tube; the scale on the tube indicates the temperature at the moment of inversion. (Helland-Hansen, 1912)

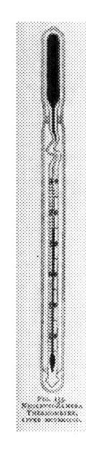

Peterson – Nansen bottle 1894

On the left it is seen open, as it is let down into the water; the lid is suspended in the upper part of the frame, and supports the cylinders as well as a weight hanging below the apparatus. When a messenger is sent down the line and strikes the water-bottle, the lid is released, and the weight draws both lid and cylinders down, clasping the apparatus together and closing it hermetically. The right - hand figure shows the water-bottle closed and ready for hauling up. (Helland-Hansen, 1912)

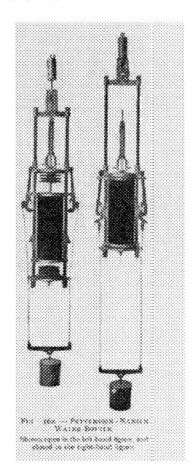

Richter reversing thermometer 1902

In many cases it is necessary to have the temperature determined with the highest possible degree of accuracy, and Richter's reversing thermometer is very satisfactory in this respect. During the " Michael Sars" Atlantic Expedition the temperature series were taken almost exclusively by the aid of these thermometers, and in most instances two thermometers were used simultaneously, so as to make quite sure of the determinations. (Helland-Hansen, 1912)

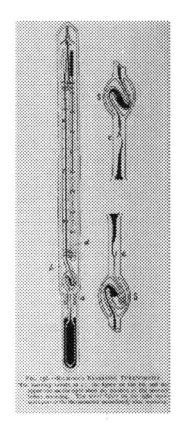

Ekman bottle 1903

V. W. Ekman has constructed an apparatus to serve as a reversing mechanism and a water-bottle at the same time ... A messenger is sent down after it and knocks out the pin ... The levers gradually draw the lids closer, and when the cylinder is wholly reversed it is held fast by a catch and encloses the water-sample hermetically. (Helland-Hansen, 1912)

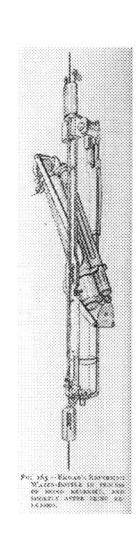

 # Ekman current meter 1903

It consisted of a propeller oriented by a vane and a mechanism that recorded the number of revolutions, a compass and a recorder that provided a statistical indication of the current directions. Inside the current meter, metal balls in a reservoir fell, one at a time, onto sectors of the compass. The number of balls in each sector provided an indication of the direction of the main currents (Manzella et al. this book)

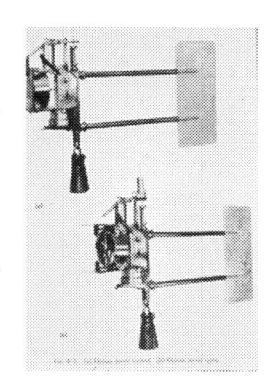

 # Photometer 1903

Investigation of the limit of light rays. A *gelatine-film was covered with a glass plate and inserted into a small envelope of thin caoutchouc, with a square opening in front through which the light is admitted.*

(Helland-Hansen, 1912)

FIG. 171.—HELLAND-HANSEN'S PHOTOMETER.
On the left, as it is sent down ; in the middle, open for exposure ; on the right, closed and ready for hauling up.

 # Plankton centrifuge 1903

During the "Michael Sars" Expedition our quantitative investigations yielded really remarkable results. Lohmann had succeeded by means of a centrifuge in determining the quantity of plankton in quite small samples of Baltic water, and we felt confident, therefore, that this excellent method ought also to prove serviceable in the open sea. (Murray, 1912)

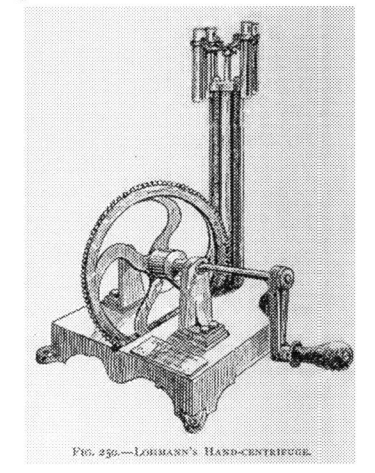

FIG. 250.—LOHMANN'S HAND-CENTRIFUGE.

Titration apparatus

The most convenient, and on the whole the most satisfactory, method of determining salinity is a chemical one, and is based on the fixed relation between the chlorine contained in a sea-water and its total salinity... When a solution of silver nitrate is added to sea-water, the chlorine is thrown down as a white precipitate of silver chloride. (Helland-Hansen, 1912)

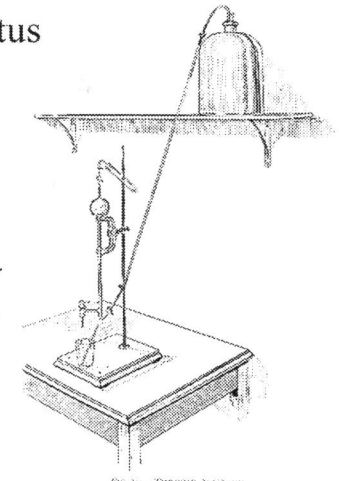

 Bathythermograph (BT) 1935

A torpedo-shaped instrument with a temperature sensor and a transducer. It detects temperature and depth down to about 300 metres. Experimented for the first time (1935) by C.G. Rossby and then improved in the framework of a collaboration among MIT, WHOI and US Navy. In 1938 the BT came into use (*Scripps Institution of Oceanography: Probing the Oceans 1936 to 1976.* San Diego, Calif: Tofua Press, 1978.
http://ark.cdlib.org/ark:/13030/kt109nc2cj/)

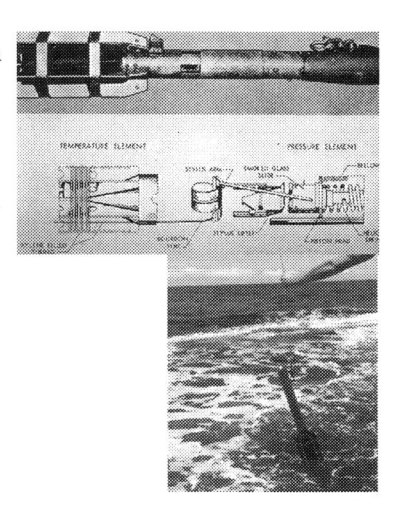

 Mechanical BathyThermograph-MBT

Bathytermograph winch (on the right) and temperature profile as derived from the smoked glass (below

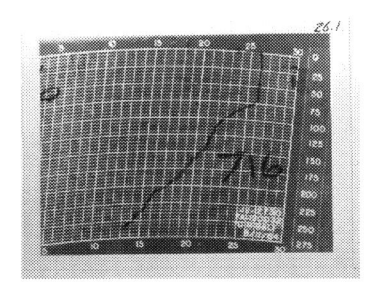

eXpendable BathyThermograph -late'50s

A torpedo-shaped instrument with a temperature sensor developed by J. M. Snodgrass in the late 1950'. The temperature data are transmitted on board a ship by means of a thin copper wire. The depth is calculated as a function of the descent time interval (Berger, *Ocean: Reflections on a Century of Exploration*, University of California Press, 2009). In figures: Sippican XBTs and launching system during the Mediterranean Forecasting System projects (1998 – 2006)

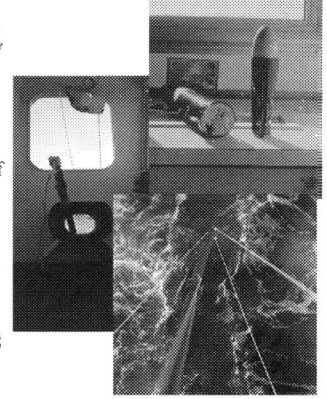

SPAR buoy. Air-sea interaction studies

60s STD systems

- Salinity(Conductivity)-Temperature-Depth profilers developed in the early 1960s.
- A "break-through" in oceanographic technology
- From thermometers and inverted bottles to the ability to acquire continuous profiles in situ
- A dramatic change in the way we investigate the oceans.

'60s CTD Kieler Hosvaldtswerke T87

- German-made CTD system representative of the most advanced technology based on thermionic valve circuits.
- Inductive conductivity cell.
- Multiplexed FM technique for sending, via electromechanical cable, the sensory signals to the on-board unit
- High voltage in the underwater unit.

 # STD Hytech Model 9006(USA)

- Designed in '62 by one of the great pioneers of oceanographic engineering, Neil Brown
- First system based on solid state components
- Unique ability in the world to provide salinity data in real time
- In situ salinity detected with electronically compensated inductive conductivity cell for the T and P effects
- Data transmitted to the on-board unit with cable telemetry (FM multiplexed)
- Complex calibration procedures
- Dimensional evolution of the C sensor during the 1960s

Aanderaa Current meter RCM4

- Spin-off activity generated by a NATO project in 1960
- The Norwegian Ivar Aanderaa was the project manager
- Innovative digital technology based on an exclusive electromechanical encoder for A to D conversion
- Current meter designed on traditional speed (Savonius rotor) and direction (large rudder) sensors
- Able to digitally store data and transmit them acoustically from a distance (including CTD parameters)
- The company was founded in 1966 and the current meter marketed as RCM4

Buoyancy Glass Sphere

- Mooring Buoyancy Units
- 20 Kg Net Buoyancy
- Low drag mooring assembly
- Used along mooring lines to balance instruments weights

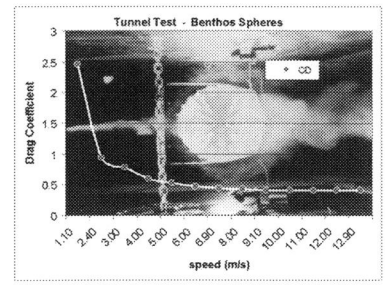

Lagrangian Drifters

- Company Albatros Marine Technology (Balearic Spain)
- Minidrifter 02
- Communications GPS DSM (Iridium Version)
- Very low Nyler Coefficient
- Diam. 10 cm
- Heigth 28cm

ALBATROS MARINE TECHNOLOGIES – MD02

Lagrangian Drifters

- **Company:**
- Clearwater (MA, USA)
- **Model:**
- CLEARSAT ARGOS/GPS MARKER
- **Communication:**
- Argos transmitter
- On-board GPS navigation
- **Dimensions:**
- The Marker Buoy is attached by means of stainless steel eyebolts at the bottom and side. Hull 40.6 cm diameter fibreglass
- **Weight:**
- 9 Kgr

CLEARWATER – CLEARSAT ARGOS/GPS MARKER

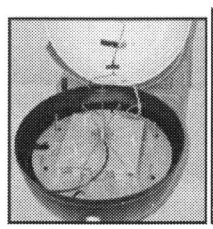

Lagrangian Drifters
MARLIN YUG – SVP-B / SVP-BT

- **Company:**
- Marlin-Yug Ltd. (Sevastopol, Ukraine)
- **Model:**
- SVP-B and SVP-BT (barometric pressure and temperature)
- **Communication:**
- Argos-2 or Iridium transmitter
- GPS receiver optional
- **Dimensions:**
- Surface float 41 cm diameter fibreglass hull
  Drogue is centered at 15m (optional up to 50m)
  Holey sock: 92cm OD, length 552cm.
  Drogue consists of 6 cylindrical sections, each 92cm long

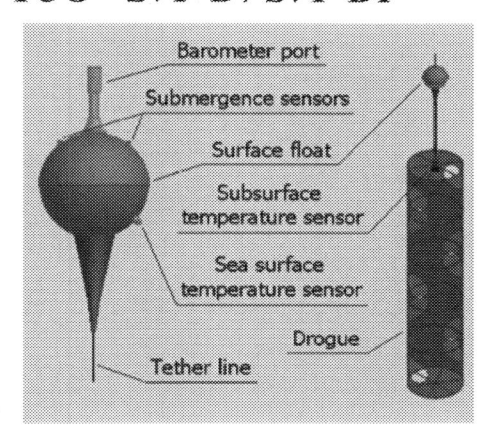

New generation Proto-current meters ('70s)

- Electromagnetic current meter Comex Equipment (UK project)
- Acoustic current meter Crouzet (FR)

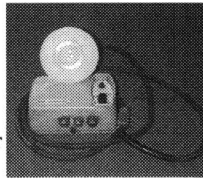

Gliders

- Autonomous underwater vehicle, propelling through buoyancy changes
- Ocean profile data: temperature, salinity, currents, and other parameters
- Able to travel far distances over long periods
- 0-200m

Autonomous Profilers

- Profiling floats measure ocean water properties
- Data transmitted to data centers via satellite

Moorings

- Multiple sensors connected to a wire anchored at the seafloor
- CTDs, acoustic doppler current profilers (ADCPs), fluorometers (chlorophyll), sediment traps, etc.
- Ocean profile at fixed stations
- long term series (>2 years)

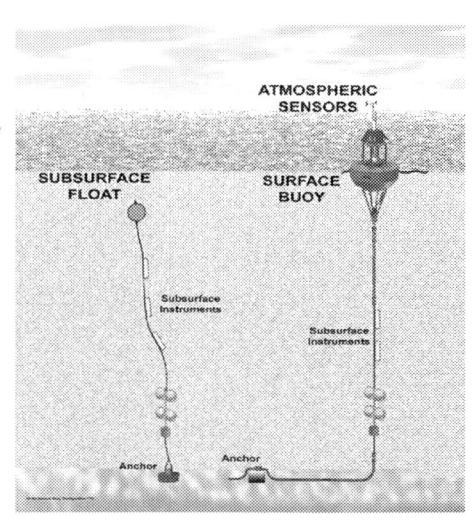

Animal Tagging

- Marine mammals tagging with CTD, geolocation, and fluorometer
- Continuous data from ocean areas unreachable to humans and robots, e.g. deep Southern Ocean
- animal behaviour and tracking data
- high resolution data transmitted in near real time through satellites

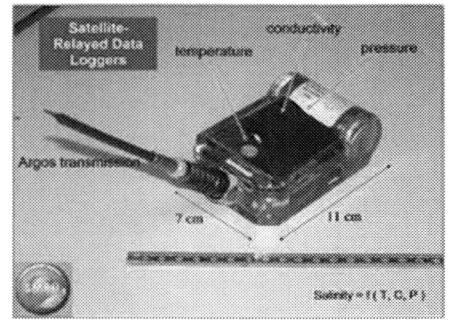

Sensors on Fishing Gear

- CTD sensors attached to fishing nets
- Mobile gears: vertical profiles and bottom transects (e.g. trawl)
- Fixed gears: vertical profile and stationary bottom measurements (e.g gill net)

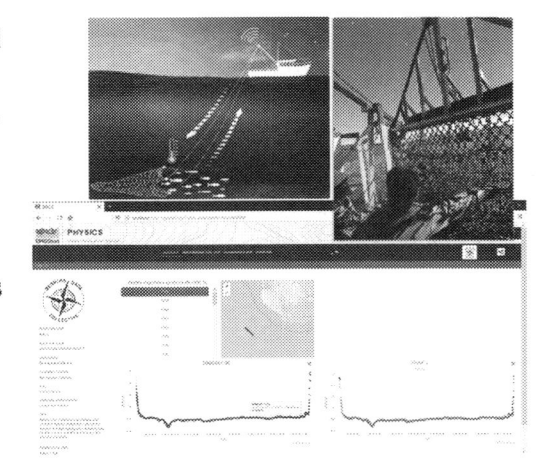

High Frequency Radar

- Ocean currents and waves in near real time
- based on the emission of electromagnetic waves and the study of the echo after reflection by the sea surface
- Large spatial scale measurements, up to 200 km

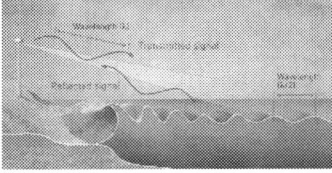

List of acronyms

AATSR	Advanced Along Track Scanning Radiometer
ABDC	Access to Biological Collection Data
ADP	Arctic Data Portal
ADU	Associated Data Unit
AfReMaS	African Register of Marine Species
AIS	Automatic Identification System
ALA	Atlas of Living Australia
AniBOS	Animal Borne Ocean Sensors
AODN	Australian Ocean Data Network
AOML	Atlantic Oceanographic and Meteorological Laboratory
API	Application Program Interface
APM	Application Performance Matrix
ASCII	American Standard Code for Information Interchange
ATSR	Along Track Scanning Radiometer
AVHRR	Advanced Very High Resolution Radiometer
BAMS	Bulletin of the American Meteorological Society
BGC-Argo	BioGeoChemical Argo
BHL	Biodiversity Heritage Library
BODC	British Oceanographic Data Centre
BOLD	Barcode of Life Data System
BOOS	Baltic Operational Oceanographic System
C3S	Copernicus Climate Change Service
CAFF	Conservation of Arctic Flora and Fauna
CAMS	Copernicus Atmospheric Service
CBD	Convention on Biological Diversity
CCHDO	Carbon Hydrographic Data Office
CCI	Climate Change Initiative
CDI	Common Data Index
CDS	Climate Data Store
CEC	Commission of the European Communities
CETAF	Consortium of European Taxonomic Facilities
CF	Climate and Forecast
CIMR	Copernicus Imaging Microwave Radiometer
CLMS	Copernicus Land Service
CMEMS	Copernicus Marine Environment Monitoring Service
CO$_2$	Carbon Dioxide
CODATA	Committee on Data
CoL	Catalogue of Life
COL-PLUS	Innovating the Catalogue of Life Systems
CoML	Census of Marine Life
COP	Conference of the Parties
CPR	Continuous Plankton Recorder
CRM	Certified Reference Materials

CRS	Coordinate Reference System
CSIRO	Commonwealth Scientific and Industrial Research Organization
CSR	Cruise Summary Reports
CSV	Comma Separated Values
CTD	Conductivity Temperature Depth sonde
CV	Curriculum Vitae
D4Science	Data infrastructure for science
DAB	Discovery and Access Broker
DATAMEQ	Data Management, Exchange and Quality
DataONE	Data Observation Network for Earth
DBCP	Data Buoy Cooperation Panel
DCDB	Data Centre for Digital Bathymetry
DG MARE	Directorate-General for Maritime Affairs and Fisheries
DG	Directorate General
DGPM	Directorate General for Maritime Policy (Portugal)
DIAS	Data and Information Access Services
DIC	Dissolved Inorganic Carbon
DIGI TWIN	digital twin "a realistic digital representation of assets, processes or systems in the built or natural environment" ("The Gemini Principles" (PDF). www.cdbb.cam.ac.uk. Centre for Digital Built Britain. 2018. Retrieved 2020-01-01.)
DIKW	Data-Information-Knowledge-Wisdom
DiSSCo	Distributed System of Scientific Collections
DIVA	Data-Interpolating Variational Analysis
DIVAnd	Data-Interpolating Variational Analysis in n dimensions
DM	Delayed Mode
DMP	Data Management Plan
DOI	Digital Object Identifier
DPSIR	Driver-Pressure-State-Impact-Response
DQ	Data Quality
DSA	Data Seal of Approval
DTM	Digital Terrain Model
DTO	Digital Twin of the Ocean
EBSA	Ecologically or Biologically Significant Area
EBV	Essential Biodiversity Variables
EBWBL	EMODnet Bathymetry released a World Base Layer Service
EC	European Commission
ECV	Essential Climate Variables
EDIOS	European DIrectory of the Ocean Observing Systems
EDMED	European Directory of Marine Environmental Data
EDMERP	European Directory of Marine Environmental Research Projects
EDMO	European Directory of Marine Organizations
EEA	European Environment Agency
EEI	Earth's Energy Imbalance
EGO	Everyone's Gliding Observatories
EIONET	European Environment Information and Observation Network
ELIXIR	European Life-Science Infrastructure
eLTER	European part of the global Long-Term Ecosystem Research

EMB	European Marine Board
EBMI	Inhaca Marine Biology Research Station
EMBRC	European Marine Biological Research Centre
EMODnet	European Marine Observation and Data Network
EMSO	European Multidisciplinary Seafloor and water column Observatory
ENA	European Nucleotide Archive
ENSO	El Niño-Southern Oscillation
ENVRI	ENVironmental Research Infrastructure
EO	Earth Observation
EoL	Encyclopaedia of Life
EOSC	European Open Science Cloud
EOV	Essential Ocean Variables
ERDAPP	Environmental Research Division's Data Access Program
ERIC	European Research Infrastructure Consortium
ERMS	European Register of Marine Species
ESA	European Space Agency
ESFRI	European Strategy Forum on Research Infrastructures
ESIP	Earth Science Information Partners
EU	European Union
EUDAT	pan-European network of research organizations, data and computing centers.
EUMETSAT	European Organization for the Exploitation of Meteorological Satellites
EUNIS	European Nature Information System
EuroArgo	European component of Argo program
EuroBioImaging-	European Research Infrastructure for Imaging Technologies in Biological and Biomedical Sciences
EurOBIS	European Ocean Biodiversity Information System
EuroFLEETS	An alliance of European marine research infrastructure to meet the evolving needs of the research and industrial communities
EuroGOOS	European component of GOOS
EuroSEA	European Ocean Observing and Forecasting System
Eurostat	European Statistical Office
EUSeaMap	Seabed habitat map for Europe
EV	Essential Variables
FAIR Principles	Findable, Accessible, Interoperable, and Reusable
FAO	Food and Agriculture Organization
FGDC	Federal Geographic Data Committee
G7	Group of Seven, Intergovernmental Organization
GACS	Global Alliance of Continuous plankton recorder Surveys
GBIF	Global Biodiversity Information Facility - An international body dedicated to providing free access to biodiversity data
GCI	GEOSS common infrastructure
GCMD	Global Change Master Directory
GCOS	Global Climate Observing System
GDAC	Global Data Assembly Centre
GEBCO	General Bathymetric Chart of the Oceans
GEF	Global Environment Facility

GEO	Global Environment Outlook
GEO BON	Group on Earth Observations Biodiversity Observation Network
GEO	Group of Earth Observation
GEOMAR	Helmholtz Centre for Ocean Research Kiel
GEOSS	Global Earth Observation System of Systems
GEOTRACES	International program which aims to improve the understanding of biogeochemical cycles and large-scale distribution of trace elements and their isotopes in the marine environment.
GES	Good Environmental Status
GGBN	Global Genome Biodiversity Network
GHRSST	Group for High Resolution SST
GIS	Geographic Information System
Globe	GLobal Oceanographic Bathymetry Explorer
GLODAP	Global Ocean Data Analysis Project
GLOSS	Global Sea Level Observing System
GO-SHIP	Global Ocean Ship-Based Hydrographic Investigations Program
GODAR	Global Oceanographic Data Archeology and Rescue
GOMON	Global Ocean Macroalgal Observing Network
GOOS	Global Ocean Observing System
GOSUD	Global Ocean Surface Underway Data
GRA	GOOS Regional Alliances
GTS	Global Telecommunication System
GTSPP	Global Temperature and Salinity Profile Program
HadISST	Hadley Center Sea Ice and Sea Surface Temperature
HarmoNIA	Harmonization and Networking for contaminant assessment in the Ionian and Adriatic Seas
HELCOM	HELsinki COMmission
HF	High Frequency
HFR	High Frequency Radar
HMAP	History of Marine Animal Populations
HTML	HyperText Markup Language
IAPB	International Arctic Buoy Program
IAPSO	International Association for the Physical Sciences of the Oceans
IBI-ROOS	Ireland Biscay Iberian Regional Operational Oceanographic System
ICEDIG	Innovation and consolidation for large scale digitization of natural heritage
ICES	International Council for the Exploration of the Sea
ICOS	Integrated Carbon Observation System
ICSU	International Council for Science
ICSU-WDS	ICSU World Data System
ICZM	Integrated Coastal Zone Management
ID	Input Data(set)
iDigBio	Integrated Digitized Biocollections
IDOE	International Decade of Ocean Exploration
IGY	International Geophysical Year
IHO	International Hydrographic Organization
IJI	International Joint Initiatives
IK	Indigenous Knowledge

IMIS	Integrated Marine Information System
IMOS	Australia's Integrated Marine Observing System
INSPIRE	INfrastructure for SPatial Information in the euRopEan Community
InSTAC	In Situ Thematic Assembling Centre
IOC	Intergovernmental Oceanographic Commission of UNESCO
IOC-ODIS	Ocean Data and Information System
IOCCp	International Ocean Carbon Coordination Project
IODE	International Oceanographic Data and information Exchange
IOOS	US Integrated Ocean Observing System
IPBES	Intergovernmental Science-Policy Platform on Biodiversity and Ecosystem Services
IPCC	Intergovernmental Panel on Climate Change
IQuOD	International Quality Controlled Ocean Database
ISC	International Science Council
ISO	International Organization for Standardization
ISSC	International Social Science Council
IT	Information Technology
ITIS-	Integrated Taxonomic Information System
IUBS	International Union of Biological Sciences
JAVA	class-based, object-oriented programming language
JCOMM	Joint technical Commission for Oceanography and Marine Meteorology
JCOMMOPS	Joint WMO-IOC Centre for in situ Ocean and Marine Meteorological Observing Program Support
JERICO	Joint European Research Infrastructure of Coastal Observatories
JERICOS3	Joint European Research Infrastructure of Coastal Observatories Science, Service, Sustainability
JGOFS	Joint Global Ocean Flux Study
JRC	Joint Research Center
JSON	JavaScript Object Notation
KBP	Kenya-Belgium Project
KMFRI	Kenya Marine and Fisheries Research Institute
LAT	Lowest Astronomical Tide
LEO	Local Environmental Observer
LifeWatch	European AgroEcology Living Lab and Research Infrastructure Network
LSID	Life Science Identifier
LTER	Long-Term Ecosystem Research
MacroBen	Integrated database on soft-bottom benthos
MANUELA	Meiobenthic and Nematode biodiversity Unravelling Ecological and Latitudinal Aspects
MAP	Mediterranean Action Plan
MARBEF	Marine Biodiversity and Ecosystem Functioning
MARPOL	International Convention for the Prevention of Pollution from Ships
MASDEA	Marine Species Database for Eastern Africa
MAT	Binary MATLAB file
MBA	Marine Biological Association
MBON	Marine Biodiversity Observation Network

MCDS	Marine Climate Data System
MDT	Mean Dynamic Topography
MEDAR	MEditerranean oceanographic Data Archeology and Rescue
MedAtlas	Mediterranean Atlas
MEDI	Marine Environmental Data Information Referral Catalogue
MEDPOL	Marine pollution assessment and control component of MAP
MEOP	Marine Mammals Exploring the Oceans Pole to Pole
MFC	Monitoring and Forecasting Center
MHHW	Mean Higher High Water
MIxS	Minimum Information about any (x) Sequence
MLLT	Mean Lower Low Tide
MODB	Mediterranean Oceanographic Data Base
MONGOOS	Mediterranean Operational Network for the Global Ocean Observing System
MSFD	Marine Strategy Framework Directive
MSP	Marine Spatial Planning
NaGISA	Natural Geography in Shore Areas
NASA	National Aeronautics and Space Administration
NASA - CMR	NASA Common Metadata Repository
NASA - GCMD	NASA Global Change Master Directory
NCEAS	National Center for Ecological Analysis and Synthesis
NCEI	NOAA's National Centers for Environmental Information
NDBC	US National Data Buoy Center
NERC	Natural Environment Research Council
NESDIS	National Environmental Satellite, Data, and Information Service
NetCDF	Network Common Data Form
NOAA	National Oceanographic and Atmospheric Administration
NODC	National Oceanographic Data Center
NoE	Network of Excellence
NOOS	North Sea Operational Oceanographic System
NRT	Near Real Time
NSBP	North Sea Benthos Project
NSBS	North Sea Benthos Survey
NVS	NERC Vocabulary Server
O&M	Observations and Measurements
O_2	Oxygen
OAI	Open Archives Initiative
OBIS	Ocean Biodiversity Information System
OBPS	Ocean Best Practices System
OceanOPS	Ocean Observations Programs Support
OceanSITES	international system of long-term, open-ocean reference stations
OCG	Observations Coordination Group
ODINAFRICA	Ocean Data and Information Network for Africa
ODIP	Ocean Data Interoperability Platform
ODIS	Ocean Data and Information System
ODSBPP	Ocean Data Standards and Best Practices Project

ODV	Ocean Data View
OECD	Organization for Economic Co-operation and Development
OGC	Open Geospatial Consortium - Define standards for sharing geographical data
OGDMTT	Ocean Glider Data Management Task Team
OISST	Optimum Interpolation Sea Surface Temperature
OMB	Office of Management and Budget
OOPC	Ocean Observations Physics and Climate Panel
OOPS	Operational Oceanographic Products and Services
OPeNDAP	Open-source Project for a Network Data Access Protocol
ORCID	Open Researcher and Contributor ID
OSGeo	Open Source GEOspatial foundation
OSPAR	OSlo-PARis Convention for the Protection of the Marine Environment of the North-East Atlantic
OSSE	Observing System Simulation Experiments
OSTIA	Operational Sea Surface Temperature and Ice Analysis
OSTP	Office of Science and Technology Policy
P2P	Pole to Pole
PACE	PArtnership for China and Europe
PANGAEA	Data Publisher for Earth and Environmental Science
pCO2	Partial pressure of carbon dioxide
pH	Power of Hydrogen
PID	Personal Identifiable Data
PIDoc	Product Information Document
PIRATA	PredIction and Research moored Array in The Atlantic
POP	Persistent Organic Pollutant
PSMSL	Permanent Service for Mean Sea Level
QA	Quality Assurance
QAS	Quality Assurance Strategy
QC	Quality Control
QF	Quality Flag
QI	Quality Indicator
RAMA	Research moored Array for African-Asian-Australian Monsoon Analysis and prediction
RDA	Research Data Alliance
RDLF	Research Data Life Cycle
RECOSCIX-WIO	Regional Cooperation in Scientific Information Exchange in the West Indian Ocean
RI	Research Infrastructures
RMP	Responsive Mode Programs
RMSD	Root Mean Square Difference
ROC	Regional Dispatch Centre
ROOS	Regional Operational Oceanographic System
RT	Real Time
RTD	Research and Technological Development
SARCE	South American Research Group on Coastal Ecosystems
SBSTTA	Subsidiary Body on Scientific, Technical and Technological Advice
SCOR	Scientific Committee on Oceanic Research

SD	Sustainable Development
SDG	Sustainable Development Goals
SDN	SeaDataNet
SeaDataCloud	Advanced SeaDataNet services
SeaDataNet	Pan-European Infrastructure for Ocean and Marine Data Management
sensorML	Standard models and an XML encoding for describing any process
SI	Système Internationale or International System
SKOS	Simple Knowledge Organization System
SLS	IOC Sea Level Station Monitoring (SLS)
SLSTR	Sea and Land Surface Temperature Radiometer
SMM	System Maturity Matrix
SOAP	Simple Object Access Protocol
SOCAT	Surface Ocean CO_2 ATlas
SONEL	Système d'Observation du Niveau des Eaux Litorales
SOOP	Ship of Opportunity Program
SOOS	Southern Oceans Observing System
SOP	Standard Operating Procedures
SOS	Sensor Observation Service
SPARQL	Sparql Protocol And Rdf Query Language
SST	Sea Surface Temperature
SWE	Sensor Web Enablement
SYNTHESIS	Synthesis of Systematic Resources
TAC	Thematic Assembling Center
TAO	Tropical Atmosphere Ocean
Tb	Brightness Temperature
TDS	THREDDS Data Server
TDWG	Biodiversity information standards (formally Taxonomic Databases Working Group)
TG-ML	MSFD Technical Group on Marine Litter
THREDDS	Thematic Real-time Environmental Distributed Data Services
TKIP	Traditional Knowledge Information Portal
TRITON	TRIangle Trans-Ocean buoy Network
TRUST	Transparency - Responsibility - User community - Sustainability — Technology
TXT	TeXT
UD	Upstream Data(set)
UHSLC	University of Hawaii Sea Level Center
UN	United Nations
UNDP	United Nations Development Program
UNEP	United Nations Environment Program
UNESCO	United Nations Educational Scientific and Cultural Organization
UNFCCC	United Nations Framework Convention on Climate Change
UNSD	United Nations Sustainable Development
URI	Universal Resource Identifier
URMO	UNESCO-IOC Register of Marine Organisms
US	United States
VLIZ	Flanders Marine Institute

VOS	Voluntary Observing Ship
VRE	Virtual Research Environment
VTS	Vessel Traffic Services
W3C	World Wide Web Consortium
WAF	Web Accessible Folder
WCS	Web Coverage Service
WDC	World Data Centre
webODV	on line Ocean Data View
WEkEO	Copernicus DIAS service
WFD	Water Framework Directive - EU Project to provide cleaner fresh water in Europe
WFS	Web Feature Service
WG	Working Group
WIPO	World Intellectual Property Organization
WISE	Water Information System for Europe (also WISE-Marine for the marine specific bit)
WMO	World Meteorological Organization
WMS	Web Map Service
WMTS	Web Map Tile Service
WOA	World Ocean Assessment
WOA	World Ocean Atlas
WOCE	World Ocean Circulation Experiment
WOD	World Ocean Database
WoRMS	World Register of Marine Species
WxS	W3C xml Schema
XBT	eXpendable BathyThermograph

Index

Printed in the United States
by Baker & Taylor Publisher Services